For over 35 years, biological scientists have come to rely on the research protocols and methodologies in the critically acclaimed *Methods in Molecular Biology* series. The series was the first to introduce the step-by-step protocols approach that has become the standard in all biomedical protocol publishing. Each protocol is provided in readily-reproducible step-by-step fashion, opening with an introductory overview, a list of the materials and reagents needed to complete the experiment, and followed by a detailed procedure that is supported with a helpful notes section offering tips and tricks of the trade as well as troubleshooting advice. These hallmark features were introduced by series editor Dr. John Walker and constitute the key ingredient in each and every volume of the *Methods in Molecular Biology* series. Tested and trusted, comprehensive and reliable, all protocols from the series are indexed in PubMed.

Plant Gametogenesis

Methods and Protocols

Edited by

Christophe Lambing

Department of Plant Sciences, University of Cambridge, Cambridge, UK;
Department of Plant Sciences, Rothamsted Research, Harpenden, UK

Humana Press

Editor
Christophe Lambing
Department of Plant Sciences
University of Cambridge
Cambridge, UK

Department of Plant Sciences
Rothamsted Research
Harpenden, UK

ISSN 1064-3745 ISSN 1940-6029 (electronic)
ISBN 978-1-0716-2252-0 ISBN 978-1-0716-2253-7 (eBook)
https://doi.org/10.1007/978-1-0716-2253-7

This Humana imprint is published by the registered company Springer Science+Business Media, LLC part of Springer Nature.
The registered company address is: 1 New York Plaza, New York, NY 10004, U.S.A.

Preface

Gametogenesis is an essential biological process by which the precursor germ cells differentiate to form mature haploid gametes. In animals, germline specification is predetermined at an early embryonic stage. In contrast, plant precursor cells are not predetermined, and germline differentiation only occurs at a late stage in plant development. All land plants have evolved from a common ancestor after the transition from ocean to land that occurred 400–500 million years ago. During plant evolution and the apparition of flowering plants, the sporophyte stage became dominant in the life cycle while the gametophyte stage was restricted to fewer cells. In flowering plants, male gametogenesis occurs in the anthers and female gametogenesis takes place in the ovules. These reproductive organs are complex with several different cell types which limit the study of gamete development.

During gametogenesis, cells experience modifications in their cellular, metabolic, and epigenetic contents. The genetic and epigenetic information present in the gametes can be inherited in the embryo. During this process, a complex control of epigenetic reprogramming exists in plants to ensure proper propagation of the epigenetic information and to maintain the silencing of repetitive transposable elements. Several techniques have been developed to study gametogenesis that include genomic, cytogenetic, and molecular biology. The recent advances in cell type isolation and genomic studies have provided information on the epigenetic and transcriptomic changes that occur during gamete formation. Moreover, complementary methods are now available to investigate cell type-specific molecular phenotypes using the whole reproductive organ. For example, RNA hybridization coupled with microscopy allows spatial and temporal detection of transcriptional activity for specific genes.

The aim of this volume is to provide protocols and detailed information on methods and analyses that will allow a readership to be informed on the most recent techniques available to study plant gametogenesis. Thorough investigation of complex biological processes often involves multidisciplinary approaches, and this is the reason I decided to combine protocols that cover a comprehensive set of methods for plant biology. This book is divided in four parts covering omics approaches, microscopy approaches, plant manipulation and genome editing, bioinformatics and data analysis. I hope this volume serves as a tool to provide the foundation for future studies and to be a source of inspiration to investigate new questions in the field.

Harpenden, UK *Christophe Lambing*

Contents

Contributors

SIDDIQUE I. ABOOBUCKER • *Department of Agronomy, Iowa State University, Ames, IA, USA*

WAHIBA AMRI-TILIOUINE • *Laboratory of Integrative Improvement of Vegetal Productions, Higher National Agronomic School (ENSA), Algiers, Algeria; Division of Biotechnology and Plant Breeding, National Institute of Agricultural Research of Algeria (I.N.R.A.A.), Algiers, Algeria*

ABDELLAH BARAKATE • *Cell and Molecular Sciences, James Hutton Institute, Dundee, UK*

JAN BARTOŠ • *Institute of Experimental Botany of the Czech Academy of Sciences, Centre of the Region Hana for Biotechnological and Agricultural Research, Olomouč, Czech Republic*

DAVID C. BAULCOMBE • *Department of Plant Sciences, University of Cambridge, Cambridge, UK*

TOMASZ BIELUSZEWSKI • *Laboratory of Genome Biology, Institute of Molecular Biology and Biotechnology, Adam Mickiewicz University, Poznan, Poland*

MONICA BORGHI • *Department of Biology, Utah State University, Logan, UT, USA*

DOHWAN BYUN • *Department of Life Sciences, Pohang University of Science and Technology, Pohang, Gyeongbuk, Republic of Korea*

YU-RU CHEN • *Department of Agronomy, Iowa State University, Ames, IA, USA*

KYUHA CHOI • *Department of Life Sciences, Pohang University of Science and Technology, Pohang, Gyeongbuk, Republic of Korea*

GREGORY P. COPENHAVER • *Department of Biology and the Integrative Program for Biological and Genome Sciences, University of North Carolina at Chapel Hill, Chapel Hill, NC, USA; Lineberger Comprehensive Cancer Center, University of North Carolina School of Medicine, Chapel Hill, NC, USA*

ALICE DARBYSHIRE • *School of Biosciences, University of Birmingham, Birmingham, UK*

LEONARDO PEREZ DE SOUZA • *Max-Planck-Institute of Molecular Plant Physiology, Potsdam-Golm, Germany*

JULIA DLUZEWSKA • *Laboratory of Genome Biology, Institute of Molecular Biology and Biotechnology, Adam Mickiewicz University, Poznan, Poland*

THOMAS DRESSELHAUS • *Department of Cell Biology and Plant Biochemistry, University of Regensburg, Regensburg, Germany*

STEFANIE DUKOWIC-SCHULZE • *Department of Cell Biology and Plant Biochemistry, University of Regensburg, Regensburg, Germany*

WOJCIECH DZIEGIELEWSKI • *Laboratory of Genome Biology, Institute of Molecular Biology and Biotechnology, Adam Mickiewicz University, Poznan, Poland*

ALISDAIR R. FERNIE • *Max-Planck-Institute of Molecular Plant Physiology, Potsdam-Golm, Germany*

TYLER FOSTER • *Department of Agronomy, Iowa State University, Ames, IA, USA*

F. CHRIS H. FRANKLIN • *School of Biosciences, University of Birmingham, Birmingham, UK*

URSULA K. FREI • *Department of Agronomy, Iowa State University, Ames, IA, USA*

BASKAR GANAPATHYSUBRAMANIAN • *Department of Mechanical Engineering, Iowa State University, Ames, IA, USA*

QUENTIN GOUIL • *Walter and Eliza Hall Institute of Medical Research, Parkville, VIC, Australia; Department of Medical Biology, The University of Melbourne, Parkville, VIC, Australia*
IAN R. HENDERSON • *Department of Plant Sciences, University of Cambridge, Cambridge, UK*
CHUNHUI HOU • *Department of Biology, School of Life Sciences, Southern University of Science and Technology, Shenzhen, China*
JIYUE HUANG • *Department of Biology and the Integrative Program for Biological and Genome Sciences, University of North Carolina at Chapel Hill, Chapel Hill, NC, USA*
MASAKO ICHIKAWA • *Agri-Bio Research Center, KANEKA Corp., Iwata, Shizuoka, Japan*
JOANNA JANKOWICZ-CIESLAK • *Plant Breeding and Genetics Laboratory, Joint FAO/IAEA Centre of Nuclear Techniques in Food and Agriculture, IAEA Laboratories Seibersdorf, International Atomic Energy Agency, Vienna, Austria*
TALUKDER Z. JUBERY • *Department of Mechanical Engineering, Iowa State University, Ames, IA, USA*
REKHA KARUMAMKANDATHIL • *Rubber Research Institute of India, Rubber Board PO, Kottayam, Kerala, India*
NORIO KATO • *Department of Biological Sciences, Tokyo Metropolitan University, Hachioji, Tokyo, Japan*
NADIA KBIRI • *Laboratory of Genome Biology, Institute of Molecular Biology and Biotechnology, Adam Mickiewicz University, Poznan, Poland*
PALLAS KUO • *Department of Plant Sciences, University of Cambridge, Cambridge, UK*
CHRISTOPHE LAMBING • *Department of Plant Sciences, Rothamsted Research, Harpenden, UK*
MERIEM LAOUAR • *Laboratory of Integrative Improvement of Vegetal Productions, Higher National Agronomic School (ENSA), Algiers, Algeria*
INNA LERMONTOVA • *Leibniz Institute of Plant Genetics and Crop Plant Research (IPK), Gatersleben, Germany*
SHAN LI • *National Key Laboratory of Rice Biology, Zhejiang Provincial Key Laboratory of Crop Genetic Resources, Institute of Crop Science, Zhejiang University, Hangzhou, China*
THOMAS LÜBBERSTEDT • *Department of Agronomy, Iowa State University, Ames, IA, USA*
PINGLI LU • *State Key Laboratory of Crop Stress Adaptation and Improvement, Key Laboratory of Plant Stress Biology, School of Life Sciences, Henan University, Kaifeng, China*
SAMUEL DANIEL LUP • *Instituto de Bioingeniería, Universidad Miguel Hernández, Elche, Spain*
CLAUDIA MARTINHO • *Department of Plant Sciences, University of Cambridge, Cambridge, UK*
TETY MARYENTI • *Department of Biological Sciences, Tokyo Metropolitan University, Hachioji, Tokyo, Japan*
JOSÉ LUIS MICOL • *Instituto de Bioingeniería, Universidad Miguel Hernández, Elche, Spain*
PAVLINA MIKULKOVA • *Central European Institute of Technology (CEITEC), Masaryk University, Brno, Czech Republic*
AOWEI MO • *School of Life Sciences, Fudan University, Shanghai, China*
LONGJIAN NIU • *Department of Biology, School of Life Sciences, Southern University of Science and Technology, Shenzhen, China*
TAKASHI OKAMOTO • *Department of Biological Sciences, Tokyo Metropolitan University, Hachioji, Tokyo, Japan*

JAMIE ORR • *Cell and Molecular Sciences, James Hutton Institute, Dundee, UK*
KIM OSMAN • *School of Biosciences, University of Birmingham, Birmingham, UK*
JANA PECINKOVA • *Central European Institute of Technology (CEITEC), Masaryk University, Brno, Czech Republic*
FERNANDO A. RABANAL • *Department of Molecular Biology, Max Planck Institute for Biology, Tübingen, Germany*
MARYIA RATNIKAVA • *Leibniz Institute of Plant Genetics and Crop Plant Research (IPK), Gatersleben, Germany*
KAREL RIHA • *Central European Institute of Technology (CEITEC), Masaryk University, Brno, Czech Republic*
EUGENIO SANCHEZ-MORAN • *School of Biosciences, University of Birmingham, Birmingham, UK*
MIRIAM SCHREIBER • *Division of Plant Sciences, The University of Dundee, James Hutton Institute, Dundee, UK*
HEÏDI SERRA • *Genetics, Diversity and Ecophysiology of Cereals, UMR 1095, INRAE, Université Clermont Auvergne, Clermont-Ferrand, France; Genetics, Reproduction and Development, Université Clermont Auvergne, CNRS, INSERM, Clermont-Ferrand, France*
DRIN SHABANI • *Department of Cell Biology and Plant Biochemistry, University of Regensburg, Regensburg, Germany*
PIERRE SOURDILLE • *Genetics, Diversity and Ecophysiology of Cereals, UMR 1095, INRAE, Université Clermont Auvergne, Clermont-Ferrand, France*
LU SUN • *State Key Laboratory of Crop Stress Adaptation and Improvement, Key Laboratory of Plant Stress Biology, School of Life Sciences, Henan University, Kaifeng, China*
RADIM SVAČINA • *Institute of Experimental Botany of the Czech Academy of Sciences, Centre of the Region Hana for Biotechnological and Agricultural Research, Olomouč, Czech Republic*
MAJA SZYMANSKA-LEJMAN • *Laboratory of Genome Biology, Institute of Molecular Biology and Biotechnology, Adam Mickiewicz University, Poznan, Poland*
YUANYUAN TAN • *The New Countryside Development Institute at Zhejiang University, Hangzhou, China*
MEENAKUMARI THUPPALE • *Rubber Research Institute of India, Rubber Board PO, Kottayam, Kerala, India*
BRADLEY J. TILL • *Plant Breeding and Genetics Laboratory, Joint FAO/IAEA Centre of Nuclear Techniques in Food and Agriculture, IAEA Laboratories Seibersdorf, International Atomic Energy Agency, Vienna, Austria; Veterinary Genetics Laboratory, University of California, Davis, Davis, CA, USA*
THOMAS KADAMPANATTU UTHUP • *Rubber Research Institute of India, Rubber Board PO, Kottayam, Kerala, India*
SONA VALUCHOVA • *Central European Institute of Technology (CEITEC), Masaryk University, Brno, Czech Republic*
CONG WANG • *State Key Laboratory of Genetic Engineering and Collaborative Innovation Center of Genetics and Development, Ministry of Education Key Laboratory of Biodiversity Sciences and Ecological Engineering, Institute of Plant Biology, School of Life Sciences, Fudan University, Shanghai, China*
WENJING WANG • *Department of Biology, School of Life Sciences, Southern University of Science and Technology, Shenzhen, China; School of Life Science and Stake Key Laboratory of Agrobiotechnology, The Chinese University of Hong Kong, Hong Kong, China*

XIAO YU WANG • *Department of Plant Sciences, University of Cambridge, Cambridge, UK*
YINGXIANG WANG • *State Key Laboratory of Genetic Engineering and Collaborative Innovation Center of Genetics and Development, Ministry of Education Key Laboratory of Biodiversity Sciences and Ecological Engineering, Institute of Plant Biology, School of Life Sciences, Fudan University, Shanghai, China*
ZHENGMING WANG • *Department of Plant Sciences, University of Cambridge, Cambridge, UK; Key Laboratory of Horticulture Biology, Ministry of Education, College of Horticulture and Forestry Sciences, Huazhong Agricultural University, Wuhan, Hubei, China*
ROBBIE WAUGH • *Division of Plant Sciences, The University of Dundee, James Hutton Institute, Dundee, UK; Cell and Molecular Sciences, James Hutton Institute, Dundee, UK*
DAVID WILSON-SÁNCHEZ • *Department of Comparative Development and Genetics, Max Planck Institute for Plant Breeding Research, Cologne, Germany*
RAMAKRISHNA YADALA • *Leibniz Institute of Plant Genetics and Crop Plant Research (IPK), Gatersleben, Germany*
CHENJIANG YOU • *State Key Laboratory of Genetic Engineering and Collaborative Innovation Center of Genetics and Development, Ministry of Education Key Laboratory of Biodiversity Sciences and Ecological Engineering, Institute of Plant Biology, School of Life Sciences, Fudan University, Shanghai, China*
LONGFEI ZHU • *Laboratory of Genome Biology, Institute of Molecular Biology and Biotechnology, Adam Mickiewicz University, Poznan, Poland*
PIOTR A. ZIOLKOWSKI • *Laboratory of Genome Biology, Institute of Molecular Biology and Biotechnology, Adam Mickiewicz University, Poznan, Poland*

Part I

Omics Approaches to Study Male and Female Germlines in Plants

Chapter 1

Measurement of Flower Metabolite Concentrations Using Gas Chromatography–Mass Spectrometry and High-Performance Liquid Chromatography–Mass Spectrometry

Monica Borghi, Leonardo Perez de Souza, and Alisdair R. Fernie

Abstract

Metabolite profiling aiming at quantifying the metabolome of flowers is emerging as a suitable tool to understand the metabolic complexity of these reproductive organs and the associations between primary and secondary metabolites which characterize them. This chapter provides a general method for the combined analyses of primary and secondary metabolites via gas chromatography–mass spectrometry (GC-MS) and high-performance liquid chromatography–mass spectrometry (LC-MS) of flower samples. We describe the preparatory steps, the procedure of metabolites' extraction and finally provide examples of data representation. The method described here can be applied to the analysis of metabolomes of entire flowers, as well as specific flower organs.

Key words Flower, Primary metabolites, Secondary metabolites, GC-MS, LC-MS

1 Introduction

To tackle the complexity of flowers and the developmental progression of gametes and floral organs at a molecular level, multiple omics approaches have been utilized. Early microarray experiments, followed by RNA sequencing, and more recently, by single-cell transcriptomics [1, 2] and ChiP sequencing [3], have certainly dominated the scene and paved the way toward the discovery of the transcriptional processes governing flower development. Proteomics [4, 5] and multielemental analysis [6] have also been applied to the study of flowers, and recently, metabolomic approaches have provided comprehensive profiles of different classes of secondary compounds present in flowers [7, 8]. Indeed, flowers are organs with a complex type of metabolism sustained mainly by importing

Christophe Lambing (ed.), *Plant Gametogenesis: Methods and Protocols*, Methods in Molecular Biology, vol. 2484, https://doi.org/10.1007/978-1-0716-2253-7_1,

carbohydrates into the floral calyx via the phloem [9]. Sucrose is the carbohydrate preferentially transported along the phloem in most of the relevant species for agricultural and physiological studies. Nevertheless, flowers also accumulate abundant amounts of carbohydrates different from sucrose. Indeed, cycles of carbohydrate hydrolysis and resynthesis frequently occur in flowers to partition and recycle compounds in various organs and tissues and sustain new metabolites' synthesis. The amino acids Asn, Asp, Gln, and Glu, which are primarily synthesized in root and leaves, are also transported to the floral calyx via the phloem and, once in flowers, are also used for the synthesis of a diverse array of metabolites. Besides, in flowers, intermediates of the tricarboxylic acid (TCA) cycle, nucleic acids, vitamins, and various secondary metabolites are also found. Indeed, the petals of many flower species accumulate Phe derivatives in a large abundance, such as flavonoids and anthocyanins. These metabolites are also found in anthers and pollen to shield male gametophytes from UV radiation. Glucosinolates, polyamines, and hydroxycinnamates are also frequently found in floral tissues [8].

This chapter describes a general protocol for the combined extraction of primary and secondary metabolites suitable to gain a holistic understanding of flower metabolism and the balance between different classes of compounds. This protocol is suitable for a wide range of flower tissues spanning from entire flowers to single organs (sepals, petals, stamens, and pistils; Fig. 1). An example where this protocol finds applications is in experiments aiming to measure the metabolome of full flowers exposed to abiotic or biotic stresses (Fig. 1a) or flowers from different species or genotypes (Fig. 1b). Indeed, drought, heat, cold, nutrient limitations, and so on are known to impact the floral metabolism dramatically [10]. Similarly, flowers infected by pathogenic microorganisms and fungi are often the subject of metabolic reprogramming [11], which can also be investigated by extracting and annotating a vast array of metabolites with the protocol presented here. Flower developmental series also represents an ad hoc application, for example, to assess changes in the metabolome as flowers progress from buds to a status of mature organs suitable for plant reproduction (Fig. 1c). Short developmental series, representing time points during a 24 h period, can also be analyzed with this methodology, for example, to investigate the strength of flowers as sink organs and the diurnal and nocturnal metabolic adjustments. Within a flower, floral organs show specific metabolic signatures [7], which can also be assessed following this protocol (Fig. 1d). In this scenario, the only limiting factor is the ability to harvest a sufficient amount of material for each of the organs, especially for species with tiny flowers, with *Arabidopsis thaliana* representing an excellent example of this kind. Similarly, floral organs can be collected from

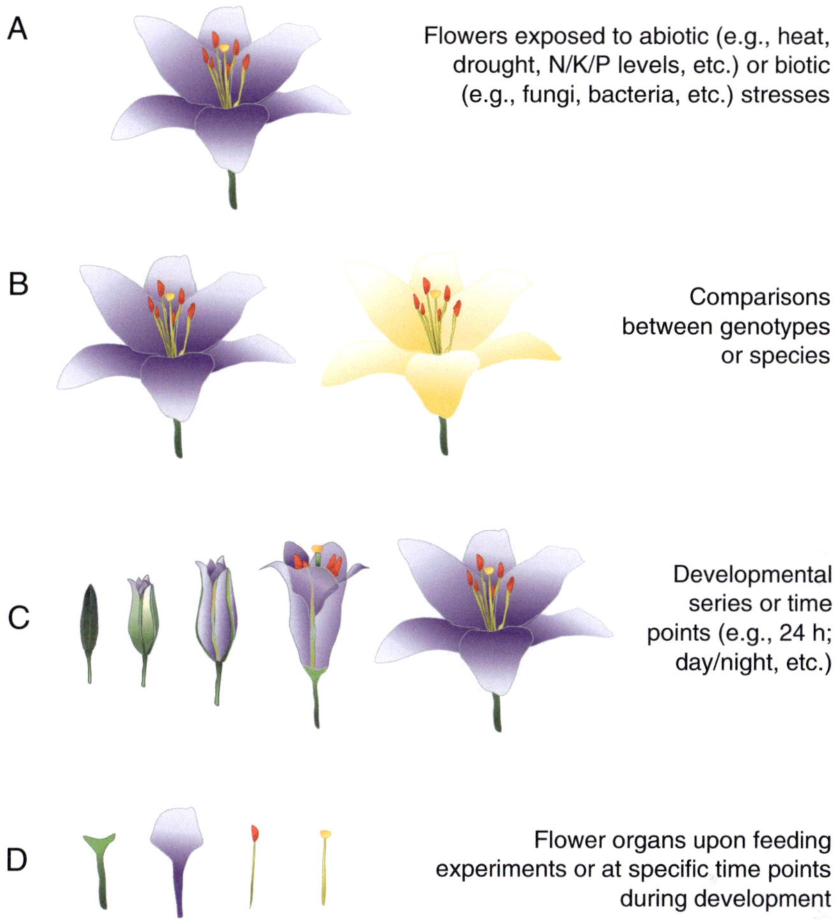

Fig. 1 Schematic representation of the types of samples suitable to be analyzed with the method described in this chapter. (**a**) Flowers exposed to abiotic and biotic stresses. (**b**) Flowers from different species or genotypes. (**c**) Flower developmental series or time points during a 24 h period. (**d**) Flower organs from feeding experiments or at time points during development

flowers fed with ^{13}C- or ^{15}N-labelled substrates to assess the metabolic fluxes and the redistribution of the stable isotope within those organs [12, 13].

2 Materials

Use ultrapure water and high-grade chemicals for the experiments.

2.1 Plant Material, Harvest, Sampling

1. Flowers or floral organs.
2. Precision balance.
3. Blade or scissor or tweezer, spatula.
4. 2.0 mL round bottom-shaped plastic tubes with screw cap.

5. Small metal beads of 3 mm or similar.
6. Mill for cryogenic grinding (*see* **Note 1**).
7. Liquid nitrogen.
8. Ice.

2.2 Metabolite Extraction

1. Methanol (MeOH) ULC/MS grade.
2. Extraction solution of MeOH containing the appropriate concentration of internal standard (IS; *see* **Note 2**).
3. Chloroform.
4. Water ULC/MS grade.
5. Centrifuge with a 2 mL tube capacity rotor.
6. Vortex mixer.
7. 2 mL or 1.7 mL snap-cap microcentrifuge tubes.
8. Speed vacuum concentrator.
9. Glass vials with screw cap for GC-MS and LC-MS analysis.

2.3 Derivatization

1. Methoxyamine hydrochloride, purity 98%.
2. Pyridine (HPLC grade).
3. Methoxyamine hydrochloride solution in pyridine: 20 mg of methoxyamine hydrochloride in 1 mL of pyridine. Dissolve by sonication for 3 min.
4. *N*-methyl-*N*-trimethylsilyl-trifluoroacetamide (MSTFA).
5. Fatty acid methyl esters (FAMES): methylcaprylate, methyl pelargonate, methylcaprate, methyllaurate, methylmyristate, methylpalmitate, methylstearate, methyleicosanoate, methyldocosanoate, lignoceric acid methylester, methylhexacosanoate, methyloctacosanoate, triacontanoic acid methylester. Dissolve esters in chloroform at the final concentration of 0.8 μL/mL (liquid esters) or 0.4 mg/mL (solid esters). Store FAMES at −20 °C.
6. MSTFA solution: add 20 μL of FAMES to 1 mL of MSTFA reagent. Prepare fresh each time.
7. Thermal mixer with blocks.
8. Sonicator (ultrasonic cleaner).

2.4 GC-TOF-MS Analysis of Primary Metabolites

1. Gas Chromatograph equipped with a time-of-flight mass spectrometer (TOF-MS) analyzer.
2. Autosampler system for the automatic injection of samples.
3. Capillary column OPTIMA 35 MS, 30 m (length), 0.25 mm (internal diameter), 0.25 μm particle size (or similar).
4. Helium carrier gas.

5. Computer and software for mass spectral data analysis (Xcalibur, MassHunter, XCMS, etc.).

Details on the procedures needed to perform profile metabolite analysis via GC-MS have been provided in Methods in Molecular Biology, vol. 860, Chapter 7 [14].

2.5 LC-MS Analysis of Secondary Metabolites

1. Ultrahigh-performance liquid chromatograph equipped with an orbitrap or TOF mass spectrometer.
2. Reversed-phase C18 UHPLC column such as Waters HSS T3 C18 100 mm (length), 2.1 mm (internal diameter), 1.8 μm particle size (or similar).
3. Distilled water ULC/MS grade.
4. Acetonitrile (ACN) ULC/MS grade.
5. Formic acid (HCOOH) ULC/MS grade.
6. Mobile phase: eluent (A) 0.1% formic acid in water, eluent (B) 0.1% formic acid in ACN.
7. Computer and software for mass spectral data analysis (Xcalibur, MassHunter, XCMS, etc.).

Details on procedures needed to perform profile metabolite analysis via LC-MS have been provided in Methods in Molecular Biology, vol. 1778, Chapter 12 [15].

3 Methods

The method described in this paragraph has been optimized to extract metabolites from flowers of Arabidopsis, tomato, petunia, and snapdragon. Given the immense diversity of metabolites of flowers of different species and changes in abundance occurring as flower progress throughout development or when exposed to biotic and abiotic stresses, the amount of flower starting material used for the extractions and volumes collected thereafter may need further adjustments.

3.1 Sample Preparation

1. Harvest flowers using clean scissors, a blade, or tweezers as appropriate for the species under investigation.
2. Place the flower material and two metal beads in a 2.0 mL round bottom-shaped plastic tubes with screw cap and immediately snap-freeze the samples in liquid nitrogen (*see* **Note 3**).
3. Homogenize the samples in a cryogenic mill at 25 Hz for 3 min. Keep the samples frozen all the time. If the cryogenic mill is not connected to a continuous flow of liquid nitrogen, freeze the holders before each use (*see* **Note 4**).
4. Quickly remove the tubes from the mill and place them in liquid nitrogen.

3.2 Sample Extraction

1. Remove the samples from the liquid nitrogen (one sample at the time), add 450 μL of MeOH extraction solution containing the appropriate IS and immediately vortex thoroughly for 10–15 s.
2. Repeat **step 1** until all samples have been processed. Keep the samples on ice at all times.
3. Homogenize the samples in the mill at 25 Hz for 3 min. As MetOH halts cellular biological processes and inactivates enzyme activity, this homogenization step can be performed at room temperature.
4. Centrifuge the samples at 20,000 × *g* for 10 min.
5. Transfer the supernatant to a 2 mL snap-cap microcentrifuge tubes (approximately 360 μL; *see* **Note 5**).
6. Add, in the following order, half volume of chloroform (e.g., 180 μL for 360 μL of supernatant) and one volume of water (e.g., 360 μL for 360 μL of supernatant) and vortex thoroughly for 10–15 s.
7. Centrifuge the samples at 20,000 × *g* for 10 min.
8. Transfer 100 μL from the supernatant to two 1.7 mL snap-cap microcentrifuge tubes. One aliquot will be utilized for derivatization and GC-MS analysis of primary metabolites, and the second aliquot for LC-MS analysis of secondary metabolites (*see* **Note 6**).
9. Dry all aliquots in the speed vacuum concentrator without heating (usually 3–5 h are needed to dry the samples completely). Dried aliquots can be stored at −80 °C.

3.3 Sample Derivatization and Data Acquisition by GC-MS

The following steps only apply to samples allocated for GC-MS analysis of primary metabolites.

1. Prepare a fresh solution of methoxyamine hydrochloride in pyridine (*see* **Note** 7).
2. Add 40 μL of methoxyamine hydrochloride solution to each sample and vortex the samples for 5 s (*see* **Note 8**).
3. Place the samples in the thermal mixer and shake them at 37 °C for 2 h at 950 rpm.
4. Prepare a fresh solution of MSTFA.
5. Retrieve the samples from the thermal mixer and spin down shortly.
6. Add 70 μL of MSTFA solution and quickly vortex the samples for 5 s.
7. Place the samples in the thermal mixer and shake at 37 °C for 30 min at 950 rpm.

8. Retrieve the samples from the thermal mixer and spin down shortly.
9. Transfer the sample to a glass vial with screw cap for GC-MS analysis.

Details on the procedures needed to perform profile metabolite analysis via GC-MS have been provided in Methods in Molecular Biology, vol. 860, Chapter 7 (*see* also **Note 9**).

3.4 Sample Preparation and Data Acquisition by LC-MS

The following steps only apply to samples allocated for LC-MS analysis of secondary metabolites.

1. Reconstitute the samples by adding 70 μL of MeOH (*see* **Note 8**).
2. Vortex the samples thoroughly until fully resuspended. If needed, briefly sonicate the samples for 3–5 min.

Details on the procedures needed to perform profile metabolite analysis via LC-MS have been provided in Methods in Molecular Biology, vol. 1778, Chapter 12 [15].

3.5 Data Visualization

After analysis of the mass spectra and normalization, the results of these measurements can be presented with a diverse array of informative graphs as listed below.

1. Bar graphs (Fig. 2a): To compare the amount of a specific compound in each of the floral organs understudy or at specific time points during development, calculate the average and plot it as a bar plot.
2. Line graphs (Fig. 2b): To visualize changes of metabolite abundance over time, calculate the metabolite average for each of the time points and plot the data as a line graph. Data collected from feeding experiments (*see* refs. 12 and 13 for comprehensive protocols on ^{13}C-labeling experiments) can also be presented as line graphs of the total ^{13}C enrichment plotted as a function of time.
3. Heatmaps (Fig. 2c): To visualize clusters of samples with similar metabolic rearrangements, calculate the correlation coefficients between metabolite pairs and plot them as a heatmap.
4. Network plots (Fig. 2d): To visualize metabolites of relevance within a group of samples, organs, or time points, calculate the correlation coefficients between metabolites or the enriched categories of metabolite species among samples and plot the data as a network.

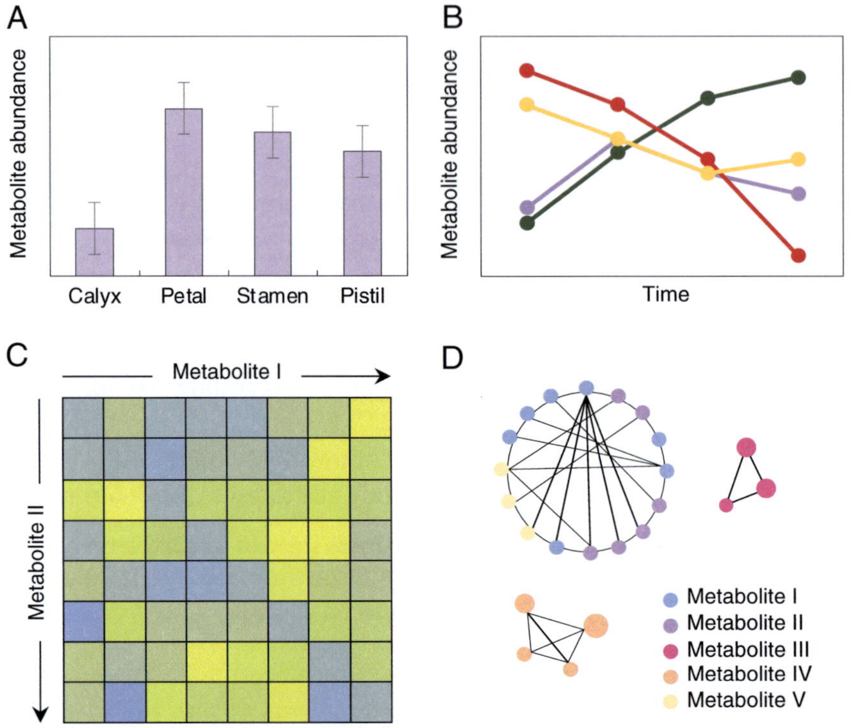

Fig. 2 Examples of data representation. (**a**) Bar graph: bars show the average abundance of a hypothetical metabolite measured in different floral organs (or in flowers of different species or at different time points during development). (**b**) Line graph: lines show the variation of metabolite abundance measured in floral organs in four consecutive time points. (**c**) Heatmap: visualization of metabolite-metabolite correlations. Color intensity is proportional to the degree of correlation, from high (yellow) to low (blue) correlated metabolites. (**d**) Network visualization of enriched metabolite groups in response to a treatment or in specific tissues. Metabolite-metabolite correlation can also be visualized on a network

4 Notes

1. A mortar and pestle can also be used, washed, rinsed, and dried prior to each use.
2. IS should be added to evaluate extraction efficiency and instrument performance. For this, a chemical compound not present in the samples and stable toward extraction and derivatization methods should be chosen. Typically, ribitol (0.2 mg/mL in water) is utilized as IS for the analysis of primary metabolites via GC-MS, and isovitexin (apigenin-6-*C*-glucoside; 5 μg/mL in 50% MeOH-water) for LC-MS analysis of secondary metabolites.
3. This protocol has been optimized to extract primary and secondary metabolites from 15 mg of flower material. Fresh flowers (e.g., Arabidopsis florets) or flower organs or parts of flower organs (e.g., portion of the corolla of petunia flowers) can all be

utilized as starting material for metabolites' extraction. For large mature flowers (e.g., petunia and snapdragon), use 15 mg of the fine powder obtained by grinding the flowers in liquid nitrogen. Long-term storage of the powder at −80 °C is also appropriate. For accurate comparative analyses of metabolites, the fresh-weight variation among samples should not exceed 10–15%.

4. Homogenization times may vary depending on the hardness, shape, and size of flowers, water content, and fibers' rigidity in the floral tissues. If longer homogenization times are needed, make sure that the samples are kept frozen at all times.
5. The amount of supernatant retrieved varies depending upon species, stage of development, treatment. Make sure to transfer the same amount of supernatant for all samples.
6. It is good practice to aliquot and dry the remaining supernatant and store it at −80 °C as a backup.
7. The methoxyamine hydrochloride solution is unstable and must be prepared fresh each time.
8. When processing samples that have been stored at −80 °C, dry all aliquots in the speed vacuum concentrator without heating for 30 min immediately before derivatization. This step will ensure that no water deriving from condensation after long-term storage at low temperatures is left in the samples.
9. Flower samples contain carbohydrates in a large abundance, which typically accumulate in the ovaries, style and stigma, and the nectaries. For a proper quantification of carbohydrates, samples' injection in the GC-MS should be performed in split and splitless modes. A 20% split mode is usually appropriate for most flower samples. However, as variation in quality and quantity of carbohydrates is often observed depending upon species and treatment, the split injection may need further optimization.

Acknowledgments

The authors gratefully acknowledge the financial support of the Max Planck Society and Utah State University startup funding to M.B.

References

1. Misra CS, Santos MR, Rafael-Fernandes M et al (2019) Transcriptomics of Arabidopsis sperm cells at single-cell resolution. Plant Reprod 32(1):29–38
2. Palovaara J, Weijers D (2020) Cell type-specific transcriptomics in the plant embryo using an adapted INTACT protocol. In: Plant metabolomics, Methods mol biol, vol 2122, pp 141–150
3. Chen D, Yan W, Fu L-Y et al (2018) Architecture of gene regulatory networks controlling

flower development in Arabidopsis thaliana. Nat Commun 9(1):1–13

4. Mergner J, Frejno M, Messerer M et al (2020) Proteomic and transcriptomic profiling of aerial organ development in Arabidopsis. Sci Data 7(1):1–11
5. Dafny-Yelin M, Guterman I, Menda N et al (2005) Flower proteome: changes in protein spectrum during the advanced stages of rose petal development. Planta 222(1):37–46
6. An D, Chen J-G, Gao Y-Q et al (2017) *AtHKT1* drives adaptation of Arabidopsis thaliana to salinity by reducing floral sodium content. PLoS Genet 13(10):e1007086
7. Li D, Heiling S, Baldwin IT et al (2016) Illuminating a plant's tissue-specific metabolic diversity using computational metabolomics and information theory. Proc Natl Acad Sci 113(47):E7610–E7618
8. Tohge T, Borghi M, Fernie AR (2018) The natural variance of the Arabidopsis floral secondary metabolites. Sci Data 5:180051
9. Borghi M, Fernie AR (2017) Floral metabolism of sugars and amino acids: implications for pollinators' preferences and seed and fruit set. Plant Physiol 175(4):1510–1152
10. Borghi M, de Souza LP, Yoshida T et al (2019) Flowers and climate change: a metabolic perspective. New Phytol 224(4):1425–1441
11. Borghi M, Fernie AR (2020) Outstanding questions in flower metabolism. Plant J 103: 1275–1288
12. de Souza LP, Fernie AR, Tohge T (2018) Carbon atomic survey for identification of selected metabolic fluxes. In: Plant metabolomics, Methods mol biol, vol 1178. Springer, pp 59–67
13. Lima VF, de Souza LP, Williams TC et al (2018) Gas chromatography–mass spectrometry-based 13 C-labeling studies in plant metabolomics. In: Plant metabolomics, Methods mol biol, vol 1178. Springer, pp 47–58
14. Osorio S, Do PT, Fernie AR (2011) Profiling primary metabolites of tomato fruit with gas chromatography/mass spectrometry. In: Plant metabolomics, Methods mol biol, vol 860. Springer, pp 101–109
15. Shimizu T, Watanabe M, Fernie AR et al (2018) Targeted LC-MS analysis for plant secondary metabolites. In: Plant metabolomics, Methods mol biol, vol 1178. Springer, pp 171–181

Chapter 2

Proteomic-, Phosphoproteomic-, and Acetylomic-Based Mass Spectrometry to Identify Tissue-Specific Protein Complexes and Phosphorylation in Plant Gametogenesis

Lu Sun, Aowei Mo, and Pingli Lu

Abstract

Using proteomics to analyze phosphorylation, acetylation, and other posttranslational modifications has been a very important method in biological research. Here we take the rice meiotic anther as an example to introduce the experimentally verified proteomic analysis methods of plant tissue-specific phosphorylation and acetylation, including total protein extraction, trypsin digestion, phosphopeptide enrichment by TiO_2 microcolumn, affinity enrichment of lysine-acetylated peptides, desalting, Nano UHPLC-MS/MS analysis, database search and data analysis, and bioinformatic analysis.

Key words Proteomic, Lysine acetylation, Phosphoproteomic, Meiosis, Posttranslational modifications

1 Introduction

Proteomic analysis has become a very common method in modern biological research. Proteomic analysis of posttranslational modifications has been widely used in various species. Phosphoproteomic analysis of pancreatic cancer cells revealed a comprehensive molecular network of radiation-induced signaling [1]. Phosphoproteomic analysis of *Arabidopsis* revealed histone H2AX phosphorylation is related to DNA damage signaling pathways [2]. Proteomic analysis of lysine acetylation in zebrafish embryos indicated that acetyl lysine might be involved in protein biosynthesis [3]. These results demonstrate that posttranslational modifications such as phosphorylation and acetylation are involved in a variety of biological processes in vivo and play a very important role. However, most of these studies have been conducted in animals and less in plants. Phosphoproteomic analysis of maize seedling leaves suggested that plants might preferentially regulate the level of posttranslational modifications (PTMs) rather than protein abundance for adapting

Christophe Lambing (ed.), *Plant Gametogenesis: Methods and Protocols*, Methods in Molecular Biology, vol. 2484, https://doi.org/10.1007/978-1-0716-2253-7_2,

to de-etiolation [4]. Proteomic analysis of lysine acetylation in strawberry leaves demonstrated that lysine acetylation is involved in multiple cellular metabolisms and cellular processes [5]. In addition, most proteomic studies on plants are focused on vegetative growth tissues and organs. Gamete development, as a key step in the process of plant sexual reproduction, related to plant reproduction, crop breeding, and yield, is a very important research field. In other species, posttranslational modification has been shown to play a role in meiosis and gametogenesis. Acetylation of the transcriptional repressor Ume6p is related to the meiotic transient transcriptomic program in yeast [6]. Phosphoproteomic analyses revealed more than 5000 proteins with over 3500 phosphorylation sites in human spermatozoa [7]. Due to the small tissue size and difficulty in obtaining materials for plant gametes, there are still few studies on this subject.

Here, we introduce an experimental method of plant tissue-specific phosphorylation and acetylation proteomics by taking rice meiotic anthers as an example, to advance related research.

2 Materials

2.1 Meiotic Stage Anther Collection

1. Rice plants (*Oryza sativa* subsp. indica cv. 9311).
2. Plastic planting bins.
3. Stereoscopic microscope.
4. Glass slides.
5. Dissecting needles.
6. Liquid nitrogen.
7. 2 ml centrifuge tubes.

2.2 Total Protein Extraction

1. Protein extraction solution: 6 M urea, 2 M thiourea, and 100 mM NH_4HCO_3, dissolve 36.024 g urea, 15.224 g thiourea, 0.7906 g NH_4HCO_3 in 100 ml of distilled water. Store at 4 °C.
2. Pestles and mortars.
3. Ultrasonic cell crusher.
4. Ultracentrifuge.
5. 2 ml centrifuge tubes.

2.3 Trypsin Digestion

1. 5 mM DTT: dissolve 0.0077 g DTT in 10 ml of distilled water. Store at 4 °C.
2. 15 mM iodoacetamide: dissolve 0.0277 g DTT in 10 ml of distilled water. Store at 4 °C.

3. Lys-C: dissolve 15 μl powder in 500 μl trypsin resuspension dilution buffer provided by the supplier. Store at −20 °C.
4. Modified Trypsin: dissolve 100 μg powder in 1 ml trypsin resuspension dilution buffer provided by the supplier. Store at −20 °C.

2.4 Phosphopeptide Enrichment by TiO_2 Microcolumn

1. TiO_2 loading buffer:1 M glycolic acid in 80% acetonitrile (ACN), 1% trifluoroacetic acid (TFA).
2. TiO_2 microcolumn: 5 mg TiO_2 beads in 200 ml of 100% acetonitrile. Stamp out a small plug of C_8 material from a 3 M Empore C_8 extraction disk using an HPLC syringe needle, and place this plug in the constricted end of a GELoader tip; Pack the TiO_2 beads on top of the disk in the GELoader tip by loading an aliquot of this TiO_2 suspension (depending on the size of the column) into the GELoader tip and applying gentle air pressure using a plastic syringe (*see* **Note 1**) [8].
3. Washing buffer: 80% ACN, 1% TFA.
4. Elution buffer 1: 2 M $NH_3{\cdot}H_2O$.
5. Elution buffer 2: 1 M $NH_3{\cdot}H_2O$ in 40% ACN.

2.5 Affinity Enrichment of Lysine-Acetylated Peptides

1. Agarose conjugated acetyl lysine antibody.
2. 3-(N-morpholino) propanesulfonic acid (MOPS) buffer: 50 mM MOPS, 50 mM NaCl, 10 mM NaH_2PO_4 H_2O, pH 7.2. Dissolve 0.5232 g MOPS, 0.146 g NaCl, 0.069 g $NaH_2PO_4{\cdot}H_2O$ in 50 ml of distilled water, adjust pH to 7.2 with NaOH.
3. 0.1% or 1% TFA: dissolve 0.1 ml or 1 ml TFA in 100 ml of distilled water. Store at room temperature.

2.6 Desalting and Nano UHPLC-MS/MS Analysis

1. Poros R3 microcolumn: Suspend 5 mg Poros Oligo R3 reversed phase material in 200 ml of 50% acetonitrile. Prepare a R3 microcolumn by stamping out a small plug of C18 material from a 3 M Empore C18 extraction disk using an HPLC syringe needle and placing this plug in the constricted end of a GELoader tip. Use R3 microcolumns of 3–6 mm (0.4–0.8 mg) depending on the amount of material (*see* **Note 2**) [9].
2. Solvent A: 0.1% formic acid (FA) in H_2O.
3. Solvent B: 0.1% FA in 95% ACN.

2.7 Database Search and Data Analysis

1. PROTEOME DISCOVERER software (Version 1.4).
2. Rice (Oryza sativa Japonica) protein sequence database TIGR.
3. MASCOT (Version 2.3.02, http://www.matrixscience.com/).
4. PPDB database (http://ppdb.tc.cornell.edu/).

5. MapMan (http://mapman.mpimpgolm.mpg.de/general/ora/ora.shtml).
6. KEGG (http://www.kegg.jp/kegg/pathway.html).
7. HMMER 2.11.1 and Pfam 30.0 HMM databases.
8. Online DAVID bioinformatics resources.
9. Web tool REVIGO.
10. STRING database (version 10.5).
11. 11.Cytoscape (version 3.5.1).
12. Ice-Logo motif builder.
13. AgriGO database.
14. NetSurfP (version 1.1).

3 Methods

3.1 Meiotic Stage Anther Collection

1. Rice plants (*Oryza sativa* subsp. indica cv. 9311) are grown in plastic planting bins in a controlled environment chamber (12 h light–12 h dark, 28 °C, and 70% relative humidity).
2. Collect florets with 3–5 mm length which contain stages 5–8 anthers determined by semi thin section ranged from premeiotic pollen mother cells to just after the release of microspores from the tetrad (*see* **Note 3**).
3. Isolate anthers on a glass slide with dissecting needles under a stereoscopic microscope and transfer them into a 2 ml centrifuge tube.
4. Immediately freeze the anthers in liquid nitrogen and store them at −80 °C until further use.

3.2 Total Protein Extraction

1. Grind rice anthers to a fine powder in liquid nitrogen with a pestle and mortar, and suspend it with protein extraction solution (600 μl protein extraction solution per 100 μg anthers powder).
2. Sonicate the suspension for 35 min (2 s sonication with 5 s intervals, keep the sample in an ice bath environment).
3. Centrifuge at 20,000 × *g* for 20 min.
4. Collect the supernatant.
5. Determine the protein concentration using the Bradford assay (*see* **Note 4**).
6. Add the extraction buffer to adjust the protein concentration to 1 mg/ml.

3.3 Trypsin Digestion

1. Reduct the proteins with 5 mM DTT at 37 °C for 1 h.
2. Alkylate the proteins with 15 mM iodoacetamide in the dark at room temperature for 1 h (*see* **Note 5**).
3. Quench the excess of iodoacetamide by adding 5 mM DTT.
4. Digest the proteins with Lys-C digestion (1 μg endoproteinase Lys-C for 100 μg protein) for 4 h at room temperature.
5. Dilute the protein solution by adding six volumes of distilled water to make sure the urea concentration in the solution is below 2 M.
6. Digest the proteins with a solution of trypsin digestion (1 μg trypsin for 50 μg protein) overnight at 37 °C.
7. Vacuum freeze-drying the sample.

3.4 Phosphopeptide Enrichment by TiO_2 Microcolumn

Use a peptide solution from 400 μg protein for phosphoproteomic analysis.

1. Resuspend the vacuum dried digest product with TiO_2 loading buffer and apply onto the TiO_2 microcolumn [8].
2. Wash four times with 20 μl of TiO_2 loading buffer and twice with 20 μl washing buffer by applying air pressure.
3. Elute the bound peptides twice with 20 μl of elution buffer 1, and with 2 μl of elution buffer 2 by applying air pressure.
4. Collect the eluate.
5. Vacuum freeze-drying the sample.

3.5 Affinity Enrichment of Lysine-Acetylated Peptides

Use a peptide solution from 400 μg protein for acetylomic analysis.

1. Dissolve the tryptic peptide mixture in 100 μl MOPS buffer.
2. Take 30 μl anti-acetyl lysine agarose beads, centrifuge at 3000 × *g* for 2 min, discard the supernatant.
3. Add 100 μl MOPS buffer to wash the beads, centrifuge at 3000 × *g* for 2 min, discard the supernatant.
4. Repeat **step 3** twice.
5. Incubate the mixture with prewashed anti-acetyl lysine agarose beads at 4 °C for more than 12 h with constant slow end-over-end rotation [9].
6. Gently wash the beads more than three times with MOPS buffer and twice with double-distilled water, every time centrifuge at 5000 rpm for 2 min.
7. Elute the bound peptides from the beads with 30 μl 0.1% TFA.
8. Collect the eluate.

3.6 Desalting

1. Add 20 μl 5% FA to equilibrate homemade Poros R3 microcolumn.
2. Repeat **step 1** again.
3. Load the eluates onto a preequilibrated homemade Poros R3 microcolumn [10].
4. Wash the R3 resin twice with 15 μl 5% FA.
5. Elute the bound peptides with 15 μl of 30% ACN follow by 15 μl of 60% ACN.

3.7 Nano UHPLC-MS/MS Analysis

1. Use an LTQ-Orbitrap Elite coupled with an Easy nLC 1000 to performed LC-MS/MS analysis (*see* **Note 6**).
2. Separate peptides with a three-step elution: 2–35% of solvent B for 3.3 h, 35–90% of solvent B for 10 min, 90% of solvent B for 5 min, 90–2% of solvent B for 2 min, and 2% of solvent B for 13 min.
3. The peptide ions are detected with the Orbitrap mass spectrometer and up to 15 of the most intense peptide ions (>5000 counts) are selected and fragmented by MS/MS using multistage activation (MSA) in the linear ion trap [11].

3.8 Database Search and Data Analysis

1. Process the raw data using PROTEOME DISCOVERER software (Version 1.4).
2. Conduct database search using an in-house Mascot server ((Version 2.3.02, Matrix Science, London, UK, http://www.matrixscience.com/) against the TIGR Rice (*O. sativa japonica*) protein sequence database (FASTA data is downloaded from ftp://ftp.plantbiology.msu.edu/pub/data/Eukaryotic_projects/O_Sativa/annotation_dbs/pseudomolecules/version 7.0/all.dir/all.pep; 2010; 66 495 sequences).
3. Specify parameters in the protein database searches as below: only tryptic peptides with up to two missed cleavage sites are allowed; 10 p.p.m. (parts per million) mass tolerance for MS and 0.6 Da for MS/MS fragment ions; carbamidomethylcysteine as a fixed modification; and protein N-acetylation, oxidized methionine, and phospho_STY (serine, threonine, and tyrosine) are permitted as variable modifications [12].
4. Identify peptides by using the following requirements: the expectation value (p) is lower than 0.05 and the peptides are ranked as No. 1 by the Mascot search engine. Decoy database searches in Mascot reveal a false positive rate lower than 1% at peptide level.
5. Identified peptides are further validated with Target Decoy PSM validator and phosphorylation sites are evaluated with phosphoRS3.0.

6. For the remaining lysine acetylated peptides, the acetylation site assignments are verified by running custom code.

3.9 Bioinformatic Analysis

1. Retrieve the best matched Arabidopsis homologs of identified rice proteins through PPDB (http://ppdb.tc.cornell.edu/).
2. Facilitate functional classification of identified rice proteins by using the best matched Arabidopsis homolog accessions through online MapMan analysis (http://mapman.mpimpgolm.mpg.de/general/ora/ora.shtml) (*see* **Note 7**).
3. Perform pathway analysis of identified rice proteins with the online software KEGG (http://www.kegg.jp/kegg/pathway.html) using the best matched Arabidopsis homolog accessions [13].
4. Perform protein domain enrichment on the identified rice proteins using the HMMER 2.11.1 and Pfam 30.0 HMM databases.
5. Perform domain enrichment using hypergeometric testing as defined in Cobbold et al. [14].
6. use the online DAVID bioinformatic resources [14] and the software Mapman v.3.6.0 to conduct functional categorization of identified rice proteins using the best matched *Arabidopsis* homolog accessions. Nonredundant and highly significant GO terms are visualized by the available web tool REVIGO [15].
7. The protein–protein interaction network is elucidated based on the STRING database (version 10.5) with a medium-confidence cutoff (0.4) and visualized with Cytoscape (version 3.5.1) [16].
8. KAC motif analysis is conducted and displayed using the ice-Logo motif builder with a *p*-value of 0.01. Set all protein sequences derived from O. sativa japonica as the analysis background [17].
9. Protein secondary structures surrounding the acetylated lysine are determined using Net-SurfP (version.1.1) [18].

4 Notes

1. For simpler samples use TiO_2 microcolumns of 3 mm (0.4 mg of TiO_2 beads). Use longer columns for more complex samples. The capacity of TiO_2 toward phosphorylated peptides is extremely high, so even a small column will retain a substantial amount of phosphorylated peptide.
2. To minimize sample loss, the length of the column selected depends on the amount of sample (approximately 1 μl bed volume of R3 material is used per 10 μg of sample).

3. Fix the rice anthers at different developmental stages in FAA (ethanol 50% (v/v), acetic acid 5.0% (v/v) and formaldehyde 3.7% (v/v)) at 4 °C for 12–24 h. Dehydrate in a graded ethanol series (50% twice, 60%, 70%, 80%, 90%, 95% for 30 min for each treatment and 100% twice, each for 1 h). Transfer the anthers to a solution of xylene and embedded in Spurr's epoxy resin using a Powertome XL cut sections (1 μm thin) and heat-fixed to glass slides. Staining with 0.1% (w/v) of toluidine blue O in 0.1 M phosphate buffer, pH 7.0 (To make 1 l of 0.1 M sodium phosphate buffer, pH 7.0: add 280 ml of 0.1 M NaH_2PO_4(-dissolve 11.998 g NaH_2PO_4 in 1000 ml of distilled water) and 720 ml of 0.1 M Na_2HPO_4 (dissolve 14.196 g Na_2HPO_4 in 1000 ml of distilled water). If the pH of the solution is too high, add additional 0.1 M NaH_2PO_4 to pH 7.0), incubate at room temperature for 30–40 min. Rinse the stained sections with pure water three times and air dried. Take photographs of the anther cross sections with a bright-field microscope with a digital camera.
4. Prepare a 100 μg/ml stock solution of BSA in water from BSA 50 mg/ml. Use a 96-well microtiter plate to prepare the reaction mix. Make triplicates for all reactions. Prepare 0, 2.5, 5, 10, 20, 30, 50 μg of BSA standard in corresponding well as described in Table 1.

   Prepare test samples in a microtiter plate well by making several dilutions with water to make a total of 160 μl volume, add 40 μl of Bradford reagent to each microtiter plate well, mix and incubate for 5 min at room temperature. Read absorbance (ABS) at 595 nm on a microtiter plate reader. If the spectrophotometer does not have a software to plot the standard curve, then plot the standard curve in Excel by plotting known BSA concentrations on *X*-axis and ABS_{595} on *Y*-axis. Obtain trendline and use the equation for the trendline and the ABS_{595} of the unknown to resolve the unknown concentration.

Table 1
Protein estimation reaction mix composition

Well	Concentration (μg/ml)	BSA (100 μg/ml) (μl)	Water (μl)	Bradford reagent (μl)
SD1	2.5	4	156	40
SD2	5	8	152	40
SD3	10	16	144	40
SD4	20	32	128	40
SD5	30	64	96	40
SD6	50	80	80	40
SD7	0	0	160	40

5. The concentration of DTT and iodoacetamide are the final concentration in the sample.
6. The state of the mass spectrometer directly affects the accuracy and clarity of the results.
7. If the gene annotation of the material species used in the experiment is relatively complete and the identified proteins are sufficient, it is best to use the identified proteins directly.

Acknowledgments

This work was supported by grants from the National Natural Science Foundation of China (No.32170356) and the National Key Research and Development Program of China (No. 2021YFA1300401).

References

1. Wiechmann S, Saupp E, Schilling D et al (2020) Radiosensitization by kinase inhibition revealed by phosphoproteomic analysis of pancreatic cancer cells. Mol Cell Proteomics 19: 1649–1663
2. Waterworth WM, Wilson M, Wang D et al (2019) Phosphoproteomic analysis reveals plant DNA damage signalling pathways with a functional role for histone H2AX phosphorylation in plant growth under genotoxic stress. Plant J 100:1007–1021
3. Kwon OK, Kim S, Lee S (2016) Global proteomic analysis of lysine acetylation in zebrafish (Danio rerio) embryos. Electrophoresis 37: 3137–3145
4. Gao Z, Shen Z, Chao Q et al (2020) Large-scale Proteomic and Phosphoproteomic Analysis of Maize SeedlingLeaves During De-etiolation. Genom Proteom Bioinf 18: 397–414
5. Fang X, Chen W, Zhao Y et al (2015) Global analysis of lysine acetylation in strawberry leaves. Front Plant Sci 6:739
6. Law M, Mallory M, Dunbrack R et al (2013) Acetylation of the transcriptional repressor Ume6p allows efficient promoter release and timely induction of the meiotic transient transcription program in yeast. Mol Cell Biol 34: 631–642
7. Urizar-Arenaza I, Osinalde N, Akimov V et al (2019) Phosphoproteomic and functional analyses reveal sperm-specific protein changes downstream of kappa opioid receptor in human spermatozoa. Mol Cell Proteomics 18: mcp.RA118.001133
8. Thingholm TE, Jorgensen TJ, Jensen ON et al (2006) Highly selective enrichment of phosphorylated peptides using titanium dioxide. Nat Protoc 1:1929–1935
9. Wu Q, Xu W, Cao L et al (2013) SAHA treatment reveals the link between histone lysine acetylation and proteome in nonsmall cell lung cancer A549 cells. J Proteome Res 12: 4064–4073
10. Gobom J, Nordhoff E, Mirgorodskaya E et al (1999) Sample purification and preparation technique based on nano-scale reversed-phase columns for the sensitive analysis of complex peptide mixtures by matrix-assisted laser desorption/ionization mass spectrometry. J Mass Spectr 34:105–116
11. Palumbo AM, Reid GE (2008) Evaluation of gas-phase rearrangement and competing fragmentation reactions on protein phosphorylation site assignment using collision induced dissociation-MS/MS and MS3. Anal Chem 80:9735–9747
12. Ye J, Zhang X, Young C et al (2010) Optimized IMAC−IMAC protocol for phosphopeptide recovery from complex biological samples. J Proteome Res 9:3561–3573
13. Ye J, Zhang Z, Long H et al (2015) Proteomic and phosphoproteomic analyses reveal extensive phosphorylation of regulatory proteins in developing rice anthers. Plant J 84:527–544
14. Cobbold SA, Santos JM, Ochoa A et al (2016) Proteome-wide analysis reveals widespread lysine acetylation of major protein complexes in the malaria parasite. Sci Rep 6:19722

15. Supek F, Bošnjak M, Škunca N et al (2011) REVIGO summarizes and visualizes long lists of gene ontology terms. PLoS One 6:e21800
16. Shannon P, Markiel A, Ozier O et al (2003) Cytoscape: a software environment for integrated models of biomolecular interaction networks. Genome Res 13:2498–2504
17. Colaert N, Helsens K, Martens L et al (2009) Improved visualization of protein consensus sequences by iceLogo. Nat Methods 6:786–787
18. Li X, Ye J, Ma H et al (2018) Proteomic analysis of lysine acetylation provides strong evidence for involvement of acetylated proteins in plant meiosis and tapetum function. Plant J 93:142–154

Chapter 3

Identifying small RNAs and Analyzing Their Association with Gene Expression Using Isolated Arabidopsis Male Meiocytes

Jiyue Huang, Chenjiang You, Cong Wang, Yingxiang Wang, and Gregory P. Copenhaver

Abstract

Meiosis is a specialized cell division that generates gametes and is essential for sexual reproduction. Studying meiosis in plants, like the model flowering plant *Arabidopsis thaliana*, contributes to our understanding of the fundamental biology of reproductive biology and has practical implications for improving economically important crop species. In this chapter, we provide a detailed protocol for capillary collection of Arabidopsis male meiocytes followed by total RNA extraction, RNA-Seq, and bioinformatics analysis of small-RNAs (sRNAs) including analysis of sRNA cluster that correlate with genomic features.

Key words Meiocyte, Capillary collection method, RNA extraction, Arabidopsis, Small-RNA Seq, RNA-Seq, Bioinformatics

1 Introduction

Meiosis is a fundamental biological process required for sexual reproduction in most eukaryotes. It produces gametes, with half the genomic complement of somatic cells, which are used during fertilization to complete a life cycle. During meiosis, parental chromosomes pair and physically interact and exchange DNA in a process called crossing over (a type of recombination). Crossovers help ensure that homologous chromosomes segregate equally into daughter cells and they shuffle parental alleles between chromosomes which generates genetic diversity. Meiosis is thought to have been present in the common ancestor of all eukaryotes

---

Jiyue Huang and Chenjiang You contributed equally with all other contributors.

Supplementary Information The online version of this chapter (https://doi.org/10.1007/978-1-0716-2253-7_3) contains supplementary material, which is available to authorized users.

Christophe Lambing (ed.), *Plant Gametogenesis: Methods and Protocols*, Methods in Molecular Biology, vol. 2484, https://doi.org/10.1007/978-1-0716-2253-7_3,

[1]. Therefore, key genes and regulatory mechanisms are shared among eukaryotes, but differences among taxa exist. Meiotic mutations that invoke checkpoint surveillance trigger cell cycle arrest or apoptosis in animals and yeast [2] but are often tolerated in plants, which makes model systems like Arabidopsis, rice, and maize ideal for investigating gene function and regulation.

Small RNAs (sRNAs) are noncoding RNAs, typically 20–30 nucleotides (nt) long, that primarily function in posttranslational silencing and epigenetic regulation mechanisms in multiple biological processes. Most sRNA functions were initially defined in somatic cells, but more recently their meiotic roles have been described. In the germ line cells of most animals, a subset of sRNAs called piwi-interacting RNAs (piRNAs) function in RNA cleavage, heterochromatin assembly and DNA methylation [3], and are essential for male meiosis in mouse [4]. There are also 21-nt phased small interfering RNAs (phasiRNAs), which are most abundant in meiotic prophase I and cleave target meiotic mRNAs required for meiotic progression and fertility, in plants [5, 6]. A recent preprint study reveals tapetal nurse cell-derived siRNAs (niRNAs) are capable of reprogramming DNA methylation in Arabidopsis meiocytes [7]. Our previous work demonstrated that meiocyte sRNAs have a distinctive genomic distribution profile compared to those produced in somatic cells [8, 9]. Moreover, by focusing on a subset of SPO11-dependent sRNAs, we proposed a putative association between sRNAs and meiotic recombination [9].

In this chapter, we present a detailed protocol for isolating Arabidopsis male meiocytes by capillary collection, extracting total RNA for RNA-seq and follow-on bioinformatics analysis of small-RNA Seq data including correlation with other genomic features. We anticipate that this method could also be easily adapted for use in other plant species including crops such as soybean [8].

2 Materials

2.1 Arabidopsis Meiocyte Isolation

1. Inflorescence from the primary shoot of *Arabidopsis thaliana* plants grown in growth room conditions at 22 °C, 16 h light and 8 h dark, and 75% humidity.
2. Dissection microscope with at least 20× magnification.
3. Inverted microscope.
4. Homemade capillary collecting system consisting of a capillary needle (*see* **item 5**), an acrylic pipette holder for 1.5 mm glass tubes, and PVC tubing (1 mm ID and 2 mm OD) with a 300 μL filtered pipette tip at one end as a disposable mouthpiece. The system is attached to a three-axis manipulator [10].
5. Capillary needles: produced from capillary glass tubes (1.5 mm outer diameter, 0.86 mm inner diameter, 15 cm length, e.g., borosilicate glass) heated, quickly stretched over a gas burner, and snapped into two pieces. Before installing the capillary needle into the acrylic pipette holder, use a pair of forceps (*see*

item 6) to enlarge the glass needle end to a tip with inner diameter of approximately 40–50 μM.

6. Fine forceps.
7. 200 μL pipette.
8. ZR BashingBead lysis tubes.
9. 1 mL Insulin Syringes.
10. Needles (27 G × 1/2).
11. Plain microscope slides.
12. Double shallow depression microscope slides.
13. Collection buffer: Dulbecco's phosphate-buffered saline (1× DPBS, 8 g/L sodium chloride; 0.2 g/L potassium phosphate, monobasic; 1.15 g/L sodium phosphate, dibasic; and 0.2 g/L potassium chloride) with 1 U/μL Recombinant RNase Inhibitor.
14. TRIzol.

2.2 Total RNA Extraction from Arabidopsis Meiocytes

1. Refrigerated microcentrifuge with a 1.5/2 mL tube rotor.
2. Geno/Grinder.
3. Nanodrop 2000.
4. Vortex.
5. 1 mL, 200 μL and 10 μL pipettes.
6. DNase and RNase free 1.5 mL Eppendorf tube.
7. RNase AWAY™ Surface Decontaminant spray bottle.
8. Chloroform.
9. TRIzol.
10. Dr. GenTLE Precipitation Carrier.
11. 100% ethanol.
12. 70% ethanol.
13. DNase and RNase free H_2O.
14. Ice bucket.

2.3 Bioinformatics Analysis of small-RNA Seq and RNA-Seq Data

1. TruSeq Small RNA Library Prep Kit.
2. TruSeq RNA Library Prep Kit.

3 Methods

3.1 Arabidopsis Meiocyte Isolation

The meiocyte isolation protocol was described in [10, 11] and is presented here with minor modifications. The workflow described in this protocol is summarized in Fig. 1.

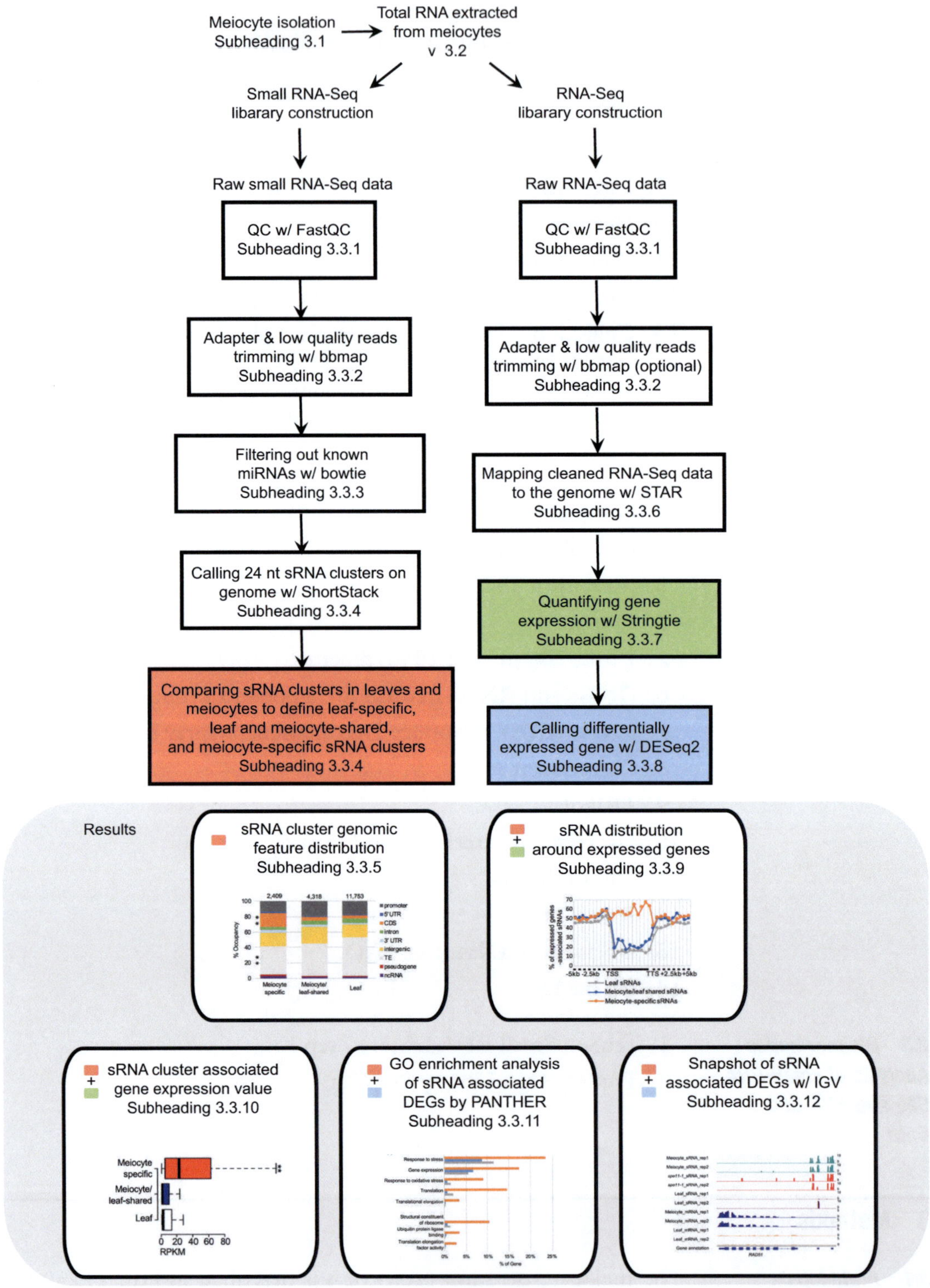

Fig. 1 Workflow for isolating Arabidopsis male meiocytes by capillary collection, extracting total RNA, performing RNA-seq and bioinformatics analyses of small-RNAs (sRNAs). Analyses include association of sRNA clusters with genomic features (using data generated from Subheading 3.3.4), the distribution of sRNA

1. Add one drop of collection buffer to a cavity of an ethanol cleaned double shallow depression slide using a 200 μL pipette and keep it on ice.
2. Add 100 μL TRIzol into a ZR BashingBead lysis tube and keep it on ice.
3. Collect an inflorescence that includes unopened floral buds from a plant using a pair of forceps and place it on a glass microscope slide under a dissection microscope (*see* **Note 1**).
4. Dissect the unopened floral buds (around 0.2–0.4 mm) using a pair of syringes with needles to isolate stage 5–7 anthers [12] (*see* **Note 2**).
5. Using the pair of syringes with needles, quickly transfer the anthers to the slide with collection buffer and continue collecting about 24 stage 5–7 anthers.
6. Physically disrupt the anthers using forceps to squeeze them in the collection buffer to release meiocytes (*see* **Note 3**).
7. Transfer the slide onto the inverted microscope stage. Add 50 μL collection buffer using a 200 μL pipette.
8. Carefully position the manipulator to insert the capillary needle into the buffer drop.
9. Use mouth-controlled air flow at one end of the PVC tubing of the capillary collecting system to gently suck the meiocyte masses into the capillary needle.
10. Collecting meiocyte masses until one-third of the capillary needle is full, then gently expel the contents into a ZR Bashing-Bead lysis tube containing 100 μL TRIzol. Keep the tube on ice.
11. Repeat the procedure (**steps 8–10**) until all meiocyte masses in the collection buffer have been collected.
12. Repeat the procedure (**steps 1–11**) until 200–300 meiocyte masses have been collected. This is enough material to construct one small RNA or RNA library.
13. Once all the meiocyte masses have been collected, add 100 μL TRIzol to the tube, label and date the sample, firmly tighten the cap, flash-freeze the tube in liquid nitrogen, and store at −80 °C for up to 1 month.

Fig. 1 (continued) around genes (associating data generated from Subheadings 3.3.4 and 3.3.7), gene expression levels of loci associated with sRNA clusters (associating data generated from Subheadings 3.3.4 and 3.3.7), GO enrichment of sRNA associated DEGs (associating data generated from Subheadings 3.3.4 and 3.3.8) and snapshot of sRNA associated DEGs (associating data generated from Subheadings 3.3.4 and 3.3.8) by a combination of homemade and publicly available tools. Graphs shown in "Results" are modified from [9]

3.2 Total RNA Extraction from Arabidopsis Meiocytes

1. Thaw the ZR BashingBead lysis tube containing the meiocyte masses on ice.
2. Load the sample into the precooling modules of Geno/Grinder, follow the manufacturer's instructions to grind sample at 1500 rpm for 1 min repeated three times (*see* **Note 4**).
3. Centrifuge the tube at 12,000 rpm for 15 s at room temperature (RT).
4. Add 700 μL TRIzol to the tube and pipette up and down 10 times to mix (with a 1 mL pipette).
5. Add 200 μL chloroform, securely cap the tube.
6. Vortex vigorously for 15 s.
7. Incubate the sample at RT for 2–3 min.
8. Centrifuge the sample for 15 min at 12,000 rpm at 4 °C.
9. Transfer the colorless upper aqueous phase into a new DNase and RNase free 1.5 mL tube.
10. Follow the manufacturer's protocol using Dr. GenTLE Precipitation Carrier to finish preparing the total RNA sample.
11. Check the purity and concentration of sample using a Nanodrop.
12. Store the sample at −80 °C.

3.3 Bioinformatics Analysis of small-RNA Seq and RNA-Seq Data

Prepare small RNA libraries following the protocol described in the TruSeq small RNA library prep reference guide. Prepare RNA libraries following the protocol described in the TruSeq RNA sample preparation v2 guide. Produce raw sequencing data using a high throughput sequencing system such as the Illumina HiSeq 3000 system (*see* **Note 5**). Example datasets prepared in this manner that can be used for the analyses described below are available in the NCBI SRA repository under the project name PRJNA510650 (https://www.ncbi.nlm.nih.gov/bioproject/PRJNA510650).

3.3.1 Quality Check of the Raw Data

Use FastQC (https://www.bioinformatics.babraham.ac.uk/projects/fastqc/) to check the dataset quality (*see* **Note 6**).

```
cd foo/bar/working_directory
 for file in 'ls *.fastq.gz'; do fastqc $file;done
```

Raw sequence data are all placed in folder "working_directory" with a compressed fastq format.

3.3.2 Adapter Trimming and Low-Quality Read Removal

BBMap (version38.82, Bushnell B., sourceforge.net/projects/bbmap/) is used to trim adapters and discard low-quality reads to achieve better mapping efficiency and accuracy.

For the 50 bp single-end small RNA read datasets:

```
bbduk.sh in=meiocyte_sRNA_rawdata_rep1.fastq.gz out=meiocy-
te_sRNA_trimmed_rep1.fastq.gz literal=TGGAATTCTCGGGTGCCAAGG
ktrim=r k=21 qtrim=rl trimq=10 mink=9 hdist=1
  bbduk.sh in=meiocyte_sRNA_rawdata_rep2.fastq.gz out=meiocy-
te_sRNA_trimmed_rep2.fastq.gz literal=TGGAATTCTCGGGTGCCAAGG
ktrim=r k=21 qtrim=rl trimq=10 mink=9 hdist=1
  bbduk.sh in=leaf_sRNA_rawdata_rep1.fastq.gz out=leaf_sRNA_-
trimmed_rep1.fastq.gz literal=TGGAATTCTCGGGTGCCAAGG ktrim=r
k=21 qtrim=rl trimq=10 mink=9 hdist=1
  bbduk.sh in=leaf_sRNA_rawdata_rep2.fastq.gz out=leaf_sRNA_-
trimmed_rep2.fastq.gz literal=TGGAATTCTCGGGTGCCAAGG ktrim=r
k=21 qtrim=rl trimq=10 mink=9 hdist=1
```

Trimming off 3′ adapters (using the option "literal=TGGAATTCTCGGGTGCCAAGG") of the sRNA reads is essential for further analysis. Using the setting "qtrim=rl trimq=10" enables quality trimming of low-quality bases of each read from both ends.

For the 150 bp paired-end RNA-Seq read datasets:

```
bbduk.sh in1=meiocyte_mRNA_rawdata_R1_rep1.fastq.gz in2=meio-
cyte_mRNA_rawdata_R2_rep1.fastq.gz out1=meiocyte_mRNA_trim-
med_R1_rep1.fastq.gz out2=meiocyte_mRNA_trimmed_R2_rep1.
fastq.gz ref=adapters.fa ktrim=r k=23 qtrim=rl trimq=10
mink=11 hdist=1 tpe tbo
  bbduk.sh in1=meiocyte_mRNA_rawdata_R1_rep2.fastq.gz in2=-
meiocyte_mRNA_rawdata_R2_rep2.fastq.gz out1=meiocyte_mRNA_-
trimmed_R1_rep2.fastq.gz out2=meiocyte_mRNA_trimmed_R2_rep2.
fastq.gz ref=adapters.fa ktrim=r k=23 qtrim=rl trimq=10
mink=11 hdist=1 tpe tbo
  bbduk.sh in1=leaf_mRNA_rawdata_R1_rep1.fastq.gz in2=leaf_mR-
NA_rawdata_R2_rep1.fastq.gz out1=leaf_mRNA_trimmed_R1_rep1.
fastq.gz out2=leaf_mRNA_trimmed_R2_rep1.fastq.gz ref=adap-
ters.fa ktrim=r k=23 qtrim=rl trimq=10 mink=11 hdist=1 tpe tbo
  bbduk.sh in1=leaf_mRNA_rawdata_R1_rep2.fastq.gz in2=leaf_mR-
NA_rawdata_R2_rep2.fastq.gz out1=leaf_mRNA_trimmed_R1_rep2.
fastq.gz out2=leaf_mRNA_trimmed_R2_rep2.fastq.gz ref=adap-
ters.fa ktrim=r k=23 qtrim=rl trimq=10 mink=11 hdist=1 tpe tbo
```

The "ref=adapters.fa" option provides adapter list used for RNA library construction. It is a multisequence fasta file which you can edit to specify the adapter used for your library preparation. "tpe" trims adapters based on where paired reads overlap. "tbo" trims both paired reads to the minimum length of either.

Both "tpe" and "tbo" are important parameters for paired-end read trimming (*see* **Note 7**).

3.3.3 Filtering Out Annotated microRNAs

1. Filter out known mature microRNAs (miRNAs) annotated in Arabidopsis from the trimmed sRNA dataset above to focus the analysis on non-miRNAs (*see* **Note 8**). To facilitate this filtering step, mature miRNA sequences from all species can be downloaded from miRbase (v22.1; http://www.mirbase.org). Arabidopsis mature miRNA sequences are extracted using the "extract_miRNAs.pl" (https://github.com/rajewsky-lab/mirdeep2/tree/master/src/) script from mirdeep2 [13]. This script also converts all "U" bases to "T" in the output.

```
extract_miRNAs.pl mature.fa ath > mature_U2T_ath.fa
```

2. Build an index from the converted mature Arabidopsis miRNA sequences using bowtie [14] to filter out known Arabidopsis miRNAs.

```
bowtie-build mature_U2T_ath.fa mature_U2T_ath
```

3. Retrieve the miRNA-free read sets and convert to a compressed format:

```
gzip -dc meiocyte_sRNA_trimmed_rep1.fastq.gz | bowtie -q -v
0 -p 12 --un meiocyte_miRNA_discarded_rep1.fastq --best ma-
ture_U2T_ath -; gzip -v meiocyte_miRNA_discarded_rep1.fastq
gzip -dc meiocyte_sRNA_trimmed_rep2.fastq.gz | bowtie -q -v
0 -p 12 --un meiocyte_miRNA_discarded_rep2.fastq --best ma-
ture_U2T_ath -; gzip -v meiocyte_miRNA_discarded_rep2.fastq
gzip -dc leaf_sRNA_trimmed_rep1.fastq.gz | bowtie -q -v 0 -p
12 --un leaf _miRNA_discarded_rep1.fastq --best mature_U2T_ath
-; gzip -v leaf_miRNA_discarded_rep1.fastq
gzip -dc leaf_sRNA_trimmed_rep2.fastq.gz | bowtie -q -v 0 -p
12 --un leaf _miRNA_discarded_rep2.fastq --best mature_U2T_ath
-; gzip -v leaf_miRNA_discarded_rep2.fastq
```

3.3.4 Generating sRNA Clusters

24-nt small RNAs, generated from dsRNA duplexes, processed by DCL3 and associated with AGO4, are one of the major functional small RNA groups in Arabidopsis. They map to clusters on genome, and mainly function in transcriptional gene silencing of viral DNA, transgenes, and transposable elements [15].

1. Extract 24-nt length reads from datasets by reformat.sh from BBMap:

```
reformat.sh in=meiocyte_miRNA_discarded.fastq_rep1.gz out=-
meiocyte_24nt_miRNA_discarded_rep1.fastq.gz minlength=24 max-
length=24
reformat.sh in=meiocyte_miRNA_discarded.fastq_rep2.gz out=-
meiocyte_24nt_miRNA_discarded_rep2.fastq.gz minlength=24 max-
length=24
reformat.sh in=leaf_miRNA_discarded.fastq_rep1.gz out=-
leaf_24nt_miRNA_discarded_rep1.fastq.gz minlength=24 max-
length=24
reformat.sh in=leaf_miRNA_discarded.fastq_rep2.gz out=-
leaf_24nt_miRNA_discarded_rep2.fastq.gz minlength=24 max-
length=24
```

2. Align 24-nt small RNA subsets on the Arabidopsis genome using ShortStack (version 3.8.5) [16]. The Arabidopsis genome sequence can be downloaded from EnsemblPlants (ftp://ftp.ensemblgenomes.org/pub/plants/release-49/fasta/arabidopsis_thaliana/dna//Arabidopsis_thaliana.TAIR10.dna.toplevel.fa.gz).

```
ShortStack --outdir meiocyte_24nt_rep1 --nohp --readfile meio-
cyte_24nt_miRNA_discarded_rep1.fastq.gz --bowtie_cores 6 --
mismatches 0 --mincov 0.5rpm --genomefile Arabidopsis_thali-
ana.TAIR10.dna.toplevel.fa
ShortStack --outdir meiocyte_24nt_rep2 --nohp --readfile meio-
cyte_24nt_miRNA_discarded_rep2.fastq.gz --bowtie_cores 6 --
mismatches 0 --mincov 0.5rpm --genomefile Arabidopsis_thali-
ana.TAIR10.dna.toplevel.fa
ShortStack --outdir leaf_24nt_rep1 --nohp --readfile
leaf_24nt_miRNA_discarded_rep1.fastq.gz --bowtie_cores 6 --
mismatches 0 --mincov 0.5rpm --genomefile Arabidopsis_thali-
ana.TAIR10.dna.toplevel.fa
ShortStack --outdir leaf_24nt_rep2 --nohp --readfile
leaf_24nt_miRNA_discarded_rep2.fastq.gz --bowtie_cores 6 --
mismatches 0 --mincov 0.5rpm --genomefile Arabidopsis_thali-
ana.TAIR10.dna.toplevel.fa
```

3. Filter out sRNA clusters with sRNA abundance less than 3 RPM (read per million):

```
awk -F "\t" '$5>=3' meiocyte_24nt_rep1/Results.txt > meiocy-
te_Results_3rpm_rep1.txt
awk -F "\t" '$5>=3' meiocyte_24nt_rep2/Results.txt > meiocy-
te_Results_3rpm_rep2.txt
```

```
awk -F "\t" '$5>=3' leaf_24nt_rep1/Results.txt > leaf_Re-
sults_3rpm_rep1.txt
awk -F "\t" '$5>=3' leaf_24nt_rep2/Results.txt > leaf_Re-
sults_3rpm_rep2.txt
```

4. Compare sRNA clusters between two biological replicates to define 24-nt sRNA clusters that overlap by more than 60% in two replicates using 1_merge_cluster.pl (Data 1) (*see* **Note 9**).

```
perl 1_merge_cluster.pl meiocyte_Results_3rpm_rep1.txt meio-
cyte_Results_3rpm_rep2.txt 0.6 > meiocyte_Results_3rpm_-
rep12_60p.txt 2> err
perl 1_merge_cluster.pl leaf_Results_3rpm_rep1.txt leaf_Re-
sults_3rpm_rep2.txt 0.6 > leaf_Results_3rpm_rep12_60p.txt 2>
err
```

5. Compare sRNA clusters from meiocyte and leaf samples to retrieve meiocyte-specific, leaf-specific sRNA, and meiocyte and leaf-shared clusters and output in BED format using 2_reduce_cluster.pl (Data 2) or 3_complimentation.pl (Data 3).
   (a) For meiocyte-specific sRNA clusters:

```
perl 2_reduce_cluster.pl meiocyte_Results_3rpm_rep12_60p.txt
leaf_Results_3rpm_rep12_60p.txt 0.6 > meiocyte_specific_clus-
ter.bed 2> err
```

   (b) For leaf-specific sRNA clusters:

```
perl 2_reduce_cluster.pl leaf_Results_3rpm_rep12_60p.txt
meiocyte_Results_3rpm_rep12_60p.txt 0.6 > leaf_specific_clus-
ter.bed 2> err
```

   (c) For meiocyte and leaf-shared sRNA clusters:

```
perl 3_complimentation.pl meiocyte_Results_3rpm_rep12_60p.txt
meiocyte_specific_cluster.txt 2 > meiocyte_leaf_shared_clus-
ter.bed 2> err
```

3.3.5 sRNA Cluster Genomic Feature Distribution

Genomic features can be extracted from the GFF (Gene feature Format) annotation file downloaded from TAIR (https://www.arabidopsis.org/download_files/Genes/Araport11_genome_release/Araport11_GFF3_genes_transposons.201606.gff.gz). We then use BLASTn (2.10.1) [17] to help us locate the genomic feature distribution of the sRNA clusters.

1. Use GenomeTools [18] to add intron position information.

```
gt gff3 -sort yes -addintrons yes Araport11_GFF3_genes_tran-
sposons.201606.gff > Araport11_GFF3_genes_transpo-
sons.201606_w_intron.gff
```

2. Generate a 500 bp promoter for all annotated genes using 4_generate_promter.pl (Data 4).

```
awk '$3~/gene/' Araport11_GFF3_genes_transposons.201606_w_in-
tron.gff
 | perl 4_generate_promter.pl - 500 > promoter.gff
```

3. Merge it with the previously generated gff file and sort using bedtools [19].

```
cat Araport11_GFF3_genes_transposons.201606_w_intron.gff pro-
moter.gff | bedtools sort -i - > sorted_Araport11_GFF3_gen-
es_transposons.201606_w_intron_promoter.gff
```

4. Retrieve intergenic region annotation in BED format.

```
bedtools complement -i sorted_Araport11_GFF3_genes_transpo-
sons.201606_w_intron_promoter.gff -g Arabidopsis_thaliana.
TAIR10.dna.toplevel.fa.fai | awk '{print $0"\tintergenic"}' >
intergenic.bed
```

"Arabidopsis_thaliana.TAIR10.dna.toplevel.fa.fai" can be generated using samtools [20] with the commend "samtools faidx Arabidopsis_thaliana.TAIR10.dna.toplevel.fa".

5. Extract the sequences of all genomic features of interest based on the genomic reference sequence.

   (a) For intergenic regions:

```
bedtools getfasta -fi Arabidopsis_thaliana.TAIR10.dna.tople-
vel.fa -bed intergenic.bed -name+ -fo intergenic.fa
```

   (b) For other genomic features:

```
bedtools getfasta -fi Arabidopsis_thaliana.TAIR10.dna.tople-
vel.fa -bed sorted_Araport11_GFF3_genes_transposons.201606_-
w_intron_promoter.gff -name+ -fo other_genomic_features.fa
```

   (c) Merge to a fasta sequence file of all genomic features of interest:

```
cat intergenic.fa other_genomic_features.fa > whole_genomic_-
features.fa
```

6. Convert each category of sRNA clusters into fasta sequence files.

```
bedtools getfasta -fi Arabidopsis_thaliana.TAIR10.dna.tople-
vel.fa -bed meiocyte_specific_cluster.bed -name+ -fo meiocy-
te_specific_cluster.fa
bedtools getfasta -fi Arabidopsis_thaliana.TAIR10.dna.tople-
vel.fa -bed leaf_specific_cluster.bed -name+ -fo leaf_speci-
fic_cluster.fa
bedtools getfasta -fi Arabidopsis_thaliana.TAIR10.dna.tople-
vel.fa -bed meiocyte_leaf_shared_cluster.bed -name+ -fo meio-
cyte_leaf_shared_cluster.fa
```

7. Determine genomic features that correlate with sRNA clusters using blastn.

```
blastn -query meiocyte_specific_cluster.fa -subject whole_
genomic_features.fa -out meiocyte_specific_cluster_blastn.out
-evalue 0.01 -outfmt "6 qseqid sseqid pident qlen length
evalue"
blastn -query leaf_specific_cluster.fa -subject whole_ geno-
mic_features.fa -out leaf_specific_cluster_blastn.out -evalue
0.01 -outfmt "6 qseqid sseqid pident qlen length evalue"
blastn -query meiocyte_leaf_shared_cluster.fa -subject whole_
genomic_features.fa -out meiocyte_leaf_shared_cluster_blastn.
out -evalue 0.01 -outfmt "6 qseqid sseqid pident qlen length
evalue"
```

8. Output sRNA cluster associated genomic feature distribution. A feature that can be mapped without mismatch and covers at least 60% of the query sRNA cluster sequence qualifies a sRNA cluster target. The blastn output files can be imported into Excel for further analysis and plotting.

```
awk -F "\t" '$3==100 && $5/$4 >= 0.6 {print $1"\t"substr($2,1,
index($2,"::")-1)}' meiocyte_specific_cluster_blastn.out |
sort | uniq | cut -f 2 | sort | uniq -c | awk '{print $2"\t"$1}'
awk -F "\t" '$3==100 && $5/$4 >= 0.6 {print $1"\t"substr($2,1,
index($2,"::")-1)}' leaf_specific_cluster_blastn.out | sort |
uniq | cut -f 2 | sort | uniq -c | awk '{print $2"\t"$1}'
awk -F "\t" '$3==100 && $5/$4 >= 0.6 {print $1"\t"substr($2,1,
index($2,"::")-1)}' meiocyte_leaf_shared_cluster_blastn.out |
sort | uniq | cut -f 2 | sort | uniq -c | awk '{print $2"\t"$1}'
```

3.3.6 Mapping Cleaned RNA-Seq Data to the Arabidopsis Genome Using STAR

STAR (Spliced Transcripts Alignment to a Reference) [21] is used to align cleaned RNA-Seq data to the Arabidopsis genome (*see* **Note 10**).

1. Generate genome index for alignment. Arabidopsis annotation file (gtf, Gene Transfer Format file) can be downloaded from TAIR: https://www.arabidopsis.org/download_files/Genes/Araport11_genome_release/Araport11_GFF3_genes_transposons.201606.gtf.gz).

```
star --runThreadN 6 --genomeSAindexNbases 12 --runMode geno-
meGenerate --genomeDir Araport11_STARindex --genomeFastaFiles
Arabidopsis_thaliana.TAIR10.dna.toplevel.fa --sjdbGTFfile
Araport11_GFF3_genes_transposons.201606.gtf
```

   Option "--genomeSAindexNbases" is calculated as log2 (GenomeLength)/2–1. For an approximately genome size of 125 Mega base in Arabidopsis, we use "--genomeSAindexN-bases 12".

2. Align cleaned RNA-Seq paired-end reads with STAR.

```
star --runThreadN 6 --runMode alignReads --genomeDir Ara-
port11_STARindex/ --sjdbGTFfile Araport11_GFF3_genes_transpo-
sons.201606.gtf --readFilesIn meiocyte_mRNA_trimmed_R1_rep1.
fastq.gz meiocyte_mRNA_trimmed_R2_rep1.fastq.gz --readFiles-
Command zcat --outSAMtype BAM SortedByCoordinate --outFileNa-
mePrefix ./meiocyte_mRNA_rep1_
star --runThreadN 6 --runMode alignReads --genomeDir Ara-
port11_STARindex/ --sjdbGTFfile Araport11_GFF3_genes_transpo-
sons.201606.gtf --readFilesIn meiocyte_mRNA_trimmed_R1_rep2.
fastq.gz meiocyte_mRNA_trimmed_R2_rep2.fastq.gz --readFiles-
Command zcat --outSAMtype BAM SortedByCoordinate --outFileNa-
mePrefix ./meiocyte_mRNA_rep2_
star --runThreadN 6 --runMode alignReads --genomeDir Ara-
port11_STARindex/ --sjdbGTFfile Araport11_GFF3_genes_transpo-
sons.201606.gtf --readFilesIn leaf_mRNA_trimmed_R1_rep1.
fastq.gz leaf_mRNA_trimmed_R2_rep1.fastq.gz --readFilesCom-
mand zcat --outSAMtype BAM SortedByCoordinate --outFileName-
Prefix ./leaf_mRNA_rep1_
star --runThreadN 6 --runMode alignReads --genomeDir Ara-
port11_STARindex/ --sjdbGTFfile Araport11_GFF3_genes_transpo-
sons.201606.gtf --readFilesIn leaf_mRNA_trimmed_R1_rep2.
fastq.gz leaf_mRNA_trimmed_R2_rep2.fastq.gz --readFilesCom-
mand zcat --outSAMtype BAM SortedByCoordinate --outFileName-
Prefix ./leaf_mRNA_rep2_
```

3.3.7 Transcript Assembly and Gene Abundance Calling by Stringtie

Stringtie [22] uses the bam files generated by STAR to assemble transcripts and call raw and normalized read count per transcript/gene in one step.

```
stringtie meiocyte_mRNA_rep1_Aligned.sortedByCoord.out.bam -o
meiocyte_mRNA_rep1.gtf -p 6 -G Araport11_GFF3_genes_transpo-
sons.201606.gff -e -A meiocyte_mRNA_rep1_gene_abund.tab
stringtie meiocyte_mRNA_rep2_Aligned.sortedByCoord.out.bam -o
meiocyte_mRNA_rep2.gtf -p 6 -G Araport11_GFF3_genes_transpo-
sons.201606.gff -e -A meiocyte_mRNA_rep2_gene_abund.tab
stringtie leaf_mRNA_rep1_Aligned.sortedByCoord.out.bam -o
leaf_mRNA_rep1.gtf -p 6 -G Araport11_GFF3_genes_transpo-
sons.201606.gff -e -A leaf_mRNA_rep1_gene_abund.tab
stringtie leaf_mRNA_rep2_Aligned.sortedByCoord.out.bam -o
leaf_mRNA_rep2.gtf -p 6 -G Araport11_GFF3_genes_transpo-
sons.201606.gff -e -A leaf_mRNA_rep1_gene_abund.tab
```

Option "-e" omits novel transcript discovery and only estimates reads aligned to reference transcripts provided by "-G". "-A" outputs a tab delimited gene abundance file, in which the last two columns display the normalized read count per gene (FPKM and TPM). An expressed gene is defined as having ≥ 1 TPM in both biological replicates. Therefore, the expressed and nonexpressed gene list (BED format files below) could be generated from the tab files of meiocyte and leaf, respectively.

```
meiocyte_expressed_gene.bed
meiocyte_nonexpressed_gene.bed
leaf_expressed_gene.bed
leaf_nonexpressed_gene.bed
```

3.3.8 Calling Differential Expressed Gene by DESeq2

Differentially expressed genes (DEGs) can be called using DESeq2 [23] following the protocol described in "using StringTie with DESeq2" step 1 to 8 on the StringTie website (http://ccb.jhu.edu/software/stringtie/index.shtml?t=manual). Two differentially expressed gene tables are generated by DESeq2 with a standard of log2 fold change ≥ 1 or ≤ -1, q value ≤ 0.05.

```
res05up_meiocyte_vs_leaf.csv
res05down_meiocyte_vs_leaf.csv
```

3.3.9 Analyzing sRNA Distribution Around Expressed Genes

The distribution of sRNA clusters around genes can be analyzed using the "5_aroundgene.pl" script (Data 5). Gene loci plus 5000 bp up- and downstream are divided into 30 equal bins, and sRNA clusters are mapped to each bin. A target bin should cover at least 60% region of the query sRNA cluster. A frequency table showing sRNA cluster distribution among all the bins is generated using output of the script and can be imported into Excel for plotting.

For meiocyte-specific sRNA clusters:

```
perl 5_aroundgene.pl meiocyte_expressed_gene.bed meiocyte_-specific_cluster.bed Arabidopsis_thaliana.TAIR10.dna.tople-vel.fa.fai 5000 10 2> err | cut -f 13-14 | sort | awk '{sum+=$2} {a[$1]+=$2} END {for (i in a) print i, a[i]/sum}' | sort -n | awk '{print $1"\t"$2}'
```

For leaf-specific sRNA clusters:

```
perl 5_aroundgene.pl leaf_expressed_gene.bed leaf_speci-fic_cluster.bed Arabidopsis_thaliana.TAIR10.dna.toplevel.fa.fai 5000 10 2> err | cut -f 13-14 | sort | awk '{sum+=$2} {a[$1]+=$2} END {for (i in a) print i, a[i]/sum}' | sort -n | awk '{print $1"\t"$2}'
```

For meiocyte and leaf-shared sRNA clusters:

```
perl 5_aroundgene.pl meiocyte_expressed_gene.bed meiocyte_-leaf_shared_cluster.bed Arabidopsis_thaliana.TAIR10.dna.to-plevel.fa.fai 5000 10 2> err | cut -f 13-14 | sort | awk '{sum+=$2} {a[$1]+=$2} END {for (i in a) print i, a[i]/sum}' | sort -n | awk '{print $1"\t"$2}'
```

3.3.10 Analyzing Gene Expression Value of sRNA Associated Genes

1. Obtain sRNA cluster associated gene list:
   (a) For meiocyte-specific sRNA clusters:

```
cat meiocyte_expressed_gene.bed meiocyte_nonexpressed_gene.bed | perl 5_aroundgene.pl - meiocyte_specific_cluster.bed Arabidopsis_thaliana.TAIR10.dna.toplevel.fa.fai 0 1 2> err | cut -f 5 | sort | uniq > meiocyte_specific_cluster_associa-te_genelist.txt
```

   (b) For leaf-specific sRNA clusters:

```
cat leaf_expressed_gene.bed leaf_nonexpressed_gene.bed | perl 5_aroundgene.pl - leaf_specific_cluster.bed Arabidop-sis_thaliana.TAIR10.dna.toplevel.fa.fai 0 1 2> err | cut -f 5 | sort | uniq > leaf_specific_cluster_associate_genelist.txt
```

(c) For meiocyte and leaf-shared sRNA clusters:

```
cat meiocyte_expressed_gene.bed meiocyte_nonexpressed_gene.
bed | perl 5_aroundgene.pl - meiocyte_leaf_shared_cluster.bed
Arabidopsis_thaliana.TAIR10.dna.toplevel.fa.fai 0 1 2> err |
cut -f 5 | sort | uniq > meiocyte_leaf_shared_cluster_asso-
ciate_genelist.txt
```

2. Retrieve gene abundance (TPM or FPKM) of these genes from the gene abundance files generated above (meiocyte_mRNA_rep1_gene_abund.tab, meiocyte_mRNA_rep2_gene_abund.tab, leaf_mRNA_rep1_gene_abund.tab and leaf_mRNA_rep2_gene_abund.tab), then plot the gene expression value of sRNA associated genes in Excel or R.

3.3.11 GO analysis of sRNA Associated DEGs

Overlap the above sRNA cluster associated gene lists (meiocyte_specific_cluster_associate_genelist.txt, leaf_specific_cluster_associate_genelist.txt and meiocyte_leaf_shared_cluster_associate_genelist.txt) with the DEG lists (res05up_meiocyte_vs_leaf.csv and res05down_meiocyte_vs_leaf.csv) to obtain the specific sRNA cluster associated DEG list. Then GO analysis can be performed by the online tool PANTHER [24] (http://www.pantherdb.org).

3.3.12 Snapshot of sRNA and sRNA-wAssociated DEGs

Load sRNA and RNA-Seq read alignment result files below into IGV [25] (2.8.2) for illustrations of the sRNA and gene loci (*see* **Note 11**). sRNA read alignment files listed below are generated by ShortStack from Subheading 3.3.4, **step 2**. RNA-Seq read alignment files are generated by STAR from Subheading 3.3.6, **step 2**.

```
meiocyte_24nt_miRNA_discarded_rep1.bam
meiocyte_24nt_miRNA_discarded_rep2.bam
leaf_24nt_miRNA_discarded_rep1.bam
leaf_24nt_miRNA_discarded_rep2.bam
meiocyte_mRNA_rep1_Aligned.sortedByCoord.out.bam
meiocyte_mRNA_rep2_Aligned.sortedByCoord.out.bam
leaf_mRNA_rep1_Aligned.sortedByCoord.out.bam
leaf_mRNA_rep2_Aligned.sortedByCoord.out.bam
```

4 Notes

1. Keep plants at ambient temperature (18–25 °C) when collecting samples. Low or high temperature could cause a stress response of the plants sampled and influence the sRNA-Seq and RNA-Seq result.

2. Two or three unopened floral buds are appropriate for meiocyte collection from a young inflorescence from the primary shoot of wild type plants grown under normal conditions. Stage 5–7 anthers that contain meiocytes [12] appear semi-transparent with a light green color. Anthers that have a yellowish color have already finished meiosis and should not be collected.
3. Squeezing the anthers with forceps will release rod-like meiocytes clusters into the collection buffer under dissection microscope. However, if you notice a lot of smaller masses, it is likely that you were using an anther containing meiocytes that had already finished meiosis. The numerous smaller masses are likely tetrad-stage developing pollen grains. In contrast, if no tissue mass is released, it is likely that you were using an anther with immature meiocytes.
4. For gDNA extraction, we suggest repeating freeze-thaw at least three times to maximize DNA yield before this step. One freeze-thaw cycle includes: put the tube in liquid nitrogen for 30 s; transfer into 37 °C water bath for 50 s; vortex for 60 s. However, we would not suggest doing so for RNA extraction, since it might increase the risk of RNA degradation.
5. Data generated by other sequencing platforms such as the Hiseq series and NextSeq series from Illumina, MinION from Oxford Nanopore, DNBSEQ-T7 from BGI and others may also be used. If other platforms or library preparation kits are used, please note that the adapter sequences may be different in Subheading 3.3.2.
6. Download links and manuals for the publicly available software described can be found in the link or reference provided for each tool.
7. Currently read trimming is considered an option, especially when the analysis is focused on quantification of gene expression. Alignment tools, like STAR or HISAT2 can "soft-trim" untrimmed reads during alignment. Total analysis time would be reduced using untrimmed reads [26].
8. Alternatively, you can also first align all the preprocessed sRNA reads, then use miRNA annotation to mask known miRNA loci. For the purpose of this protocol, in which we focus on the relationship between non-miRNAs and genes, we prefer to filter out miRNAs first to simplify the procedure.
9. All home-made perl scripts, such as "1_merge_cluster.pl" can be found in Data 1.
10. The choice of mapping tool is often a matter of personal preference and relies on the computational resources that are available. If you are only interested in gene expression quantification, we recommend Salmon [27] or Kallisto [28], which

are much faster and require less memory. If you are also interested in novel gene or transcript discovery, we recommend using STAR or HISAT2 [29].

11. Alternatively, convert a sorted bam file to a binary tiled data (tdf) file by igvtools from IGV interface [25], or to a bigWig (bw) file by bamCoverage from deeptools [30] will reduce memory load and provide smoother visualization.

Acknowledgments

G.P.C. and J.H. were supported by a grant from the US National Science Foundation (IOS-1844264). C.Y. was supported by National Natural Science Foundation of China (31900445). Y.W. and C.W. were supported by National Natural Science Foundation of China (31925005). The authors have no conflict of interest.

References

1. Loidl J (2016) Conservation and variability of meiosis across the eukaryotes. Annu Rev Genet 50:293–316
2. Yang X, Makaroff CA, Ma H (2003) The Arabidopsis MALE MEIOCYTE DEATH1 gene encodes a PHD-finger protein that is required for male meiosis. Plant Cell 15(6):1281–1295
3. Ozata DM, Gainetdinov I, Zoch A et al (2019) PIWI-interacting RNAs: small RNAs with big functions. Nat Rev Genet 20(2):89–108
4. Greenberg MVC, Bourc'his D (2019) The diverse roles of DNA methylation in mammalian development and disease. Nat Rev Mol Cell Biol 20(10):590–607
5. Zhang YC, Lei MQ, Zhou YF et al (2020) Reproductive phasiRNAs regulate reprogramming of gene expression and meiotic progression in rice. Nat Commun 11(1):6031
6. Jiang P, Lian B, Liu C et al (2020) 21-nt phasiRNAs direct target mRNA cleavage in rice male germ cells. Nat Commun 11(1):5191
7. Long J, Walker J, She W et al (2021) Nurse cell-derived small RNAs define paternal epigenetic inheritance in Arabidopsis. bioRxiv:2021.2001.2025.428150
8. Huang J, Wang C, Li X et al (2020) Conservation and divergence in the meiocyte sRNAomes of Arabidopsis, soybean, and cucumber. Plant Physiol 182(1):301–317
9. Huang J, Wang C, Wang H et al (2019) Meiocyte-specific and AtSPO11-1-dependent small RNAs and their association with meiotic gene expression and recombination. Plant Cell 31(2):444–464
10. Wang Y, Cheng Z, Lu P et al (2014) Molecular cell biology of male meiotic chromosomes and isolation of male meiocytes in Arabidopsis thaliana. Methods Mol Biol 1110:217–230
11. Chen C, Retzel EF (2013) Analyzing the meiotic transcriptome using isolated meiocytes of Arabidopsis thaliana. In: Pawlowski WP, Grelon M, Armstrong S (eds) Plant meiosis: methods and protocols. Humana Press, Totowa, NJ, pp 203–213
12. Sanders PM, Bui AQ, Weterings K et al (1999) Anther developmental defects in *Arabidopsis thaliana* male-sterile mutants. Sex Plant Reprod 11(6):297–322
13. Friedlander MR, Mackowiak SD, Li N et al (2012) miRDeep2 accurately identifies known and hundreds of novel microRNA genes in seven animal clades. Nucleic Acids Res 40(1): 37–52
14. Langmead B, Trapnell C, Pop M et al (2009) Ultrafast and memory-efficient alignment of short DNA sequences to the human genome. Genome Biol 10(3):R25
15. Borges F, Martienssen RA (2015) The expanding world of small RNAs in plants. Nat Rev Mol Cell Biol 16(12):727–741
16. Axtell MJ (2013) ShortStack: comprehensive annotation and quantification of small RNA genes. RNA 19(6):740–751

17. Camacho C, Coulouris G, Avagyan V et al (2009) BLAST+: architecture and applications. BMC Bioinformatics 10:421
18. Gremme G, Steinbiss S, Kurtz S (2013) GenomeTools: a comprehensive software library for efficient processing of structured genome annotations. IEEE/ACM Trans Comput Biol Bioinform 10(3):645–656
19. Quinlan AR, Hall IM (2010) BEDTools: a flexible suite of utilities for comparing genomic features. Bioinformatics 26(6):841–842
20. Li H, Handsaker B, Wysoker A et al (2009) The Sequence Alignment/Map format and SAMtools. Bioinformatics 25(16):2078–2079
21. Dobin A, Davis CA, Schlesinger F et al (2013) STAR: ultrafast universal RNA-seq aligner. Bioinformatics 29(1):15–21
22. Pertea M, Pertea GM, Antonescu CM et al (2015) StringTie enables improved reconstruction of a transcriptome from RNA-seq reads. Nat Biotechnol 33(3):290–295
23. Love MI, Huber W, Anders S (2014) Moderated estimation of fold change and dispersion for RNA-seq data with DESeq2. Genome Biol 15(12):550
24. Mi H, Muruganujan A, Ebert D et al (2019) PANTHER version 14: more genomes, a new PANTHER GO-slim and improvements in enrichment analysis tools. Nucleic Acids Res 47(D1):D419–D426
25. Robinson JT, Thorvaldsdottir H, Winckler W et al (2011) Integrative genomics viewer. Nat Biotechnol 29(1):24–26
26. Liao Y, Shi W (2020) Read trimming is not required for mapping and quantification of RNA-seq reads at the gene level. NAR Genom Bioinform 2(3):lqaa068
27. Patro R, Duggal G, Love MI et al (2017) Salmon provides fast and bias-aware quantification of transcript expression. Nat Methods 14(4):417–419
28. Bray NL, Pimentel H, Melsted P et al (2016) Near-optimal probabilistic RNA-seq quantification. Nat Biotechnol 34(5):525–527
29. Kim D, Langmead B, Salzberg SL (2015) HISAT: a fast spliced aligner with low memory requirements. Nat Methods 12(4):357–360
30. Ramirez F, Ryan DP, Gruning B et al (2016) deepTools2: a next generation web server for deep-sequencing data analysis. Nucleic Acids Res 44(W1):W160–W165

Chapter 4

CTAB DNA Extraction and Genotyping-by-Sequencing to Map Meiotic Crossovers in Plants

Pallas Kuo, Ian R. Henderson, and Christophe Lambing

Abstract

Reciprocal DNA crossovers between chromosomes form new allelic combinations and contribute to the formation of novel genetic diversity. Crossovers are formed during meiosis of germ cells and these recombination events have influenced plant genome evolution and are used during breeding to create improved plant varieties. Meiotic crossovers are not uniformly formed across the genome but instead occur in regions with low nucleosome density. The recombination landscape differs between the model plant organism *Arabidopsis thaliana* and crops such as rice and maize. Genotyping-by-sequencing is a technique that can detect crossover location and provide information on the recombination landscape genome-wide. This technique can be used to compare crossover position between ecotypes, species, and mutant lines to gain information on factors controlling meiotic recombination. In this protocol, we describe the steps to purify DNA from plant tissue, prepare 96 DNA libraries in parallel and perform quality control before next-generation sequencing.

Key words Crossover, Meiosis, CTAB DNA extraction, Genotyping-by-sequencing, DNA sequencing, Recombination, Plant

1 Introduction

During meiosis, the two sets of chromosomes align to each other's and reciprocally exchange genetic information in a process called DNA crossover. Crossover breaks genetic linkage to form new allelic combinations that can be inherited [1]. Meiotic recombination has an important role in creating genetic diversity and contributes to plant genome evolution. This process can be utilised during breeding to create improved crop varieties by breaking unbeneficial linkage between genes on the same chromosome [2].

Meiotic crossovers are not uniformly distributed throughout the genome. Instead, crossovers are formed in regions of low-nucleosome density while crossover formation is repressed in the heterochromatic regions enriched in silencing epigenetic factors [3, 4]. In crops such as maize, the genome composition is

Christophe Lambing (ed.), *Plant Gametogenesis: Methods and Protocols*, Methods in Molecular Biology, vol. 2484, https://doi.org/10.1007/978-1-0716-2253-7_4,

made of ~80% of transposons and the majority of the chromatin adopts a heterochromatin structure that prevents formation of meiotic crossovers. It is estimated that about 20% of genes are located in heterochromatin regions in maize and barley [2]. This limits the introduction of genetic diversity and the potential for future crop improvement.

Genotyping-by-sequencing (GBS) is a technique that provides information on crossover rate genome-wide. GBS consists of generating a hybrid between two ecotypes from which the information of single nucleotide polymorphisms (SNPs) discriminating the two genotypes are known. Large numbers of the progeny from the F_1 hybrid are sequenced at low coverage and the sequencing data are then mapped to a reference genome where the SNP data are used to detect shifts between genotypes along a given chromosome which reflect the presence of crossovers [5]. The crossover landscape inferred from this analysis can be generated with low- or high-sequence coverage [6] and from a small or a large F_2 population [7] without impacting the data, which confirm the effectiveness of the technique.

GBS has been widely used to report the meiotic crossover landscape between plant species [8, 9], ecotypes [10], sexes [8], and between wild-type and mutant lines [4, 10–14]. This technique has advanced our understanding of meiosis by confirming the role of several genes on the control of meiotic recombination, by investigating changes in crossover landscape between lines with different genetic background and by providing information on local factors influencing crossover locations. In this protocol, we describe a method (1) to extract DNA from plant tissue, (2) to prepare DNA library for 96 samples in parallel, and (3) to perform quality control and quantification of the DNA libraries prior to next-generation sequencing.

2 Materials

2.1 Plant Growth, Tissue Collection, and DNA Extraction

1. Borosilicate glass beads (3 mm).
2. TissueLyser.
3. Mortar and pestle.
4. CTAB buffer: 140 mM sorbitol, 220 mM Tris–HCl pH 8.0, 22 mM EDTA, 800 mM NaCl, 0.1% (v/v) N-Lauryl sarcosine, 0.8% (w/v) cetyltrimethylammonium bromide (CTAB).
5. Chloroform–isoamyl alcohol (24:1).
6. Isopropanol.
7. 70% ethanol (v/v in ddH_2O).
8. RNase A solution: 5 μl of 100 mg/ml of RNase A in 5 ml of nuclease-free H_2O.

9. 3 M sodium acetate (pH 5.2). pH is adjusted with glacial acetic acid.
10. Absolute ethanol.
11. DNA concentration system such as Qubit.

2.2 DNA Library Preparation

1. Shearase enzyme (1 U/μl).
2. 0.5 M EDTA (pH 8.0). Adjust the pH with NaOH.
3. AMPure beads or equivalent.
4. Freshly made 80% ethanol (v/v in ddH_2O).
5. Elution buffer: 10 mM Tris–HCl, pH 8.0.
6. SYBR Gold or an equivalent high-fluorescence DNA dye.
7. T4 DNA ligase buffer.
8. dNTPs.
9. T4 DNA polymerase (3 U/μl).
10. T4 polynucleotide kinase (10 U/μl).
11. DNA Polymerase I, Large Klenow fragment (5 U/μl).
12. Klenow Fragment ($3' \rightarrow 5'$ exo-) (5 U/μl).
13. Adaptors P1 and P2 (*see* **Note 1**).
14. Quick Stick ligase or an equivalent enzyme.
15. PEG solution: 8 mM PEG, 2.5 M NaCl.
16. KAPA Hot Start mix or an equivalent enzyme.
17. Primers P1 and P2 (*see* **Note 1**).

3 Methods

CTAB DNA extraction has been optimised for Arabidopsis, Brassica, and wheat. To work on other plant species, we recommend testing CTAB DNA extraction on few samples at first.

3.1 Tissue Collection and DNA Extraction

1. Grow Arabidopsis in controlled growth room under long-day conditions (16 h light at 150 μmol light intensity–8 h dark) at 20 °C. Grow Brassica in glasshouses with controlled environment under long-day conditions (16 h light at 400 μmol light intensity–8 h dark) at 20 °C. Grow wheat in glasshouses with controlled environment under long-day conditions (16 h light at 400 μmol light intensity–8 h dark) at 22 °C.
2. Collect four leaves with a length of 4 cm each from an individual Arabidopsis plant. The plant should be ~5 weeks old at the stage from rosette development to bolting. Transfer the leaves in an Eppendorf tube and add four glass beads. For Brassica, collect four leaves with a length of 4 cm each from an individual

plant. The plant should be ~4 weeks old at the stage from rosette development to bolting. For wheat, collect two leaves with a length of 7 cm each from an individual plant. The plant should be ~3 month old at tillering stage.

3. Snap freeze the tubes and store at −80 °C until the day of the experiment.
4. For Arabidopsis: on the day of the experiment, place the tubes in a TissueLyser at 30 Hz for 3 min (*see* **Note 2**).

   For Brassica and wheat: on the day of the experiment, transfer the leaf tissue to a mortar containing liquid nitrogen and grind the tissue to a fine powder with a pestle.
5. Add 800 μl of CTAB buffer and resuspend the leaf tissue by inverting the tube 10 times.
6. Incubate at 65 °C for 20 min under agitation at 700 rpm. Invert the tubes occasionally to ensure that no aggregate forms at the bottom of the tube.
7. Add 800 μl of Chloroform: Isoamyl alcohol (24:1) solution to the tube.
8. Vigorously vortex the tube for 10 s to emulsify the solution.
9. Centrifuge at 13,000 rpm for 5 min at 4 °C.
10. Transfer 700 μl of the aqueous phase to an Eppendorf tube.
11. Add an equal volume of isopropanol and vortex briefly.
12. Incubate for 15 min at room temperature.
13. Centrifuge at 13,000 rpm for 25 min at 4 °C.
14. Discard the supernatant.
15. Add 500 μl of 70% ethanol and centrifuge at 13,000 rpm for 5 min at 4 °C.
16. Discard the supernatant.
17. Air-dry the samples for 20 min or spin the sample in a SpeedVac for 4 min at room temperature.
18. Resuspend the DNA in 100 μl of RNase A solution.
19. Incubate the sample at 37 °C for 30 min.
20. Add 1/10 volume of 3 M sodium acetate and 2.5 volume of ice-cold absolute ethanol.
21. Incubate at −20 °C for an hour.
22. Centrifuge at 13,000 rpm for 25 min at 4 °C.
23. Add 500 μl of 70% ethanol and centrifuge at 13,000 rpm for 5 min at 4 °C.
24. Air-dry the samples for 20 min or spin the sample in a SpeedVac for 4 min at room temperature.
25. Resuspend the DNA with 20 μl of nuclease-free water.

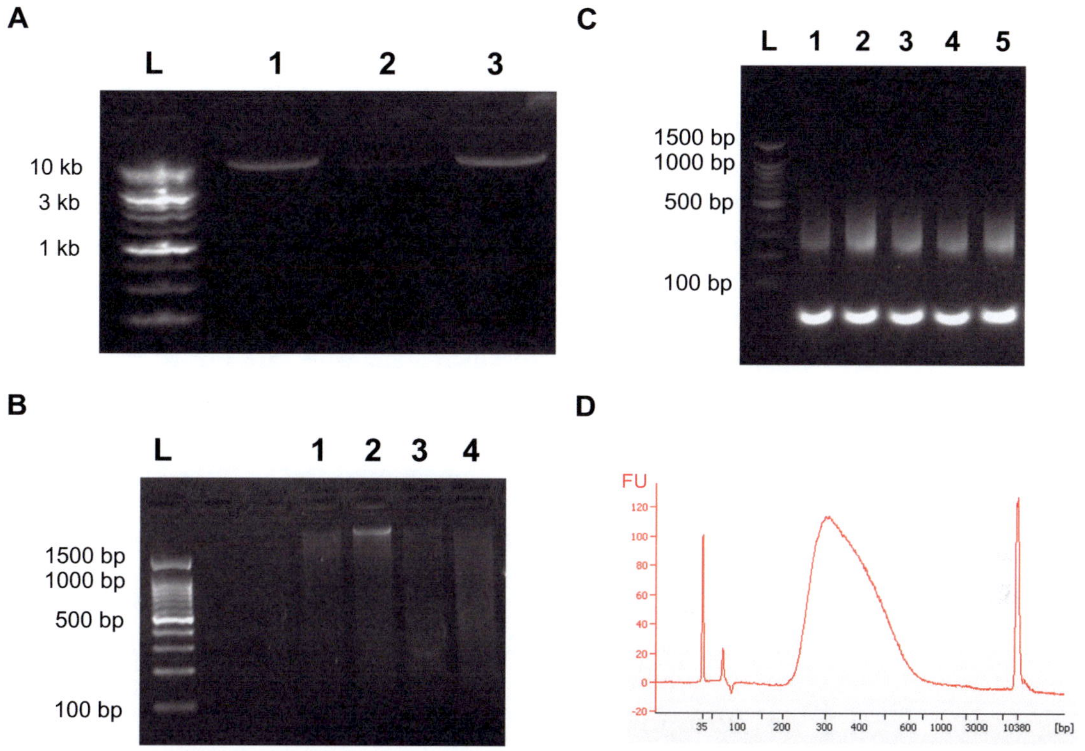

Fig. 1 DNA extraction and library preparation for genotyping-by-sequencing. (**a**) CTAB genomic DNA is loaded on a 1% agarose gel with SYBR Gold or an equivalent high fluorescence DNA dye. "L" represents the DNA ladder. "1-3" represent independent genomic DNA samples. (**b**) Sheared genomic DNA is loaded on a 1% agarose gel with SYBR Gold or an equivalent high fluorescence DNA dye. "L" represents the DNA ladder. "1-4" represent independent genomic DNA samples after shearing. (**c**) PCR-amplified DNA libraries are loaded on a 1.5% agarose gel with SYBR Gold or an equivalent high fluorescence DNA dye. DNA libraries are mostly composed of fragments between 200 bp and 600 bp. "L" represents the DNA ladder. "1-5" represent independent DNA library samples. (**d**) Bioanalyzer electropherogram of a DNA library sample. The DNA library is located between the lower (35 bp) and upper peak markers (10,380 bp). A small peak is detected below 100 bp and it represents primer dimers

26. Quantify the DNA concentration by Qubit (*see* **Note 3**).
27. Normalise each sample to a concentration of 15 ng/μl with nuclease-free water.
28. Load 1 μl of each sample on a 1% agarose gel with SYBR Gold or an equivalent high fluorescence DNA dye to check quality and intensity (Fig. 1a).

3.2 DNA Library Preparation

The protocol below is designed to produce a pool of 96 DNA libraries.

3.2.1 DNA Fragmentation

1. Prepare a reaction for DNA fragmentation as follows: 150 ng of DNA, 3 μl of 5× Shearase buffer, 0.4 ml of Shearase, ddH_2O up to 15 μl (*see* **Note 4**).

2. Incubate the samples at 37 °C for 20 min follows by an incubation at 65 °C for 5 min to inactivate the Shearase activity.
3. Add 1.6 μl of 0.5 M EDTA and mix briefly.
4. Spin down the solution and add 1.8 volumes of AMPure beads (*see* **Note 5**).
5. Pipette the solution up and down and incubate at room temperature for 5 min.
6. Place the 96-well plate on a magnetic stand and incubate at room temperature for 5 min.
7. Gently remove the supernatant with care not to remove the beads (*see* **Note 6**).
8. Add 500 μl of 80% ethanol.
9. Incubate for 30 s at room temperature.
10. Remove the supernatant.
11. Repeat **steps 8–10**.
12. Air-dry the samples for 5 min at room temperature.
13. Remove the plate from the magnetic stand and add 20 μl of elution buffer.
14. Resuspend the beads.
15. Gently spin down the solution but without aggregating the beads at the bottom of the tubes.
16. Incubate for 2 min at room temperature.
17. Place the 96-well plate on a magnetic stand and incubate for 5 min at room temperature.
18. Transfer the supernatant in a new 96-well plate. The supernatant contains the fragmented DNA (*see* **Note 7**) (Fig. 1b).

3.2.2 End Repair

1. Prepare a reaction for DNA end repair as follows: 18 μl of fragmented DNA, 3 μl of 10× DNA ligase buffer, 1.2 μl of 10 mM dNTPs, 1 μl of T4 DNA polymerase, 1 μl of T4 polynucleotide kinase, 0.25 μl of DNA Polymerase I, Large Klenow fragment, and 5.5 μl of water.
2. Incubate the samples at 20 °C for 30 min.
3. Spin down the solution and add 1.8 volumes of AMPure beads.
4. Pipette the solution up and down and incubate at room temperature for 5 min.
5. Place the 96-well plate on a magnetic stand and incubate at room temperature for 5 min.
6. Gently remove the supernatant with care not to remove the beads.
7. Add 500 μl of 80% ethanol.

8. Incubate for 30 s at room temperature.
9. Remove the supernatant.
10. Repeat **steps 7–9**.
11. Air-dry the samples for 5 min at room temperature.
12. Remove the plate from the magnetic stand and add 12 μl of elution buffer.
13. Resuspend the beads.
14. Gently spin down the solution but without aggregating the beads at the bottom of the tubes.
15. Incubate for 2 min at room temperature.
16. Place the 96-well plate on a magnetic stand and incubate for 5 min at room temperature.
17. Transfer the supernatant in a new 96-well plate.

3.2.3 A-Tailing

1. Prepare a reaction for DNA A-tailing as follows: 10 μl of DNA, 1.5 μl of 10× NEB buffer 2 (or equivalent buffer), 1.5 μl of 2 mM dATP, 0.5 μl of Klenow Fragment (3′ → 5′ exo-), 1.5 μl of nuclease-free water.
2. Incubate the samples at 37 °C for 30 min.
3. Spin down the solution and add 1.8 volumes of AMPure beads.
4. Pipette the solution up and down and incubate at room temperature for 5 min.
5. Place the 96-well plate on a magnetic stand and incubate at room temperature for 5 min.
6. Gently remove the supernatant with care not to remove the beads.
7. Add 500 μl of 80% ethanol.
8. Incubate for 30 s at room temperature.
9. Remove the supernatant.
10. Repeat **steps 7–9**.
11. Air-dry the samples for 5 min at room temperature.
12. Remove the plate from the magnetic stand and add 11 μl of elution buffer.
13. Resuspend the beads.
14. Gently spin down the solution but without aggregating the beads at the bottom of the tubes.
15. Incubate for 2 min at room temperature.
16. Spin down the samples to bring the beads at the bottom of the tubes.

3.2.4 Adaptor Ligation

1. Prepare a reaction for adaptor ligation as follows: 11 μl of DNA, 0.6 μl of 10 μM Adaptor P1, 0.6 μl of 10 μM Adaptor P2, 5 μl of 4× QuicK Stick ligation buffer, 1 μl of Quick Stick ligase, 1.8 μl of nuclease-free water.
2. Incubate the samples at 20 °C for 30 min follows by an incubation at 65 °C for 10 min.

3.2.5 Pooling Samples

1. Pool 8 samples from the same lane. The 96-well plate has 12 lanes and therefore 12 tubes of 8 pooled-samples are prepared.
2. Add 1.8 volume of PEG solution and mix by pipetting up and down 10 times.
3. Incubate the solution for 5 min at room temperature.
4. Place the tubes on a magnetic stand and incubate at room temperature for 5 min.
5. Gently remove the supernatant with care not to remove the beads.
6. Add 500 μl of 80% ethanol.
7. Incubate for 30 s at room temperature.
8. Remove the supernatant.
9. Repeat **steps 6–8**.
10. Air-dry the samples for 5 min at room temperature.
11. Remove the tubes from the magnetic stand and add 33 μl of elution buffer.
12. Resuspend the beads.
13. Incubate for 2 min at room temperature.
14. Place the tubes on a magnetic stand and incubate for 5 min at room temperature.
15. Transfer the supernatant to a new tube.

3.2.6 Size Selection to Enrich for DNA Fragments Between 300 bp and 600 bp

1. Add 0.55 volume of AMPure beads and mix by pipetting up and down 10 times.
2. Incubate for 5 min at room temperature.
3. Place the tubes on a magnetic stand and incubate at room temperature for 5 min.
4. Transfer the supernatant to a new tube.
5. Add 0.23 volume of AMPure beads and mix by pipetting up and down 10 times.
6. Incubate for 5 min at room temperature.

7. Place the tubes on a magnetic stand and incubate at room temperature for 5 min.
8. Add 200 μl of 80% ethanol.
9. Incubate for 30 s at room temperature.
10. Remove the supernatant.
11. Repeat **steps 8–10**.
12. Air-dry the samples for 5 min at room temperature.
13. Remove the tubes from the magnetic stand and add 20 μl of elution buffer.
14. Resuspend the beads.
15. Incubate for 2 min at room temperature.
16. Place the tubes on a magnetic stand and incubate for 5 min at room temperature.
17. Transfer the supernatant to a new tube.

3.2.7 PCR Amplification and DNA Purification

1. Prepare a PCR reaction as follows: 12 μl of DNA, 25 μl of KAPA Hot Star mix, 1.5 μl of 10 mM Primer P1, 1.5 μl of 10 mM Primer P2, 10 μl of nuclease-free water.
2. Run the PCR with the following conditions: 95 °C for 3 min, (95 °C for 20 s, 65 °C for 30 s, 72 °C for 30 s) × 12 cycles, 72 °C for 5 min (*see* **Note 8**) (Fig. 1c).
3. Add 1 volume of AMPure beads and mix by pipetting up and down 10 times.
4. Incubate for 5 min at room temperature.
5. Place the tubes on a magnetic stand and incubate at room temperature for 5 min.
6. Add 200 μl of 80% ethanol.
7. Incubate for 30 s at room temperature.
8. Remove the supernatant.
9. Repeat **steps 6–8**.
10. Air-dry the samples for 5 min at room temperature.
11. Remove the tubes from the magnetic stand and add 10 μl of elution buffer.
12. Resuspend the beads.
13. Incubate for 2 min at room temperature.
14. Place the tubes on a magnetic stand and incubate for 5 min at room temperature.
15. Transfer the supernatant to a new tube. The supernatant contains the DNA library.

3.3 DNA Concentration and Quality Control of Library Preparation

1. Measure the DNA concentration for each pool of eight samples by Qubit.
2. Perform quality control of each pool of eight samples by BioAnalyzer (Fig. 1d). If primer dimers are detected with high concentration, repeat **steps 3–15** of Subheading 3.2.7.
3. Determine the molarity for the 12 pooled samples.
4. Combine the 12 pooled of 8 samples with equal molarity and prepare the correct molarity according to the recommendations provided by the supplier for next-generation sequencing.

4 Notes

1. Adaptors and primers should be designed to match the requirements of the sequencing provider.
2. Keep the tissue cold at all time. If necessary, place the tube in liquid nitrogen.
3. DNA concentration generally ranges between 15 and 80 ng/μ l.
4. Use a 96-well plate to prepare 96 DNA libraries in parallel. All 8 samples from each lane of the 96-well plate should have similar signal intensity on the agarose gel. The 8 samples from the same lane will be pooled together after the step of adapter ligation (Subheading 3.2.4) and it is important that all the samples have similar DNA concentration to ensure equal representation during DNA library sequencing.
5. AMPure beads are stored at 4 °C. Beads should be left at room temperature for 30 min before the start of the experiment.
6. To prevent the collection of beads, ~3 μl of the supernatant is left in the tube. If the beads are accidentally collected, transfer the beads back in the tube and incubate for 5 min on a magnetic stand at room temperature. Then, gently discard the supernatant without disrupting the beads.
7. To check the quality of the DNA fragmentation, 1 μl of sheared DNA is loaded on a 1% agarose electrophoresis gel and stained with SYBR Gold or an equivalent high fluorescence DNA dye.
8. A test PCR can be performed to make sure that the quality of the DNA and the amplification are adequate. PCR products are loaded on a 1.5% agarose gel with SYBR Gold or an equivalent high fluorescence DNA dye. The predicted size of PCR products is between 300 and 600 bp.

Acknowledgments

This work was supported by a BBSRC grant-aided support as part of the Institute Strategic Programme designing Future Wheat Grant (BP/P016855/1).

References

1. Mercier R, Mezard C, Jenczewski E et al (2015) The molecular biology of meiosis in plants. Annu Rev Plant Biol 66:297–327
2. Taagen E, Bogdanove AJ, Sorrells ME (2020) Counting on crossovers: controlled recombination for plant breeding. Trends Plant Sci 25:455–465
3. Choi K, Zhao X, Tock AJ et al (2018) Nucleosomes and DNA methylation shape meiotic DSB frequency in *Arabidopsis thaliana* transposons and gene regulatory regions. Genome Res 28:532–546
4. Underwood CJ, Choi K, Lambing C et al (2018) Epigenetic activation of meiotic recombination near *Arabidopsis thaliana* centromeres via loss of H3K9me2 and non-CG DNA methylation. Genome Res 28:519–531
5. Rowan BA, Patel V, Weigel D et al (2015) Rapid and inexpensive whole-genome genotyping-by-sequencing for crossover localization and fine-scale genetic mapping. G3 5: 385–398
6. Serra H, Lambing C, Griffin CH et al (2018) Massive crossover elevation via combination of HEI10 and recq4a recq4b during Arabidopsis meiosis. Proc Natl Acad Sci U S A 115: 2437–2442
7. Rowan BA, Heavens D, Feuerborn TR et al (2019) An ultra high-density *Arabidopsis thaliana* crossover map that refines the influences of structural variation and epigenetic features. Genetics 213:771–787
8. Kianian PMA, Wang M, Simons K et al (2018) High-resolution crossover mappings reveals similarities and differences of male and female recombination in maize. Nat Commun 9:2370
9. Wijnker E, James GV, Ding J et al (2013) The genomic landscape of meiotic crossovers and gene conversions in Arabidopsis thaliana. elife 2:e01426
10. Blackwell AR, Dluzewska J, Szymanska-Lejman M et al (2020) MSH2 shapes the meiotic crossover landscape in relation to interhomolog polymorphism in Arabidopsis. EMBO J 39:e104858
11. Lambing C, Kuo PC, Tock AJ et al (2020) ASY1 acts as a dosage-dependent antagonist of telomere-led recombination and mediates crossover interference in *Arabidopsis*. Proc Natl Acad Sci U S A 117:13647–13658
12. Lawrence EJ, Gao H, Tock AJ et al (2019) Natural variation in TBP-ASSOCIATED FACTOR 4b controls meiotic crossover and germline transcription in Arabidopsis. Curr Biol 29: 2676–2686
13. Yelina NE, Lambing C, Hardcastle TJ et al (2015) DNA methylation epigenetically silences crossover hot spots and controls chromosomal domains of meiotic recombination in Arabidopsis. Genes Dev 29:2183–2202
14. Kurzbauer MT, Janisiw MP, Paulin LF et al (2021) ATM controls meiotic DNA double-strand break formation and recombination and affects synaptonemal complex organization in plants. Plant Cell 33(5):1633–1656

Chapter 5

Interrogating Global Chromatin Interaction Network by High-Throughput Chromosome Conformation Capture (Hi-C) in Plants

Wenjing Wang, Longjian Niu, and Chunhui Hou

Abstract

High-throughput chromosome conformation capture (Hi-C) enables the global quantification of chromatin interaction frequency in eukaryotic nuclei. This method is based on in situ Hi-C, in which chromatin is cross-linked with formaldehyde, then digested with restriction enzyme. Biotin-labeled nucleotide is incorporated before the spatially adjacent DNA ends are ligated, making it possible to enrich specifically the chimeric ligation products for deep sequencing. In this chapter, we describe a modified in situ Hi-C protocol for the global chromatin interaction analysis in plants.

Key words 3D genome organization, Chromatin interaction, In situ Hi-C, Chromosome conformation capture, High-throughput sequencing, Plants

1 Introduction

Chromosome conformation capture (3C)-based techniques measure the chromatin interaction frequency between DNA fragments that are physically proximal in the three-dimensional (3D) space [1, 2]. Hi-C has been developed to utilize the next generation sequencing (NGS) technology with the purpose to comprehensively interrogate the chromatin interactions, profiling all fragment contacts in the DNA library [3]. Original Hi-C has been modified to minimize condition-dependent variations that introduce inaccuracy in the quantification of chromatin interaction frequency [4–6]. When applied on plant [7–9], Hi-C reproducibility is often compromised because of the presence of cell wall and other plant-specific substances in plant tissues. Here, we describe a protocol of in situ Hi-C (Fig. 1) for the construction of DNA library from plant tissue, that has been successfully applied on many different species of plants.

Christophe Lambing (ed.), *Plant Gametogenesis: Methods and Protocols*, Methods in Molecular Biology, vol. 2484, https://doi.org/10.1007/978-1-0716-2253-7_5,

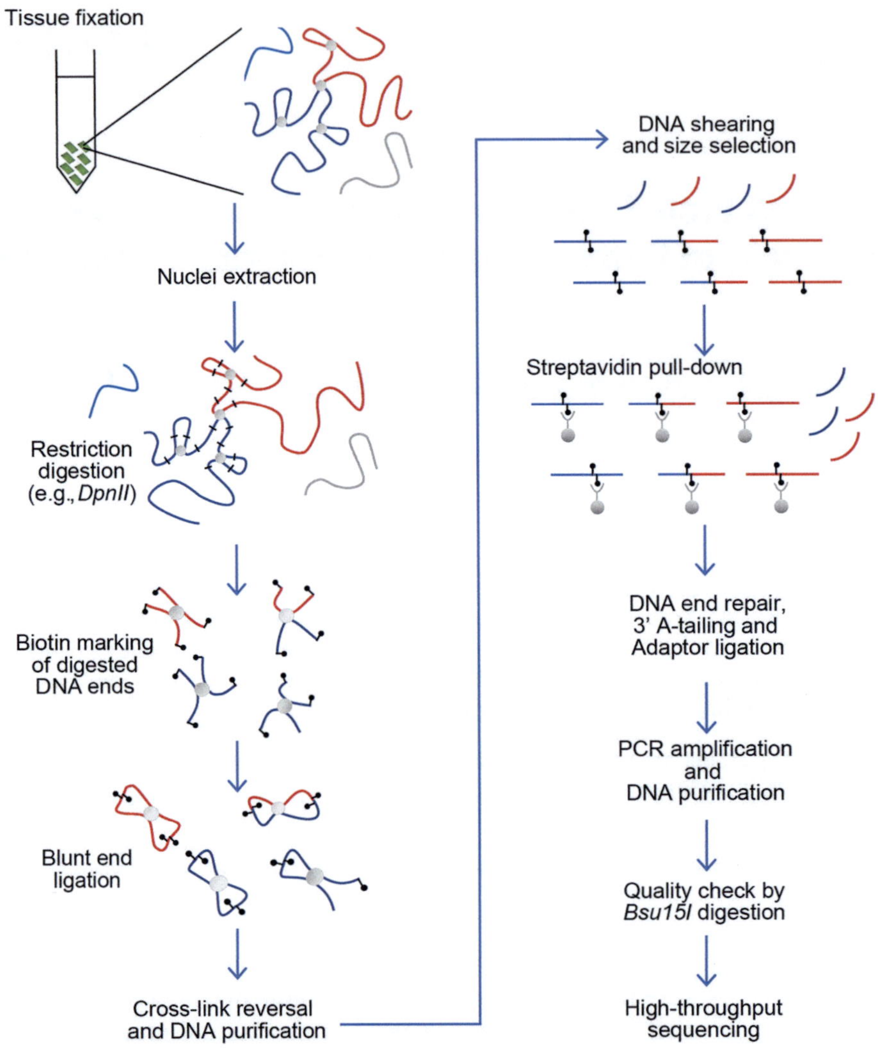

Fig. 1 Hi-C DNA ligate library construction

2 Materials

Prepare all solutions with molecular biology grade water or double-distilled water (ddH_2O).

Adaptor and PCR primers' sequences are shown in Table 1.

2.1 Cross-Linking

1. 1× PBS.
2. 37% (w/v) formaldehyde solution.
3. 2.5 M glycine: dissolve 9.35 g glycine in 50 ml ddH_2O and filter-sterilize using Steriflip column.

Table 1
Sequences of Adaptor and PCR primers

Y-Adaptors	
Y-Adaptor-F	5′ phosphate group-GATCGGAAGAGCACACGTCTGAACTCCAGTCAC-3′
Y-Adaptor-R	5′-TACACTCTTTCCCTACACGACGCTCTTCCGATCT-3′
	The underlined sequences are complementary
PCR primers	
Universal primer	5′-AATGATACGGCGACCACCGAGATCTACACTCTTTCCCTACACGAC-3′
Index primer	5′-CAAGCAGAAGACGGCATACGAGAT/CGTGATGT/GACTGGAGTTCA GACGTGT(/index sequence/)-3′

2.2 Nuclei Extraction

1. Mortar.
2. Liquid nitrogen.
3. 1× ice-cold lysis buffer: 10 mM Tris-HCl (pH 8.0), 10 mM NaCl, 0.2% Igepal, 1× protease inhibitors. Adjust the pH of Tris-HCl stock solution with sodium hydroxide (NaOH).
4. 4′,6-diamidino-2-phenylindole (DAPI) stain solution.

2.3 Restriction Digestion

1. 0.5% (w/v) sodium dodecyl sulfate (SDS).
2. 10% (v/v) Triton X-100: mix 1 ml 100% Triton X-100 with 9 ml ddH_2O.
3. 10 U/μl DpnII restriction enzyme.
4. 10× NEB buffer 3.1.

2.4 Biotin Marking and Ligation

1. 10 mM dATP, dGTP, dTTP.
2. 0.4 mM Biotin-dCTP.
3. 5 U/μl DNA Polymerase I Large (Klenow) Fragment.
4. 10× T4 DNA ligase buffer with 10 mM ATP.
5. 10% (v/v) Triton X-100.
6. 50 mg/ml bovine serum albumin (BSA).
7. 400 U/μl T4 DNA ligase.

2.5 Cross-Link Reversal and DNA Purification

1. 10% (w/v) SDS.
2. 20 mg/ml proteinase K.
3. 20 mg/ml RNase A.
4. Phenol–chloroform–isoamyl alcohol (25:24:1).
5. 3 M sodium acetate (pH 5.2), adjust pH with glacial acetic acid.

6. Ice-cold 100% ethanol.
7. 70% (v/v) ethanol.
8. 1× TE buffer (pH 8.0), adjust pH with hydrochloric acid (HCl).

2.6 DNA Shearing and Size Selection

1. Sonicator.
2. PCR purification kit.
3. 1.8% agarose gel in 0.5× TBE: 89 mMol/l Tris; 89 mMol/l boric acid; 2 mMol/l EDTA (pH 8.5). Adjust the pH of EDTA stock solution with NaOH.
4. Agarose gel electrophoresis apparatus.
5. Gel purification kit.

2.7 Streptavidin Pull-Down

1. Streptavidin Dynabeads.
2. Magnet.
3. 1× Tween Wash Buffer (TWB): 5 mM Tris-HCl (pH 7.5), 0.5 mM EDTA, 1 M NaCl, 0.05% Tween 20. Adjust the pH of Tris-HCl stock solution with HCl.
4. 2× binding buffer (BB): 10 mM Tris-HCl (pH 7.5), 1 mM EDTA, 2 M NaCl. Adjust the pH of Tris-HCl stock solution with HCl.
5. 1× T4 DNA ligase buffer.

2.8 End Repair

1. Magnet.
2. 10× T4 DNA ligase buffer with 10 mM ATP.
3. 25 mM dNTP mix.
4. 3 U/μl T4 DNA polymerase I.
5. 10 U/μl T4 polynucleotide kinase (PNK).
6. 5 U/μl DNA Polymerase I Large (Klenow) Fragment.
7. 1× Tween Wash Buffer (TWB).
8. 1× NEBuffer 2.1.

2.9 3′ A-Tailing

1. Magnet.
2. 10× NEBuffer 2.1.
3. 10 mM dATP.
4. 5 U/μl Klenow exo-.
5. 1× Tween Wash Buffer (TWB).
6. 1× Quick ligation reaction buffer.

2.10 Adaptor Ligation

1. Magnet.
2. 1× Quick ligation reaction buffer.

3. Illumina paired end adaptors.
4. DNA Quick ligase.
5. 1× Tween Wash Buffer (TWB).
6. 1× PCR buffer.

2.11 PCR Amplification

1. AMPure XP beads.
2. Magnet.
3. 1.8% agarose gel in 0.5×TBE.
4. Agarose gel electrophoresis apparatus.
5. Universal primer and index primer.
6. 2.5 mM dNTP.
7. 5× PCR buffer.
8. DNA polymerase.
9. 70% ethanol.
10. 1× TE buffer (pH 8.0).
11. *BSU15I* restriction enzyme.

3 Methods

3.1 Cross-Linking (See Note 1)

1. Cut the plant tissue into small pieces with a surgical scalpel (*see* **Note 2**).
2. Transfer plant tissue into a 50-ml tube. Add 36 ml 1×PBS at room temperature.
3. Add 1 ml 37% (w/v) formaldehyde to reach a final concentration of 1%. Incubate at room temperature for 30 min. The successfully cross-linked material should become translucent (*see* **Notes 3** and **4**).
4. Stop the cross-linking reaction by adding 1/10 reaction volume of 2.5 M glycine. Invert the tube several times to mix thoroughly. Incubate for 5 min at room temperature.
5. Centrifuge briefly to pellet the plant tissue. Wash the cross-linked material two times with 20 ml ddH_2O. The cross-linked material can be safely stored at −80 °C for up to 6 months.

3.2 Nuclei Extraction

1. Remove extra solution.
2. Weigh and take 3–5 g fixed material to a mortar. Grind the material to powder in liquid nitrogen.
3. Thaw the frozen material by adding 20 ml ice-cold lysis buffer containing protease inhibitors (*see* **Notes 5** and **6**).
4. Rotate at 30 rpm at 4 °C for 30 min.
5. Put a Miracloth upon a Sefar Nitex mesh to make a filter. Put it into a funnel. Filter the lysed sample through the Miracloth.

6. Centrifuge the sample at 2500 × *g* at 4 °C for 5 min to pellet the nuclei. Discard the supernatant carefully (*see* **Note 7**).
7. Resuspend the nuclei pellet with 1 ml ice-cold lysis buffer.
8. Take 10 μl of sample and stain with 10 μl DAPI, to confirm the purity and integrity of the extracted nuclei.
9. Save 50 μl of the sample to check the quality and integrity of genomic DNA.
10. Centrifuge at 2500 × *g* at 4 °C for 5 min. Discard the supernatant carefully.

3.3 Digestion

1. Suspend nuclei with 50 μl 0.5% (w/v) SDS. Mix carefully to avoid bubbles.
2. Incubate at 62 °C for 10 min without shaking (*see* **Note 8**).
3. Put onto ice immediately.
4. Add 25 μl 10% (v/v) Triton X-100 and 140 μl ddH_2O. Rotate at 30 rpm at 37 °C for 20 min (*see* **Note 9**).
5. Add 25 μl NEB buffer 3.1, digest chromatin with 10 μl 10 U/μ l DpnII restriction enzyme. Rotate at 30 rpm at 37 °C for at least 4 h.
6. Incubate the reaction at 65 °C for 20 min to inactivate DpnII without shaking.
7. Save 25 μl of sample and reverse cross-link as described in Subheading 3.5 with reaction volume adjusted accordingly.
8. Extract and purify DNA for electrophoresis as described in Subheading 3.5 with reaction volume adjusted accordingly.
9. Load purified DNA and DNA isolated from **step 9** in Subheading 3.2 on 0.8% agarose gel (0.5× TBE) to check the digestion efficiency.

3.4 Biotin Marking and Ligation

1. Prepare the master mix as follows.

1.5 μl	10 mM dATP
1.5 μl	10 mM dGTP
1.5 μl	10 mM dTTP
37.5 μl	0.4 mM Biotin-dCTP
10 μl	5 U/μl DNA Polymerase I Large (Klenow) Fragment

2. Add 52 μl master mix to the digested sample. This step fills the overhangs of digested fragments and marks the DNA ends with biotin-labeled nucleotides.
3. Rotate at 30 rpm at 37 °C for 45 min. Then, put onto ice immediately.

4. Prepare the ligation master mix as follows.

125 μl	10× T4 DNA ligase buffer with 10 mM ATP
125 μl	10% Triton X-100
3 μl	50 mg/ml BSA
10 μl	400 U/μl T4 DNA ligase
685 μl	ddH_2O

5. Add 948 μl ligation master mix to the sample. Rotate at 30 rpm at 16 °C for 4 h and then at room temperature for 1 h.

3.5 Cross-Linking Reversal and DNA Purification

1. Add 130 μl 10% SDS and 20 μl 20 mg/ml proteinase K. Incubate at 65 °C overnight (*see* **Note 10**).
2. Put on ice for 5 min. Add 20 μl 20 mg/ml RNase A. Incubate at 37 °C for at least 30 min.
3. Add 1.4 ml phenol–chloroform–isoamyl alcohol (25:24:1) to the reaction. Vortex vigorously for 30 s.
4. Centrifuge at 14,000 × *g* at room temperature for 5 min.
5. Transfer the aqueous phase to a new 2-ml tube.
6. Add 1/10 volume of 3 M sodium acetate (pH 5.2). Invert several times to mix.
7. Add 2 volumes of ice-cold absolute ethanol. Invert several times to mix. Incubate at −80 °C for 1 h.
8. Centrifuge at 14,000 × *g* at 4 °C for 5 min. Discard the supernatant.
9. Add 1 ml 70% ethanol. Centrifuge at 14,000 × *g* at 4 °C for 5 min. Discard the supernatant.
10. Dry the pellet at room temperature.
11. Dissolve the DNA in 100 μl 1× TE buffer.
12. Take 10 μl of sample and run in 0.8% agarose gel (0.5× TBE). Compare with the DNA purified from **step 9** in Subheading 3.2 and **step 8** in Subheading 3.3, to check the efficiency of ligation (Fig. 2).

3.6 DNA Shearing and Size Selection

1. Sonicate the DNA to a 300–500 bp size range using a sonicator. Set the parameters as below (*see* **Note 11**):

100 μl	Volume of DNA Library
10	Fill Level
15	Duty Cycle
500	PIP

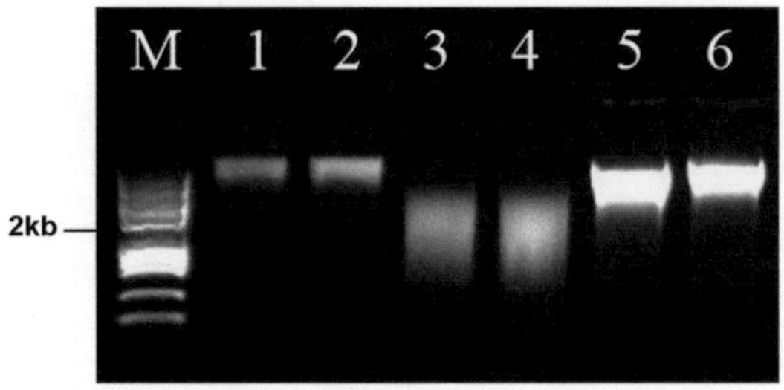

Fig. 2 Agarose gel electrophoresis of genomic, digested, and religated DNA. M, 1 kb DNA ladder. Lanes 1 and 2, genomic DNA. Lanes 3 and 4, cross-linked chromatin digested with DpnII. Lanes 5 and 6, DpnII digested DNA ligated with T4 DNA ligase. 0.8% agarose gel in 0.5× TBE

2. Purify the sonicated DNA with a PCR purification kit, elute the DNA with 20 μl water or 1× TE.
3. Run the DNA in a 1.8% agarose gel (0.5× TBE).
4. Recover DNA fragments within the 300–500 bp size range with a gel purification kit.
5. Elute the DNA with 100 μl water.

3.7 Streptavidin Pull-Down

1. Wash 100 μl Streptavidin Dynabeads with 200 μl 1× TWB in a low-retention tube (*see* **Note 12**).
2. Incubate and rotate at 30 rpm at room temperature for 5 min.
3. Put the tube onto a magnet for 5 min. Remove the supernatant with a pipette.
4. Resuspend the beads with 100 μl 2× BB.
5. Add 100 μl DNA sample from **step 5** in Subheading 3.6 to the beads. Incubate and rotate the suspension at 30 rpm at room temperature for 15 min.
6. Put the tube onto a magnet for 5 min. Remove the supernatant with a pipette.
7. Resuspend the beads with 200 μl 1× TWB. Incubate and rotate the suspension at 30 rpm at 55 °C for 2 min.
8. Put the tube onto a magnet for 5 min. Remove the supernatant with a pipette.
9. Resuspend the beads with 200 μl 1× T4 DNA ligase buffer. Transfer the beads into a new low-retention tube.
10. Put the tube onto a magnet for 5 min. Remove the supernatant with a pipette.

3.8 End Repair

1. To repair the recovered sonicated DNA fragments, suspend the beads with 100 μl master mix prepared as follows.

10 μl	10× T4 DNA ligase buffer with 10 mM ATP
2 μl	25 mM dNTP mix
4 μl	3 U/μl T4 DNA polymerase
5 μl	10 U/μl T4 polynucleotide kinase (PNK)
1 μl	5 U/μl DNA Polymerase I Large (Klenow) Fragment
78 μl	ddH_2O

2. Incubate and rotate the suspension at 30 rpm at 20 °C for 30 min.
3. Place the tube onto a magnet for 5 min. Remove the supernatant with a pipette.
4. Resuspend the beads with 200 μl 1× TWB. Incubate and rotate the suspension at 30 rpm at 55 °C for 2 min.
5. Place the tube onto a magnet for 5 min. Remove the supernatant with a pipette.
6. Resuspend the beads with 200 μl 1× NEBuffer 2. Transfer the beads into a new low-retention tube.
7. Place the tube onto a magnet for 5 min. Remove the supernatant with a pipette.

3.9 Deoxynucleotide A-Tailing

1. Resuspend the beads with 100 μl master mix prepared as follows.

80 μl	ddH_2O
10 μl	10× NEBuffer 2.1
5 μl	10 mM dATP
5 μl	5 U/μl NBE Klenow exo-

2. Incubate and rotate the suspension at 30 rpm at 37 °C for 30 min for 3′ A addition.
3. Place the tube onto a magnet for 5 min. Remove the supernatant with a pipette.
4. Resuspend the beads with 200 μl 1× TWB. Incubate and rotate the suspension at 30 rpm at 55 °C for 2 min.
5. Place the tube onto a magnet for 5 min. Remove the supernatant with a pipette.
6. Resuspend the beads with 200 μl 1× Quick ligation reaction buffer. Transfer the beads to a new low-retention tube.
7. Put the tube onto a magnet for 5 min. Remove the supernatant with a pipette.

3.10 Adaptor Ligation

1. Resuspend the beads with 110 μl master mix prepared as follows.

100 μl	1× Quick ligation reaction buffer
5 μl	Illumina paired end adaptor (Y-Adaptor) (25 μM)
5 μl	NEB DNA Quick ligase (400 U/μl)

2. Incubate and rotate the reaction at 30 rpm at room temperature for 30 min.
3. Place the tube onto a magnet for 5 min. Remove the supernatant with a pipette.
4. Resuspend the beads with 200 μl 1× TWB. Incubate the suspension at 55 °C for 2 min, to remove unligated adaptors and to inactivate ligase.
5. Place the tube onto a magnet for 5 min. Remove the supernatant with a pipette.
6. Resuspend the beads with 200 μl 1× PCR buffer. Transfer the beads into a new low-retention tube.
7. Place the tube onto a magnet for 5 min. Remove the supernatant with a pipette.
8. Resuspend the beads with 20 μl 1× PCR buffer.

3.11 PCR Amplification and Sequencing

1. Prepare the reaction as follows.

1.5 μl	On-bead DNA
0.7 μl	PCR Primers (25 μM each)
2 μl	2.5 mM dNTP
5 μl	5× PCR buffer (TransStart FastPfu Fly DNA Polymerase Kit)
0.5 μl	FastPfu Fly DNA polymerase (2.5 U/μl)
15.3 μl	ddH_2O

2. Set the PCR program as follows.

1 cycle	98 °C	30 s
9/12/15 cycles	98 °C 62 °C 72 °C	10 s 30 s 60 s
1 cycle	72 °C	7 min

3. Run PCR products from tested cycles (9/12/15) in 1.8% agarose gel (0.5× TBE).

4. Estimate the amplified DNA amount and choose an optimal PCR cycle number; carry out more PCR reactions to produce Hi-C library DNA for high-throughput sequencing.
5. Add 0.8 volume of AMPure XP beads to PCR reaction.
6. Vortex the reaction for 30 s, then incubate at room temperature for 10 min.
7. Place the tube onto a magnet for 5 min. Remove the supernatant with a pipette.
8. Wash the beads once with 150 μl 70% ethanol.
9. Dry the beads at room temperature.
10. Resuspend the beads with 12 μl 1× TE buffer. Incubate at room temperature for 10 min to elute the DNA.
11. Place the beads onto a magnet for 5 min. Transfer the supernatant to a new low-retention tube.
12. Quantify the concentration of Hi-C DNA library on a Qubit 4.0 Fluorometer.
13. Check the size range of Hi-C DNA library by running 1–2 μl of DNA on Agilent 2100 Bioanalyzer.
14. For quality control, digest 50 ng of Hi-C DNA library with *Bsu15I*, which recognizes and digests DNA sequence 5′-ATCGAT-3′, a new site created at the ligation junction from filled DpnII-digested ends (Fig. 3).
15. Hi-C DNA library is ready to be sequenced paired-endly on a high-throughput sequencing platform.

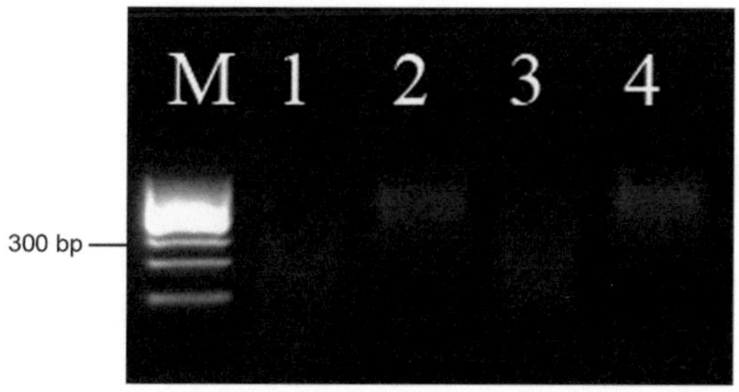

Fig. 3 Hi-C library quality confirmation. Hi-C DNA ligates digested with restriction enzyme *Bsu15I*. M, 100 bp DNA ladder. Lanes 1 and 3, amplified Hi-C library DNA digested with *Bsu15I*. Lanes 2 and 4, amplified Hi-C library DNA as control. 1.8% agarose gel in 0.5× TBE

4 Notes

1. For unfixed frozen materials, cross-linking procedure is different. Before cross-linking and cell lysis, the frozen materials undergo a freeze substitution process, during which a precrosslinking solution containing formaldehyde of low concentration is added [10].
2. It is recommended to select fresh and soft tissue with uniform texture to improve cross-linking efficiency.
3. Some plant species contain polysaccharides, polyphenols, and some readily oxidizable substances. Adding β-mercaptoethanol during cross-linking prevents the sample from being oxidized.
4. Be careful with formaldehyde. Avoid direct contact with skin.
5. To avoid the degradation of cross-linked DNA–protein complexes by inhibiting endogenous proteases.
6. Perform the nuclei extraction on ice to better maintain the integrity of nuclei.
7. Treat the fragile nuclei carefully and gently.
8. Incubation with SDS removes uncrosslinked proteins and increases chromatin accessibility.
9. SDS can be quenched by Triton X-100 to prevent the denaturation of restriction enzyme.
10. Proteinase K is used to degrade cross-linked DNA–protein complexes.
11. Parameters need to be optimized for the equipment used.
12. Streptavidin Dynabeads capture the ligation products through the interaction with biotin. Low-retention tubes reduce the loss of DNA.

Acknowledgments

We gratefully acknowledge financial support from the National key R&D Program of China (2018YFC1004500), the National Natural Science Foundation of China (31571347 to C.H.), and Southern University of Science and Technology (No.G02226301 and Y01501821 to C.H.).

References

1. Dekker J, Rippe K, Dekker M et al (2002) Capturing chromosome conformation. Science 295(5558):1306–1311
2. Louwers M, Splinter E, Van Driel R et al (2009) Studying physical chromatin interactions in plants using Chromosome Conformation Capture (3C). Nat Protoc 4(8): 1216
3. Lieberman-Aiden E, Van Berkum NL, Williams L et al (2009) Comprehensive mapping of

long-range interactions reveals folding principles of the human genome. Science 326(5950):289–293

4. Belton JM, McCord RP, Gibcus JH et al (2012) Hi–C: a comprehensive technique to capture the conformation of genomes. Methods 58(3):268–276
5. Rao SSP, Huntley MH, Durand NC et al (2014) A 3D map of the human genome at kilobase resolution reveals principles of chromatin looping. Cell 159(7):1665–1680
6. Niu L, Shen W, Huang Y et al (2019) Amplification-free library preparation with SAFE Hi-C uses ligation products for deep sequencing to improve traditional Hi-C analysis. Commun Biol 2(1):1–8
7. Wang C, Liu C, Roqueiro D et al (2015) Genome-wide analysis of local chromatin packing in *Arabidopsis thaliana*. Genome Res 25(2):246–256
8. Dong P, Tu X, Chu PY et al (2017) 3D chromatin architecture of large plant genomes determined by local A/B compartments. Mol Plant 10(12):1497–1509
9. Dong Q, Li N, Li X et al (2018) Genome-wide Hi-C analysis reveals extensive hierarchical chromatin interactions in rice. Plant J 94(6): 1141–1156
10. Zheng W, Yang Z, Ge X et al (2021) Freeze-Substitution-Hi-C, a convenient and cost-effective method for capturing the natural 3D chromatin conformation from frozen samples. J Genet Genomics 48(3):237–247

Part II

Microscopy Approaches to Study Gametogenesis in Plants

Chapter 6

Cytogenetic Techniques for Analyzing Meiosis in Hexaploid Bread Wheat

Kim Osman, F. Chris H. Franklin, and Eugenio Sanchez-Moran

Abstract

This chapter describes several cytogenetic procedures developed for investigating meiotic recombination in pollen mother cells (PMCs) of hexaploid wheat (*Triticum aestivum*) using standard fluorescence microscopy. Two basic methods are used to prepare slides for microscopy. In the cytological technique, wheat anthers are excised, fixed and used to prepare chromosome spreads which can be visualized following the application of a fluorescent DNA stain. In the immunocytological technique, fresh anthers are used to prepare chromosome spreads for analyzing the localization of meiotic proteins by applying specific antibodies followed by fluorescently tagged secondary antibodies. Both methods can be combined with the use of DNA probes to label specific chromosome regions such as telomeres, centromeres, and rDNA sequences in a procedure known as fluorescence in situ hybridisation (FISH). In addition, the cytological technique can be used in conjunction with S-phase incorporation of the DNA base analog, 5-bromo-2′-deoxyuridine (BrdU), and a modified immunolocalization procedure for a convenient meiotic time course assay. Although these protocols were developed for *T. aestivum* cv. Cadenza, they are directly applicable to other varieties and we have used them successfully for several other hexaploid cultivars and the tetraploid *Triticum turgidum* cv. Kronos.

Key words Wheat, *Triticum aestivum*, Hexaploid, Meiosis, Recombination, Chromosomes, Immunolocalization, FISH, BrdU, Cytology

1 Introduction

Meiotic recombination generates genetic variation and establishes the physical links between homologous chromosomes necessary for accurate segregation at the division stages. The allohexaploid, *Triticum aestivum* (bread wheat), is ideal for cytogenetic analysis of meiosis in a polyploid crop species, possessing relatively large chromosomes which, due to the influence of the *Ph* loci, exhibit diploid-like behavior for pairing and recombination purposes [1–4]. Like other cereals, wheat suffers from a severe bias in meiotic recombination such that most genetic crossover (CO) events, observed cytologically as chiasmata, occur near the chromosome ends (distal

Christophe Lambing (ed.), *Plant Gametogenesis: Methods and Protocols*, Methods in Molecular Biology, vol. 2484, https://doi.org/10.1007/978-1-0716-2253-7_6,

regions). This, combined with the fact that CO frequency per chromosome pair is generally low (typically 1–3), imposes severe limitations on the development of new varieties through plant breeding, as genes situated in the interstitial and centromere-proximal regions of chromosomes rarely recombine and the problem of linkage drag when introgressing desirable traits is exacerbated. A detailed understanding of the meiotic process and the control of CO formation is therefore essential in order to fully benefit from recent advances in wheat genomics and other technological developments [5, 6]. Our laboratory has developed a range of cytological techniques for investigating CO formation in *Arabidopsis thaliana*, later adapting them for use in crop species such as brassica and barley [7–9]. Here, we describe several procedures, suitable for standard fluorescence microscopy, which we developed for use in wheat. These have recently been used to carry out a detailed analysis of the meiotic program in cv. Cadenza [10] to provide a reference framework for functional studies of meiotic genes via mutant analysis [11] and ongoing initiatives to manipulate CO frequency/distribution for the benefit of plant breeding.

Two basic procedures are used to prepare slides for analysis, both involving enzymatic digestion of the PMC cell wall to allow spreading of the meiotic chromosomes. In the cytological technique, spreads are prepared from fixed anthers and are heat treated with acetic acid to dissociate cells during the spreading procedure. This method is highly convenient as floral material at the correct stage of meiosis can be harvested, fixed, and stored for analysis several weeks later. It is routinely used for assessing chiasma/CO frequency and distribution at metaphase I and for examining the integrity of the division stages (Fig. 1a, b). The technique can be further refined by the application of specific DNA probes (FISH analysis) and it can also be used in time course assays to assess S-phase incorporation of the thymidine analog, BrdU, in conjunction with immunolocalization of ASY1 and ZYP1 (which mark the chromosome axes and synaptonemal complex respectively) (Fig. 1c). However, unlike anti-ASY1 and anti-ZYP1, many meiotic antibodies do not work particularly well for spreads prepared from fixed anthers, so the immunocytological technique is routinely used for investigating the localization of proteins during prophase I (Fig. 1d–f). In this procedure, Lipsol detergent is used to clear cytoplasm from the PMC spreads, exposing epitopes on the chromosome-associated proteins, whilst reducing background due to nonspecific binding of the antibodies. This technique can also be combined with the use of DNA probes (Immuno-FISH) (Fig. 1g, h). Although the immunocytological procedure is less convenient than the cytological technique and needs to be carried out using fresh anthers when they are ready, it is a powerful tool for performing detailed analyses of meiotic recombination.

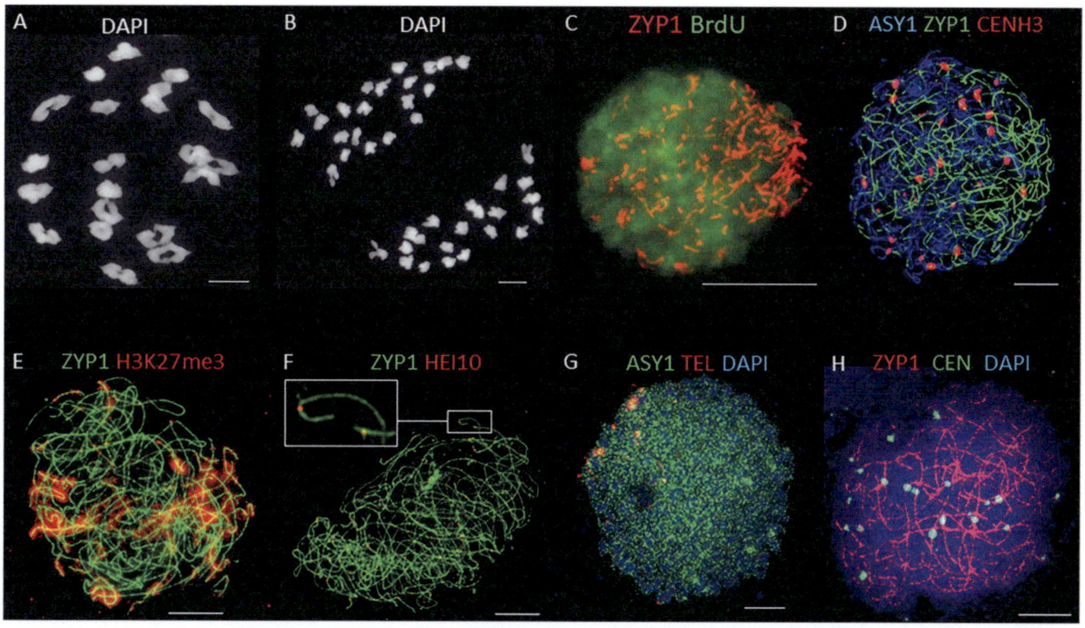

Fig. 1 Chromosome spreads from PMCs of hexaploid wheat. (**a**) and (**b**) Basic cytological spreads with DAPI-stained DNA showing (**a**) metaphase I with 21 bivalents (homologous chromosome pairs held together by COs/chiasmata) and (**b**) segregation of homologs at the first meiotic division. (**c**) Zygotene nucleus from time course experiment showing BrdU-labeling 30 h after injection. The polymerising SC is marked by anti-*At*ZYP1 antibody. (**d–f**) Immunocytological spreads showing: (**d**) mid-zygotene nucleus with unsynapsed axes marked by anti-*At*ASY1 antibody, the SC marked by anti-*At*ZYP1 antibody and the centromeric regions marked by anti-*Ta*CENH3 antibody; (**e**) pachytene nucleus showing enrichment of H3K27me3 (marked by anti-H3K27me3 antibody) at distal chromosome regions and the SC marked by ZYP1; (**f**) Late pachytene nucleus showing HEI10, essential for class I COs [12], marked by anti-*Hv*HEI10 antibody and the SC marked by ZYP1. (**g**) and (**h**) Immuno-FISH showing (**g**) clustered telomeres (marked by FISH) in the "bouquet" arrangement at early leptotene as ASY1 linearises along the chromosome axes; (**h**) centromeric regions (marked by FISH) with the SC marked by ZYP1 and DNA stained with DAPI in late prophase I. Bar = 10 μM

2 Materials

2.1 Excision of Anthers and Preparation of Slides (General)

1. *Triticum aestivum*, cv. Cadenza, plants are grown in multipurpose compost in a controlled-environment growth chamber with photoperiod 16 h, temperature 20 °C, and relative humidity 60%.
2. Binocular dissecting microscope with a graticule.
3. Petri dish (9 cm diameter).
4. Whatman filter paper (to fit inside petri dish).
5. Two pairs of fine watchmakers forceps.
6. Fine mounted needle.
7. Brass rod.
8. Microscope slides and coverslips (*see* **Notes 1** and **2**).

9. Antifade mounting medium (e.g., Vectashield).
10. Coplin jars, or similar, for slide washing.
11. Hot plate.
12. Humidified container for slide incubation (e.g., plastic sandwich box containing damp tissues and two glass rods or Pasteur pipettes for supporting slides).
13. Ice.

2.2 Preparation of Slides: Basic Cytological Technique

1. Watch glass.
2. Fixative: 3 volumes absolute ethanol:1 volume glacial acetic acid. Make fresh each day and keep on ice.
3. Citrate buffer, pH 4.5: 10 mM citrate buffer, pH 4.5. To prepare 10 ml of buffer, pH 4.5 mix 445 μl of 100 mM sodium citrate, 555 μl of 100 mM citric acid and 9 ml of sterile deionized water. Stock solutions should be autoclaved and stored at 4 °C. The citrate buffer is prepared as required.
4. Digestion medium stock: 1% cellulase, 1% pectolyase, 10 mM citrate buffer, pH 4.5. Dissolve enzymes in buffer and store in 333 μl aliquots at −20 °C.
5. Working digestion medium: 0.3% cellulase, 0.3% pectolyase, 10 mM citrate buffer, pH 4.5. Dilute a freshly thawed 333 μl aliquot of digestion medium stock with 667 μl of 10 mM citrate buffer, pH 4.5. Use immediately.
6. 70% acetic acid: dilute glacial acetic acid with sterile deionized water.
7. DAPI DNA stain: 5 μg/ml of 4′,6-diamidino-2-phenylindole (DAPI) in sterile deionized water. Prepare a 1 mg/ml stock solution of 4′,6-diamidino-2-phenylindole (DAPI) in sterile deionized water and store in aliquots at - 20 °C. For use, dilute to 5 μg/ml in Vectashield (*see* Subheading 2.1, **item 9**) and store protected from light at 4 °C.

2.3 Preparation of Slides: Immunocytological Technique

1. Cavity slide.
2. Parafilm cut into slide-sized rectangles.
3. Aluminum foil.
4. Immuno digestion medium: 0.4% cytohelicase, 1.5% sucrose, 1% polyvinylpyrrolidone (MW 40,000). Dissolve in sterile deionized water and store at −20 °C in aliquots of 100 μl or 200 μl.
5. Spreading medium: 1% Lipsol. Dilute in sterile deionized water.
6. Paraformaldehyde fixative: 4% paraformaldehyde, pH 8.0. Heat 100 ml sterile deionized water to 60 °C in a microwave and add four drops of 1 M NaOH. Weigh out 4 g

paraformaldehyde (EM grade) in a fume hood, add to the prewarmed water–NaOH and mix on a magnetic stirrer until dissolved (approx. 1 h). Filter through Whatman paper, allow to cool and, if necessary, adjust the pH to 8.0 with 1 M NaOH or 1 M HCl. The fixative can be stored for up to 1 week at 4 °C.

7. PBST wash solution: 137 mM NaCl, 2.7 mM KCl, 10 mM Na_2HPO_4, 1.8 mM KH_2PO_4, pH 7.4, 0.1% Triton X-100. We use tablets to prepare 1× PBS but it can also be prepared as a 10× stock (1.37 M NaCl, 27 mM KCl, 100 mM Na_2HPO_4, 18 mM KH_2PO_4 in deionized water), autoclaved and stored at room temperature. When diluted to 1× PBS, the pH should be approximately 7.4. If necessary, adjust with HCl or NaOH as appropriate. Add the 0.1% Triton X-100 before use.
8. Blocking buffer: 3% Bovine Serum Albumin (BSA) in PBST. Filter sterilize and store at 4 °C.
9. Primary antibodies: dilute in blocking buffer as appropriate (e.g., 1:100, 1:500, 1:1000).
10. Secondary antibodies (e.g., anti-rat, anti-rabbit, anti-guinea pig, or anti-mouse) conjugated to FITC, Cy3, or Alexa Fluor fluorophores. Stock solutions are prepared, aliquoted and stored according to the manufacturers' instructions. Before use, they are diluted in blocking buffer as appropriate.

2.4 Fluorescence In Situ Hybridization (FISH) Analysis

1. Probe labeling: use a nick translation labeling kit according to the manufacturer's instructions. Use either digoxigenin-11-dUTP or biotin-16-dUTP as nucleotide conjugates for labeling DNA. *See* Subheading 2.5 for probes.
2. 20× SSC buffer: 3 M NaCl, 300 mM trisodium citrate, pH 7.0. For 1 l, dissolve 175 g NaCl and 88.2 g trisodium citrate ($2H_2O$) in 900 ml of deionized water. Adjust to pH 7 with HCl, and make up to 1 l. Autoclave for storage at room temperature. Dilute to 2× SSC or 4× SSC as required.
3. Pepsin digestion medium: 0.01% pepsin, 10 mM HCl. Prewarm 100 ml of 10 mM HCl to 37 °C. Just before use, add 10 mg pepsin.
4. Paraformaldehyde fixative: *see* Subheading 2.3, **item 6**.
5. Alcohol Series: 70%, 85%, and 100% absolute ethanol. Prepare 100 ml of each, using sterile deionized water for dilution.
6. Hybridization mix: 71% formamide, 14% dextran sulfate (MW 500,000), 2.9× SSC, pH 7. Add 5 ml deionized formamide (molecular biology grade) (*see* **Note 3**) and 1 ml of 20× SSC (*see* **item 2**) to 1 g dextran sulphate (MW 500,000) and make up to 7 ml with sterile deionized water. Dissolve at 65 °C, cool to room temperature and, if necessary, adjust the pH to 7 with HCl or NaOH as appropriate. Store as 500 μl aliquots at −20 °C.

7. Probe mixture: 50–100 ng probe DNA, 50% formamide, 10% dextran sulphate (MW 500,000), 2× SSC, pH 7. Prepare 20 μl per slide as follows: 14 μl hybridization mix (*see* Subheading 2.4, **item 6**), 0.5–2.0 μl each labeled probe (*see* Subheading 2.5) and add sterile deionized water to 20 μl.
8. Vulcanizing rubber solution (e.g., as used for repairing bicycle tires).
9. Water bath set at 45 °C.
10. Post-hybridization wash solutions: 50% deionized formamide (*see* **Note 3**), 2× SSC, pH 7 (× 3); 2× SSC (× 1) and 4× SSC, 0.05% Tween 20 (× 1). These five wash solutions should be made up in Coplin jars or equivalent and can be reused several times; store at 4 °C between use and replace solutions monthly.
11. 4T wash buffer: 4× SSC, 0.05% Tween 20. For 1 l, combine 200 ml of 20× SSC with 800 ml of sterile deionized water and add 500 μl of Tween 20.
12. Digoxigenin specific antibody fluorophore conjugate: antibody fluorophore conjugate (amount as specified by the manufacturer), 0.5% digoxigenin blocking reagent, 4× SSC, 0.05% Tween 20. Prepare the digoxigenin blocking solution by resuspending 0.5% of a commercial digoxigenin blocking reagent in 4T wash buffer (*see* Subheading 2.4, **item 11**), centrifuge at 19,000 × *g* for 5 min and store the supernatant in 1 ml aliquots at −20 °C. Antibodies should be diluted (to manufacturer's specification) in the digoxigenin blocking solution just prior to use. Use for the detection of digoxigenin-labeled probes. Suitable fluorophores include FITC (green) or Rhodamine (red).
13. Streptavidin–fluorophore conjugate: streptavidin–fluorophore conjugate (amount as specified by the manufacturer), 0.5% dried skimmed milk, 4× SSC, 0.05% Tween 20. Prepare the biotin blocking solution by resuspending 0.5% dried skimmed milk in 4T wash buffer (*see* Subheading 2.4, **item 11**), centrifuge at 19,000 × *g* for 5 min and store the supernatant in 1 ml aliquots at −20 °C. Streptavidin conjugates should be diluted (to manufacturer's specification) in the biotin blocking solution just prior to use. Use for the detection of biotin-labeled probes. Suitable fluorophores include Cy3 (red) or FITC (green).

2.5 DNA Probes for FISH

1. Telomere probe: Use the oligonucleotide sequences 5′-$(TTTAGGG)_5$-3′ and 5′-$(CCCTAAA)_5$-3′ [13] at 0.3 μM each as both primer and template in a 50 μl primary PCR reaction. Perform PCR as follows: 94 °C, 1 min; 55 °C, 30 s; 72 °C, 1 min (8 cycles); then 94 °C, 1 min; 60 °C, 30 s; 72 °C, 1 min 30 s (24 cycles). Use 3 μl of the primary reaction as a template in a

secondary PCR reaction with dTTP replaced by dUTP-biotin/digoxigenin (*see* Subheading 2.4, **item 1**) to directly label the probe.

2. Centromere probe: Use the oligonucleotide sequences 5′-AAGAGGGGGGAATTTCGTGG-3′ and 5′-TTGAAGCCGCCATAAGGAGT-3′ [14] at 0.4 μM each in a 50 μl primary PCR reaction to amplify a 192 bp direct repeat sequence from the centromeric region of hexaploid wheat using leaf genomic DNA as template. Perform PCR as follows: 94 °C, 3 min (1 cycle); 94 °C, 1 min; 54 °C, 1 min; 72 °C, 45 s (35 cycles); 72 °C, 5 min (1 cycle). Use 3 μl of the primary reaction as a template in a secondary PCR reaction with dTTP replaced by dUTP-biotin/digoxigenin (*see* Subheading 2.4, **item 1**) to directly label the probe.
3. 5*S* rDNA probe: Use plasmid pTa794 containing the 5*S* rDNA repeat of *T. aestivum* as a 410 bp cloned insert [15]. Label the plasmid DNA with dUTP-biotin/digoxigenin using a nick–translation kit according to the manufacturer's instructions (*see* Subheading 2.4, **item 1**).
4. 45*S* rDNA probe: Use plasmid pTa71 containing the 18*S*–25*S* rDNA genes and spacer regions from *T. aestivum* as a 9 kb cloned insert [16]. Label the plasmid DNA with dUTP-biotin/digoxigenin using a nick-translation kit according to the manufacturer's instructions (*see* Subheading 2.4, **item 1**).

2.6 BrdU Meiotic Time Course

1. Eye protection and gloves.
2. 850 W microwave.
3. Microwave-safe plastic slide rack.
4. 2× microwave-safe plastic boxes, just large enough to accommodate slide rack.
5. BrdU solution: 10 mM BrdU, 1× PBS, 0.1% Tween 20. Dissolve 10 mM BrdU in 1× PBS (*see* Subheading 2.3, **item 7**). Store as 500 μl aliquots at −20 °C. Thaw immediately before use and add Tween 20 to 0.1%. BrdU solution is included in 5-bromo-2′-deoxyuridine labeling and detection kits but as larger volumes are often required, it can be more cost-effective to purchase it separately and prepare it in the laboratory.
6. Anti-BrdU primary and secondary antibodies: antibody (amount as specified by the manufacturer), 1× PBS, 0.1% Triton X-100. Dilute antibodies in PBST (*see* Subheading 2.3, **item 7**) just prior to use. Antibodies are available in 5-bromo-2′-deoxyuridine labeling and detection kits or can be purchased separately.
7. Hypodermic needle for BrdU injection (25 g × 1 in.) and a 1 ml plastic syringe.

8. Citrate buffer, pH 6: 10 mM citrate buffer, pH 6. Dissolve 2.94 g of trisodium citrate in 800 ml of deionized water, pH to 6 with a few drops of 1 M citric acid and make up to 1 l with water. Autoclave for storage at room temperature for several weeks.
9. PBST wash solution: *see* Subheading 2.3, **item** 7.
10. Fixative: *see* Subheading 2.2, **item 2**.

3 Methods

3.1 Excision of Anthers for Slide Preparation

1. Harvest a flower spike (*see* **Notes 4** and **5**) by excising it from the sheath of enclosing tissue using a small pair of sharp scissors. Place on damp Whatman paper inside a petri dish to prevent it drying out.
2. Calibrate the graticule of a dissecting microscope so that 10 bars = 1 mm. Use two pairs of watchmakers forceps to excise anthers of the required length (as a rough guide, 0.6–1.4 mm should cover the entire range of meiotic stages in Cadenza).
3. For the immunocytological technique, freshly harvested anthers should be used directly (*see* Subheading 3.3). For the cytological technique, anthers should first be fixed (*see* Subheading 3.2).

3.2 Preparation of Slides: Basic Cytological Technique

1. Place freshly dissected anthers (*see* Subheading 3.1) into 1 ml of fresh ice-cold fixative solution in a 1.5 ml microfuge tube. After 30 min, use a fine pipette to carefully remove fixative and replace with fresh solution. Store anthers at 4 °C and carry out a final change of fixative on the following day. Anthers should be left in fixative for ~1 week before preparing slides; they can be stored at 4 °C for several months but better quality chromosome spreads are obtained if slides are prepared within 1 month of fixation.
2. Use a Pasteur pipette or 1 ml disposable tip with the end excised to transfer a suspension of anthers in fixative to a watch glass. Using a dissecting microscope, carefully remove fixative and replace with 0.5 ml citrate buffer, pH 4.5 and leave for 2 min. Carry out two further washes in citrate buffer, pH 4.5.
3. Replace final citrate buffer, pH 4.5 wash with working digestion medium and place watch glass in a humidified container at 37 °C for 90–105 min.
4. Replace digestion medium with 0.5 ml of cold citrate buffer, pH 4.5 or sterile deionized water to stop the enzymatic reaction.

5. Add 2 μl of sterile deionized water to the middle of a slide and transfer up to 5 anthers to the water droplet (handle anthers gently—they are fragile after the digestion step). Quickly macerate by tapping gently with the end of a brass rod, ensuring that the suspension does not dry out.
6. Add 10 μl of 70% acetic acid, place slide on a hot plate at 45 °C and stir gently with a fine, mounted needle, spreading the material to an area of ~1 cm diameter. After ~30 s (or sooner if the material starts to dry out) add a further 10 μl of 70% acetic acid and continue stirring for up to 1 min.
7. Remove the slide from the hot plate and quickly add ~150 μl of ice-cold fixative as a ring around the material.
8. Place the slide, material side up, at a shallow angle on a paper towel (e.g., with one end resting on a pencil) and allow to air-dry.
9. The slides are now ready for basic cytology. They can either be used directly for microscopy by applying 15 μl of DAPI DNA stain and a 22 × 50 mm coverslip (7 μl of stain for a 22 × 22 mm coverslip) or they can be taken forward for FISH analysis or time course experiments (*see* Subheadings 3.4 and 3.5).

3.3 Preparation of Slides: Immunocytological Technique

1. Prewarm a humidified container by placing on a 37 °C hot plate.
2. Put 20–30 freshly dissected anthers of the appropriate size (*see* Subheading 3.1) into 20 μl of immuno digestion medium in the well of a cavity slide. Place inside the humidified container on the hot plate and incubate for 4 min.
3. Remove the cavity slide from the container and use the flat end of a brass rod to apply gentle pressure to the anthers to extrude PMCs into the digestion medium. Quickly use a fine needle to remove anther debris from the PMC suspension, taking care not to allow the suspension to dry out (add more immuno digestion medium if necessary). Return the slide to the container and incubate for a further 3 min.
4. Add 10 μl of 1% Lipsol to the middle of a microscope slide. Cut off the end of a disposable tip and use it to pipet 2 μl of the PMC suspension into the drop of Lipsol on the slide. Use the side of a fine needle to gently spread the suspension to an area the size of a coverslip.
5. Immediately add 12 μl of cold paraformaldehyde fixative to the spread in a fume hood.
6. Use the remainder of the PMC suspension to make more slides—by working quickly it is usually possible to prepare up to four slides from 20 μl of digestion mix (*see* **Note 6**).

7. Allow the slides to dry for 1–2 h in the fume hood.
8. Wash slides in PBST (2 × 5 min).
9. Apply blocking buffer to each slide as follows: pipet 100 μl of blocking buffer onto the middle of a Parafilm rectangle placed at the edge of a lab bench; remove slide from the Coplin wash jar and gently tap the edge on paper tissue to allow excess solution to drain away; carefully lower the slide with the PMC spread side downward on to the Parafilm so that the blocking buffer forms a thin layer over the surface of the slide and the Parafilm adheres to it, excluding air bubbles.
10. Place the slide, Parafilm side upward, in a humidified container and leave at room temperature for 30 min.
11. Use forceps to carefully remove Parafilm from slides, allowing excess blocking buffer to drain onto tissue paper. Apply fresh Parafilm loaded with the primary antibodies at the required dilution (*see* Subheading 2.3, **items 8** and **9**).
12. Incubate in the humidified container at 4 °C overnight (*see* **Note 7**).
13. Remove Parafilm and wash slides in PBST (3 × 5 min at room temperature).
14. Apply fresh Parafilm loaded with secondary antibodies at the required dilution (*see* Subheading 2.3, **item 10**) and incubate in the humidified container at 37 °C for 30 min. During this, and all subsequent steps, care should be taken to protect the light-sensitive fluorophores (e.g., wrap the humidified chamber in aluminum foil during incubation).
15. Wash the slides as in **step 13**.
16. Drain excess solution from the slides and apply antifade mounting medium, with or without DAPI DNA stain as preferred (*see* Subheading 2.1, **item 9** and Subheading 2.2, **item 7**) and coverslips.

3.4 Fluorescence In Situ Hybridization Analysis

3.4.1 Preparation of Slides

1. Slides prepared using the basic cytological technique (*see* Subheading 3.2) can be used directly for FISH. If slides have previously been used for microscopy, use a tissue to carefully remove excess immersion oil and soak in absolute ethanol until coverslips float off.
2. If FISH is to be combined with immunolocalization, prepare slides following the immunocytological technique up to the addition of a primary antibody (*see* Subheading 3.3, **steps 1–13**), then apply a secondary antibody conjugated to biotin and incubate at 37 °C for 30 min.

3.4.2 Fluorescence In Situ Hybridization

1. Wash slides in 2× SSC for 10 min.
2. Digest with pepsin digestion medium at 37 °C for 90 s (do not over-digest as this can result in the loss of material from the slide).
3. Wash in 2× SSC for 10 min.
4. Fix the material in paraformaldehyde fixative in a fume hood for 10 min.
5. Wash briefly in sterile deionized water.
6. Dehydrate the preparations by passing the slides through an alcohol series (2 min washes in 70%, 85% and 100% ethanol made up in deionized water).
7. Drain the slides on paper tissue.
8. Apply 20 μl of probe mixture to each slide (*see* Subheading 2.4, **item 7**), add a 22 × 22 mm coverslip and seal the edges with vulcanizing rubber solution. Place on a 75 °C hotplate for 4 min to denature the DNA.
9. Transfer the slides to a prewarmed humidified container and place at 37 °C overnight to allow DNA renaturation and binding of the probe(s) to the target sequence(s).
10. Prewarm the posthybridization washes in a 45 °C water bath in a fume hood. Remove the rubber solution and coverslips with fine forceps and pass the slides through the five washes in the order they are listed in Subheading 2.4, **item 10** (5 min each wash).
11. Transfer the slides to 4T wash buffer at room temperature for 5 min.
12. Detect digoxigenin-labeled probes by loading 100 μl of an anti-digoxigenin secondary antibody fluorophore conjugate on to Parafilm and applying to slides as in Subheading 3.3, **step 9**. Incubate in a humidified container at 37 °C in the dark for 30 min.
13. Remove Parafilm and wash slides in 4T wash buffer in the dark (3 × 5 min).
14. Repeat **steps 12** and **13** with a streptavidin–fluorophore conjugate to detect biotin-labeled probes/primary antibodies.
15. Drain off excess solution on paper tissue (do not allow slides to dry), apply antifade mounting medium, with or without DAPI DNA stain as preferred (*see* Subheading 2.1, **item 9** and Subheading 2.2, **item 7**) and a coverslip.

3.5 BrdU Meiotic Time Course

A meiotic time course assay based on the incorporation of the thymidine analog, 5-bromo-2′-deoxyuridine (BrdU), into newly synthesized DNA was developed for Arabidopsis [17] and later adapted for use in barley [9]. For wheat, we administer BrdU by direct injection into the floral tissues of the growing plant, as described for barley [9, 18], but we use a different detection method, based on a modification of [19] which allows convenient monitoring of BrdU incorporation in conjunction with immunolocalization of chromosome axis/SC proteins in fixed material.

1. Use a hypodermic needle and syringe to inject up to 500 μl of BrdU solution into the cavity above a developing flower spike (taken as Time 0) (*see* **Note 8**).
2. Harvest spikes at defined time points, excise anthers in the correct size range for meiosis and fix and prepare slides as described for the basic cytological technique but do not add DAPI stain or coverslips (see Subheading 3.2).
3. Place slides in a microwave-safe slide rack.
4. Place enough 10 mM citrate buffer, pH 6 in a microwave-safe plastic box to just cover the slide rack (but do not add the rack yet).
5. Heat the buffer in a microwave until it just begins to bubble (in our 850 W microwave this takes approximately 4 min for 350 ml of buffer).
6. Wait until the bubbles subside, then immediately place the rack of slides in the hot citrate buffer, return the box to the microwave and heat for a total of 45 s using a series of short pulses: 4 × 10 s, 1 × 5 s (the aim is to heat the slides without allowing the buffer to boil) (*see* **Note 9**).
7. Immediately transfer the rack of slides to a second plastic box containing PBST wash solution at room temperature and incubate for at least 10 min.
8. Detect BrdU and axis/SC protein(s) by immunolocalization as described in Subheading 3.3, **steps 9–16** (*see* **Note 10**).
9. Slides are viewed with a fluorescence microscope fitted with appropriate filters (e.g., DAPI, blue; FITC, green; and Texas Red) and equipped with an image capture and analysis system.

4 Notes

1. It is advisable to preclean glass slides to remove any residues from the manufacturing process which can sometimes prevent adhesion of plant material. Wash sequentially in 100% acetone (10 min), sterile deionized water (5 min) and 100% ethanol (5 min). Leave to air-dry, then wrap in cling film to protect from dust until ready for use.
2. Any good quality slides can be used but for the immunocytological technique we prefer polylysine slides as they prevent "lifting" of the PMC nucleus periphery which we sometimes observe with other types of slide.

3. We deionize formamide by mixing 200 ml with 10 g Amberlite ion-exchange resin on a magnetic stirrer at room temperature for at least 1 h. Filter out the resin and store formamide at 4 °C.
4. Before beginning analysis, it is advisable to first become familiar with identifying material at the optimum stage for meiosis. In our growth conditions, it takes Cadenza ~10 weeks from seed sowing to reach the correct stage of inflorescence development, but this time will vary between cultivars and growth environments. The developing flower spike can usually be detected as it begins to grow up into the leaf sheath cavity below the flag leaf, by applying gentle pressure between finger and thumb. With experience, it is possible to use this method to assess the optimum spike size for harvest (*see* **Note 5**).
5. Reference [20] provides a useful guide to staging reproductive development in Cadenza. The middle of a flower spike contains anthers at the most advanced stage of development, with more immature anthers being found toward the top and bottom of the spike. In our growth conditions, Cadenza spikes of 6–7 cm provide a useful range of anther sizes (0.6–1.4 mm), covering most of the meiotic stages. If the focus of the analysis is on early/late meiosis, slightly smaller/bigger spikes can be taken to optimize the number of anthers at the required stage
6. This is useful for analyzing a specific stage of meiosis with several different antibody combinations.
7. If preferred, it is usually possible to incubate primary antibodies at 37 °C for 30 min but it is advisable to optimize incubation conditions for individual antibodies. We routinely perform incubation at 4 °C overnight but have found that for some antibodies, extending the incubation time to 2 days improves signal strength.
8. Successful injection into the cavity above the developing flower spike is best achieved by holding the needle/syringe pointing downward, as close to the vertical as possible. It is advisable to wear gloves and eye protection because droplets of solution often begin to leak from the sheaf of leaves surrounding the spike before the full volume of 500 μl can be injected.
9. Preparing slides for immunolocalization using the microwave/citrate buffer treatment facilitates BrdU detection by improving access of the BrdU primary antibody to the DNA.
10. For time course experiments it is useful to combine BrdU detection with ASY1/ZYP1 immunolocalization for accurate meiotic staging. In three-color immunolocalization, ASY1 and ZYP1 can either be used singly (in combination with DAPI DNA stain) or together (by omitting DAPI from the mounting medium).

Acknowledgments

We would like to thank Steve Price and Karen Staples (University of Birmingham, UK) for technical and horticultural support and Stefan Heckmann (IPK, Gatersleben, Germany) for the kind gift of anti-*Hv*HEI10 antibody. This work was supported by the Biotechnology and Biological Sciences Research Council (BB/N002628/1 "Releasing natural variation in bread wheat by modulating meiotic crossovers").

References

1. Riley R, Chapman V (1958) Genetic control of the cytologically diploid behaviour of hexaploid wheat. Nature 182:713–715
2. Sears ER, Okamoto M (1958) Intergenomic chromosome relationships in hexaploid wheat. In: Proc. X Int. Congr. Genet. Montreal, pp 258–259
3. Martín AC, Rey MD, Shaw P et al (2017) Dual effect of the wheat Ph1 locus on chromosome synapsis and crossover. Chromosoma 126: 669–680
4. Serra H, Svačina R, Baumann U et al (2021) Ph2 encodes the mismatch repair protein MSH7-3D that inhibits wheat homoeologous recombination. Nat Commun 12:803
5. IWGSC (2018) Shifting the limits in wheat research and breeding using a fully annotated reference genome. Science 361:eaar7191
6. Adamski M, Borrill P, Brinton J et al (2020) A roadmap for gene functional characterisation in crops with large genomes: lessons from polyploid wheat. eLife 9:e55646
7. Armstrong SJ, Caryl AP, Jones GH et al (2002) Asy1, a protein required for meiotic chromosome synapsis, localizes to axis-associated chromatin in Arabidopsis and Brassica. J Cell Sci 115:3645–3655
8. Osman K, Higgins JD, Sanchez-Moran E et al (2011) Pathways to meiotic recombination in Arabidopsis thaliana. New Phytol 190:523–544
9. Higgins JD, Perry RM, Barakate A et al (2012) Spatiotemporal asymmetry of the meiotic program underlies the predominantly distal distribution of meiotic crossovers in barley. Plant Cell 24:4096–4109
10. Osman K, Algopishi U, Higgins JD et al (2021) Distal bias of meiotic crossovers in hexaploid bread wheat reflects spatio-temporal asymmetry of the meiotic program. Front Plant Sci 12, 631323
11. Krasileva KV, Vasquez-Gross HA, Howell T et al (2017) Uncovering hidden variation in polyploid wheat. Proc Natl Acad Sci U S A 114:E913–E921
12. Chelysheva L, Vezon D, Chambon A et al (2012) The Arabidopsis HEI10 is a new ZMM protein related to Zip3. PLoS Genet 8: e1002799
13. Ijdo JW, Wells RA, Baldini A et al (1991) Improved telomere detection using a telomere repeat probe (TTAGGG)n generated by PCR. Nucleic Acids Res 19:4780
14. Ito H, Nasuda S, Endo TR (2004) A direct repeat sequence associated with the centromeric retrotransposons in wheat. Genome 47: 747–756
15. Gerlach WL, Dyer TA (1980) Sequence organization of the repeating units in the nucleus of wheat which contain 5S rRNA genes. Nucleic Acids Res 8:4851–4865
16. Gerlach WL, Bedbrook JR (1979) Cloning and characterization of ribosomal RNA genes from wheat and barley. Nucleic Acids Res 7:1869–1885
17. Armstrong SJ, Franklin FCH, Jones GH (2003) A meiotic time-course for Arabidopsis thaliana. Sex Plant Reprod 16:141–149
18. Ahn YJ, Cuacos M, Ayoub MA et al (2020) In planta delivery of chemical compounds into barley meiocytes: EdU as compound example. Methods Mol Biol 2061:381–402
19. Chelysheva L, Grandont L, Vrielynck N et al (2010) An easy protocol for studying chromatin and recombination protein dynamics during Arabidopsis thaliana meiosis: immunodetection of cohesins, histones and MLH1. Cytogenet Genome Res 129:143–153
20. Fernández-Gómez J, Talle B, Tidy AC et al (2020) Accurate staging of reproduction development in Cadenza wheat by non-destructive spike analysis. J Exp Bot 71:3475–3484

Chapter 7

Super-resolution Chromatin Visualization Using a Combined Method of Fluorescence In Situ Hybridization and Structured Illumination Microscopy in *Solanum lycopersicum*

Pallas Kuo, Alice Darbyshire, and Christophe Lambing

Abstract

Chromatin organization influences gene and transposon expression, and regulates various cellular processes. Higher order chromatin structure has been widely studied using genomic approaches and microscopy image analyses. Chromosome conformation capture and sequencing the junction of DNA fragments enables the study of both chromatin interaction and chromosome folding. However, certain cell types are embedded in other cell types which complicate the process of studying them using high-throughput genomic approaches. To overcome this limitation, high-resolution microscopy techniques are now available to investigate chromatin organization in single cells. In this chapter, we provide a detailed protocol to prepare chromosome spreading from tomato nuclei, to label genomic loci by fluorescence in situ hybridization, and to visualize these locations at high resolution with Structured Illumination microscopy.

Key words Meiosis, Recombination, Chromatin, Solanum lycopersicum, Fluorescence in situ hybridization, Super-resolution, SIM microscopy

1 Introduction

Chromatin is a dynamic nucleoprotein structure where genomic DNA is organized, and where multiple cellular processes take place. Chromatin is generally organized in two compartments: gene-rich euchromatin, with open chromatin structure; or densely compacted heterochromatin, which is enriched in repeat elements such as transposons [1]. Chromatin adopts specific structures that are associated with defined epigenetic marks to regulate gene and transposon expression [2]. For instance, expression of transposons can be deleterious for the genome, and therefore transposons are typically repressed [3]. The repression of transposons over the constitutive heterochromatic regions is therefore associated with enrichment of the repressive H3K9 methylation mark [3].

Christophe Lambing (ed.), *Plant Gametogenesis: Methods and Protocols*, Methods in Molecular Biology, vol. 2484, https://doi.org/10.1007/978-1-0716-2253-7_7,

The degree of chromatin compaction differs greatly between the stages of a cell and also between cell types. Take meiosis for an example. Meiosis is a specialized cell division taking place in germ cells and the degree of chromatin compaction in meiosis differs significantly from somatic cells. Meiocytes undergo a large number of programmed DNA double strand breaks (DSBs) [4], and meiotic chromatin is formed in arrays of loops that are organized along the chromosome axis [5]. This linear structure of chromatin organization is thought to facilitate the formation of DSBs and to control their repair outcomes [4, 5]. Chromatin compaction can be determined via the associations with certain epigenetic marks, and it can also be distinguished via the use of DNA dyes such as 4′,6-diamidino-2-phenylindole (DAPI). When observed with an epifluorescent microscopy, somatic nuclei are generally compact, and contain a number of DAPI-dense regions known as "chromocenters." In contrast, meiocytes at early prophase I stage appear largely linear in structure, and the chromocenters are generally comprised of constitutive heterochromatic regions [6]. These observations suggest that chromatin interaction is different between somatic and meiotic cells.

Heterochromatic region size is different between plant species, and larger genomes tend to have a higher proportion of transposon elements and a more compact chromatin structure [7]. The varying amount of heterochromatin between species suggests that chromatin interaction can also be different between species. Several techniques have been developed to analyze chromatin organization. For instance, chromosome conformation capture techniques are commonly used to detect the spatial interaction between two distinct chromatin fragments or domains [8]. These techniques are particularly effective to provide information on long-range chromatin interactions, such as in between an enhancer of transcription, and the promoter of a gene and/or topologically associating domains (TADs), which are self-interacting genomic regions [2, 8]. The presence of TADs has been identified in mammals including humans [1] and mice [9]. In plants, TADs were reported in rice, but not in Arabidopsis and wheat [10–14]. Therefore, the study of chromatin interactions in species where TADs are not detectable rely on other techniques. Chromosome spreading, followed by DNA staining and/or immunostaining, for example, reveals protein–chromatin interaction over the entire genome. However, this technique does not give detailed information on specific chromatin regions. To circumvent this limitation, fluorescence in situ hybridisation (FISH) uses a probe to hybridize to a specific genomic sequence which is labelled with a fluorescence dye for visualization. This strategy has provided crucial insights in centromere and telomere pairing in plants [15, 16]. However, conventional widefield fluorescence microscopy has a limited resolution and a range of super-resolution microscopes is now available to use at observing biological processes at a subcellular level [17, 18]. Combining FISH with super-resolution microscopy can provide important new information on chromatin organization at an unprecedented resolution. In this protocol, we describe the preparation of chromosome spreading and detection of chromatin with FISH probes on tomato nuclei. We combine this technique with Structured Illumination Microscopy (SIM) to visualize chromatin at a resolution of ~120 nm in *xy*- and 300 nm in *z*-dimension.

2 Materials

2.1 Plant Growth and Tissue Collection

1. Grow *Solanum lycopersicum* cv M82 in a controlled growth room under long day conditions (16 hrs a day at 300 μmol light intensity–8 hrs night at dark) at 23 °C.
2. Fixative solution: 3:1 (v/v) ethanol–glacial acetic acid.
3. 15 ml falcon tubes and pipettes.

2.2 Chromosome Spread

1. Citrate buffer: 4.45 mM citric acid, 5.55 mM sodium citrate.
2. Enzyme mix: 1% (w/v) cellulase, 1% (w/v) pectolyase, 1% (w/v) cytohelicase dissolved in citrate buffer.
3. Fixative solution: 3:1 (v/v) ethanol–glacial acetic acid.
4. 60% glacial acetic acid solution diluted in deionized water.
5. Staining solution: 4′,6-diamidino-2-phenylindole at a concentration of 10 μg/ml in anti-fading Vectashield mounting medium.
6. Pipettes, jar, moist boxes, fine forceps, brass rod, microscope slides and coverslips.
7. Binocular magnifier.
8. Incubator.
9. Hotplate.
10. Hair dryer.

2.3 Fluorescence In Situ Hybridisation

1. 20× SSC: 3 M NaCl, 300 mM Trisodium citrate, pH = 7.0 adjusted with 1 M HCl.
2. DNase-free RNase A: 100 μg/ml of DNase-free RNase A in 2×SSC.
3. Pepsin solution: 10 mg of pepsin in 100 ml of 10 mM HCl.
4. 4% paraformaldehyde solution: 4 g of paraformaldehyde dissolved in 100 ml of deionized water.
5. Ethanol solution at 70%, 85% and 100%. Diluted with deionized water.

6. Hybridisation solution: 1 g of dextran sulfate, 5 ml of deionized formamide, 1 ml of 20× SSC, fill up to 7 ml with deionized water.
7. Formamide solution: 50% of formamide, 2× SSC.
8. 4T: 4× SSC, 0.05% (v/v) of Tween 20.
9. Pipettes, moist box, glue, coverslip, jars.
10. Incubator.
11. Heating block.

2.4 Structured Illumination Microscopy

This protocol was optimized using a ZEISS Elvra 7 Microscope with Lattice SIM. Alternative microscopes can be used to achieve super-resolution images, but image acquisition may differ between microscope specifications.

1. Microscope with Lattice SIM Capacity.
2. Carl Zeiss Immersol Immersion Oil (518F, RF 1.518).
3. ZEISS ZEN Black Software.
4. Slide cleaning solution: 70% of ethanol in deionized water.
5. Cotton swabs.

3 Methods

3.1 Fixation of Tissue and Chromosome Spreading

1. Collect plant tissues (*see* **Note 1**) with a forceps and incubate them in fixative solution for 4 h at 4 °C. After a day, replace with fresh fixative solution and repeat this step for two more times. Keep the samples at all time at 4 °C. A discoloration of the tissue should be visible at this point.
2. Transfer the fixed tissue to a jar and add fresh fixative solution. If the tissue is exclusively somatic (e.g., leaf), proceed to the next step. If the aim of the experiment is on meiosis, dissect the inflorescences and keep the desired floral buds before transferring them into the jar. Discard the rest of the inflorescence.
3. Remove the fixative solution and incubate the tissue with citrate buffer for 2 min.
4. Remove the citrate buffer and add a second volume of citrate buffer. Incubate for 2 min. Repeat one more time.
5. Remove the citrate buffer and add a solution of enzyme mix. Place the jar in a moist box and incubate the box at 37 °C for 1.5 h (*see* **Note 2**).
6. Remove the enzyme solution and replace with citrate buffer.
7. Transfer one bud or a section of somatic tissue to a drop of water on the center of a microscopic slide. Gently tap the tissue with a brass rod until a white suspension become visible.

8. Add two times 5 μl of 60% acetic acid and gently mix the suspension.
9. Place the slide on a hotplate and incubate at 48 °C for 1 min.
10. Add 85 μl of ice-cold fixative solution around the drop of acetic acid and gently move the slide to mix the two solutions.
11. Dry the slide upside down with a hair dryer.

3.2 Fluorescence In Situ Hybridization

All the steps are performed at room temperature, unless otherwise mentioned.

1. Incubate the microscopic slides in a jar containing a solution of 2×SSC for 10 min.
2. Add a solution of RNase A on the slide and cover the solution with Parafilm. Place the slides in a moist box and incubate for 1 hr at 37 °C.
3. Remove the parafilm and incubate the slides in 2×SSC for 5 min.
4. Transfer the slides to a pepsin solution and incubate for 90 s at 37 °C (*see* **Note 3**).
5. Transfer the slides to a solution of 2×SSC for 10 min.
6. Transfer the slides to a solution of 4% paraformaldehyde for 10 min.
7. Briefly rinse the slides in deionized water.
8. Transfer the slides respectively to a solution of 70%, 85%, and 100% ethanol for 2 min each time.
9. Leaves the slides to air-dry for 2 min.
10. Add 20 μl of the first probe solution (probe diluted in hybridisation solution) on the slides and add a cover glass (*see* **Note 4**).
11. Seal the cover glass with glue.
12. Place the slides on a hot plate at 75 °C for 4 min.
13. Incubate the slides in a moist box at 37 °C for 1–2 days (*see* **Note 5**).
14. Gently remove the glue and the cover glass.
15. Incubate the slides in a solution of formamide at 45 °C for 5 min (*see* **Note 6**).
16. Discard the formamide solution and repeat **step 15** two more times.
17. Transfer the slides in a solution of 2×SSC at 45 °C for 5 min.
18. Transfer the slides in a solution of 4T at 45 °C for 5 min.
19. Transfer the slides in a solution of 4T for 5 min.

20. Add the second antibody solution (volume according to the manufacturer's recommendations) and incubate the slides in a moist box at 37 °C for 30 min (*see* **Notes 7** and **8**).
21. Transfer the slides in a solution of 4T for 5 min.
22. Repeat **step 21** for two more times.
23. Add a solution of DAPI and add a cover glass. Proceed with image capture or store the slides in the dark at 4 °C.

3.3 Image Capture Using Structured Illumination Microscopy

1. Clean the prepared slide with slide cleaning solution on a clean cotton swab. Take care to ensure no dirt or old immersion oil remaining on the coverslip.
2. Apply a small amount of immersion oil onto the coverslip and place on the stage in the microscope.
3. Identify cells for capture by utilizing the "Locate" tab within the ZEN Black software. Use the FSet77HE filter block which will allow viewing of red and green channels (*see* **Note 9**). Locate cells using the eyepieces.
4. Once cells have been identified, capture cells using the "Acquire" tab within the ZEN Black software. First, ensure that "Technique" is set to "Lattice SIM," and all appropriate lasers are turned on. Ensure the filter is set to LBF (quad band). Under the "Channels" drop-down, add and remove new "Tracks." Within each track, assign a laser and a pseudocolor for viewing the image. Under the "Live" or "Continuous" tab, view the target cell. Set focus, and exposure level/laser power (under the "Channels" drop-down, selecting one track to edit at a time). Capture the image when ideal settings for that cell are obtained. The captured image will be the unprocessed, raw image.
5. Under "Processing" in the ZEN Black software, select the "SIM" option on the menu. Load in the unprocessed image and select "Apply" to run the SIM program. The image obtained after processing is the final image (Fig. 1), saved by

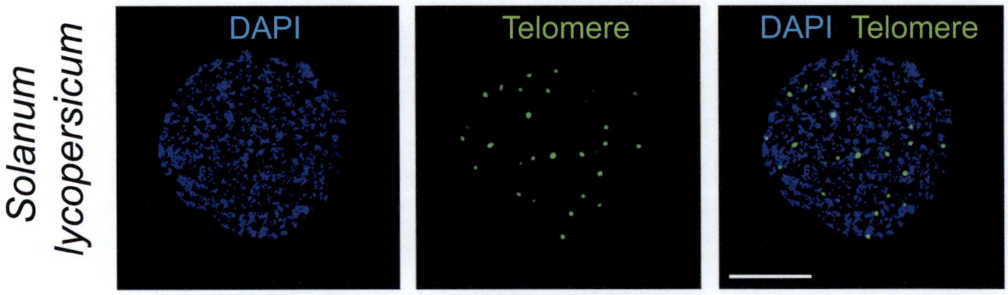

Fig. 1 Fluorescence in situ hybridisation of telomeres in *Solanum lycopersicum* leaf nucleus. Chromosome spread from *Solanum lycopersicum* leaves were hybridized with a telomeric probe conjugated with FITC dye. Chromatin was stained with DAPI. Scale bar = 10 μM

default as a .czi file. Any further processing can be conducted using ZEN Black or Blue software, or by loading the image into FIJI/Image J using the "Bio-Formats Importer" plugin.

4 Notes

1. To study meiosis, collect unopened floral buds from young plants (8 week-old plants for *Solanum lycopersicum* cv M82).
2. The optimal incubation time for enzyme digestion may differ between tissue types and it is recommended to test different conditions. Digestion of leaf and bud tissue is achieved after an incubation of 1.5 hrs.
3. The optimal duration for the digestion step with pepsin may vary between tissue types.
4. The optimal dilution for each probe should be determined.
5. The appropriate incubation time should be tested for each probe. Several factors influence the incubation time: the dilution of the probe, the size of the region targeted (small regions targeted by the probe may require longer incubation time than larger regions) and the level of nonspecific background.
6. **Steps 15** and **16** in Subheading 3.3 should be performed in a fume hood.
7. Antibody dilution should be tested and determined for each antibody.
8. If the probes are conjugated with a fluorescence dye, skip **steps 20–22** in Subheading 3.2.
9. Filter blocks used will depend on the microscope setup. This information will be available from the manufacturer/sales representative.

Acknowledgments

We would like to thank Alessandro di Maio (University of Birmingham) for his advice and knowledge of SIM. This work was supported by a BBSRC grant-aided support as part of the Institute Strategic Programme designing Future Wheat Grant (BP/P016855/1) and the BBSRC grant 1644151.

References

1. Lieberman-Aiden E, van Berkum NL, Williams L et al (2009) Comprehensive mapping of long-range interactions reveals folding principles of the human genome. Science 326:289–293
2. Gibcus JH, Dekker J (2013) The hierarchy of the 3D genome. Mol Cell 49:773–782
3. Law JA, Jacobsen SE (2010) Establishing, maintaining and modifying DNA methylation

patterns in plants and animals. Nat Rev Genet 11(3):204–220

4. Mercier R, Mézard C, Jenczewski E et al (2015) The molecular biology of meiosis in plants. Annu Rev Plant Biol 66:297–327
5. Kleckner N (2006) Chiasma formation: chromatin/axis interplay and the role(s) of the synaptonemal complex. Chromosoma 115(3): 175–194
6. Pecinka A, Chevalier C, Colas I et al (2020) Chromatin dynamics during interphase and cell division: similarities and differences between model and crop plants. J Exp Bot 71(17): 5205–5222
7. Lambing C, Franklin FCH, Wang CJR (2017) Understanding and manipulating meiotic recombination in plants. Plant Physiol 173: 1530–1542
8. Sati S, Cavalli G (2017) Chromosome conformation capture technologies and their impact in understanding genome function. Chromosoma 126:33–44
9. Dixon JR, Selvaraj S, Yue F et al (2012) Topological domains in mammalian genomes identified by analysis of chromatin interactions. Nature 485:376–380
10. Liu C, Cheng YJ, Wang JW et al (2017) Prominent topologically associated domains differentiate global chromatin packing in rice from Arabidopsis. Nat Plants 3:742–748
11. Golicz AA, Bhalla PL, Edwards D et al (2020) Rice 3D chromatin structure correlates with sequence variation and meiotic recombination rate. Commun Biol 2:235
12. Dong Q, Li N, Li X et al (2018) Genome-wide Hi-C analysis reveals extensive hierarchical chromatin interaction in rice. Plant J 94: 1141–1156
13. Feng S, Cokus SJ, Schubert V et al (2014) Genome-wide Hi-C analyses in wild-type and mutants reveal high-resolution chromatin interactions in Arabidopsis. Mol Cell 55:694–707
14. Concia L, Veluchamy A, Ramirez-Prado JS et al (2020) Wheat chromatin architecture is organized in genome territories and transcription factories. Genome Biol 21(1):104
15. Armstrong SJ, Franklin FC, Jones GH (2001) Nucleolus-associated telomere clustering and pairing precede meiotic chromosome synapsis in Arabidopsis thaliana. J Cell Sci 114:4207–4217
16. Zhang J, Pawlowski WP, Han F (2013) Centromere pairing in early meiotic prophase requires active centromeres and precedes installation of the synaptonemal complex in maize. Plant Cell 25:3900–3909
17. Lambing C, Osman K, Nuntasoontorn K et al (2015) Arabidopsis PCH2 mediates meiotic chromosome remodeling and maturation of crossovers. PLoS Genet 11:e1005372
18. Darrier B, Arrieta M, Mittmann SU et al (2020) Following the formation of synaptonemal complex formation in wheat and barley by high-resolution microscopy. Methods Mol Biol 2061:207–215

Chapter 8

Application of Chemical Inhibitors in Live Cell Imaging of Plant Meiosis Using Light Sheet Fluorescence Microscopy

Sona Valuchova, Pavlina Mikulkova, Jana Pecinkova, and Karel Riha

Abstract

Live imaging combined with the application of chemical inhibitors is a powerful research tool that enables researchers to precisely time the inhibition of cellular processes and study the consequences of these perturbations. This approach is usually applied to in vitro cultivated cells that are easily accessible to chemical treatments and microscopic observations. Here we describe a method for live cell imaging of Arabidopsis meiocytes embedded within floral organs combined with the application of a chemical drug at desired timepoints during meiosis. We describe a customized solution for the Zeiss Z.1 light sheet microscope, including sample preparation and data processing, and demonstrate its utility for the analysis of meiotic progression upon spindle inhibition.

Key words Live imaging, Meiosis, Reproduction, Light sheet fluorescence microscopy, LSFM, Germline, Oryzalin, Arabidopsis

1 Introduction

Flowering plants represent 90% of land plants and are adapted to most land environments. Their evolutionary advantage derives at least partly from protecting reproductive cells within flower organs. However, this anatomical protection makes plant sexual reproduction less accessible to microscopy techniques and hinders real-time observation of cellular events governing germline differentiation. Many processes involved in sexual reproduction, including germ cell differentiation, meiosis, gametophyte maturation, fertilization, and embryogenesis, take place underneath multiple cell layers that form the sexual organs. One way to overcome this challenge is through in vitro cultivation of excised reproductive cells such as ovules, pollen mother cells, and zygotes, in microscopic chambers [1–3]. However, viability of excised cells in cultivation media is a

Christophe Lambing (ed.), *Plant Gametogenesis: Methods and Protocols*, Methods in Molecular Biology, vol. 2484, https://doi.org/10.1007/978-1-0716-2253-7_8,

limiting factor. Arabidopsis has relatively small flowers with reproductive cells at a depth that is accessible to confocal microscopy [4]. Hence, at least in Arabidopsis, it is possible to subject the plant germline to microscopic observation in the context of an entire floral bud. This approach has been applied for live imaging of male meiosis in pollen mother cells by confocal microscopy for up to 30 h [5]. Phototoxicity and growth of the sample out of the field of view are the main obstacles for imaging over longer periods of time. Light sheet fluorescence microscopy (LSFM) proved to be a good solution to these problems [6, 7]. It uses a thin sheet of laser to illuminate a sample in a cross-section and capture all emitted fluorescence. The sample can be continuously imaged while passing through the laser sheet, which allows for very fast three-dimensional image acquisition of objects several millimeters in size, while minimizing exposure of cells to the laser. In some setups, a sample can be rotated in front of the objective to acquire images from multiple angles for more faithful 3D reconstruction. In plants, LSFM has mainly been used for root imaging [7–9], but this technology has also recently been applied for live-imaging of developing Arabidopsis flowers [10]. This highly versatile approach enables various aspects of plant sexual reproduction to be visualized at subcellular resolution, including chromosome segregation in male and female meiosis, rapid movements of pachytene chromosomes, and asymmetric pollen mitosis.

Chemical genetics is an experimental approach whereby small chemical compounds are used to perturb specific molecular functions to study particular processes in living systems [11]. Real time imaging combined with chemical inhibitor treatment is a powerful research tool that permits essential cellular functions to be distributed at precise times and the resulting perturbations to be recorded with high spatiotemporal resolution. This approach is often used to study diverse aspects of cell cycle progression and has also been applied in meiosis research [12, 13]. Here we present a live cell imaging protocol of Arabidopsis meiosis by LSFM that includes the application of a chemical drug at precise timepoints during meiotic progression. We describe in detail the preparation of samples for imaging male and female meiocytes and their treatment with oryzalin, a dinitroaniline herbicide that disrupts polymerization of microtubules [14]. We present two different methods of applying the drug into the microscopic chamber: (1) buffer exchange that brings the inhibitor rapidly to the sample, but results in sample drift, and (2) buffer addition that retains sample position throughout the experiment. Using live cell imaging of meiotic spindle and chromosome segregation we demonstrate that both methods result in efficient delivery of the drug to meiotic cells.

2 Materials

2.1 Seeds and Soil

1. Seeds of *Arabidopsis thaliana* ecotype Columbia-0 harboring HTA10:RFP to visualize chromatin and pRSP5A::G3GFP: TUA5 for detection of microtubules.
2. 12% sodium hypochlorite (NaClO)
3. 35% HCl
4. Peat soil (TS3 type).
5. Vermiculite.
6. Parasitic nematodes, Steinernema.
7. Filter paper.
8. Yellow sticky labels.

2.2 List of Solutions

1. 5MS liquid chamber medium: ½ MS + 5% sucrose. Weigh 2.45 g of Murashige and Skoog Medium including vitamins and MES buffer and 50 g of sucrose and transfer to a volume of 900 ml of demineralized water. Adjust pH to 5.7 with 1 M of KOH solution. Autoclave and make 50 ml aliquots
2. 5MS solid mounting: ½ MS + 5% sucrose + 1% low gelling agarose. Take 50 ml of freshly prepared and pH-adjusted 5MS liquid media and transfer it to a glass bottle. Add 0.5 g of low gelling temperature agarose. Autoclave and aliquot into Eppendorf tubes prior to solidification. Store aliquots at 4 °C (*see* **Note 1**)
3. Oryzalin stock solution (100 mM). Dissolve 100 mg oryzalin in 3 ml DMSO (dimethylsulfoxide) and store at −20 °C. Prepare final solution fresh in liquid 5MS: 33 μM for the buffer exchange method or a working solution of 305.25 μM for the buffer addition method. Keep the oryzalin solution at room temperature to avoid temperature shock for the sample.

2.3 Sample Preparation

1. 2× tweezers
2. Microscopic ruler (Blank Slide with 0.1 and 0.01 mm scale).
3. Sterile 1.5 ml Eppendorf tube.
4. Pipetman volume 10–100 μl + appropriate sterile tips.
5. Stereomicroscope.

2.4 Sample Mounting

1. Glass capillary (size 4).
2. Piston (size 4).
3. Thermoblock.

2.5 Microscope

Microscope and all accessories are obtained from Zeiss (*see* **Note 2**).

1. Capillary holder (Sample holder for capillary for Light Sheet Z.1),
2. Capillary holder inserts (blue, for capillary size 4),
3. Light sheet Z.1 microscope.
4. Light Sheet Z.1 Detection Optics 20×/1.0 (W Plant-Apochromat 20×/1.0 Corr DIC M27 75mm_4909000228).
5. Lightsheet Z.1 Illumination Optics 10×/0.2.
6. Chamber type: Sample Chamber with carriage Lightsheet Z.1.
7. Peltier block S (Heating Block Sample Chamber Lightsheet Z.1).
8. Incubation Temperature Sensor Light Sheet Z.1.
9. PC with good storage and processing capabilities with ZEN software for Light Sheet.
10. Syringe A (volume 50 ml).
11. Syringe B (volume 20 ml).
12. Syringe tube Hose (sterile).
13. Luer lock connectors.

2.6 Software

1. ZEN lightsheet + deconvolution.
2. FIJI [15].
3. Imaris 9.2.1.

3 Methods

3.1 Seed Sterilization

1. Place ~30 μl of seeds into open eppendorf tubes organized in a plastic box.
2. Insert box into a desiccator under a fume hood.
3. Pour 100 ml of 12% sodium hypochlorite in a glass beaker inside the desiccator.
4. Add 10 ml of 35% HCl to the hypochlorite solution (*see* **Note 3**).
5. Immediately close the desiccator and incubate the seeds with the chlorine gas that is produced for 5 h.
6. Open the desiccator within the fume hood, close the box with seeds to prevent contamination, and transfer it to a laminar flow hood.
7. Open the box and air dry the seeds for at least for 15 min in laminar flow hood before use.

3.2 Soil Preparation

1. Mix the peat soil (TS3 type) with vermiculite in a 3:1 ratio.
2. Fill pots with soil and soak the soil with water.
3. Add two small spoons of nematodes into 10 l of water and stir.
4. Water pots with nematode solution.
5. Leave the pots in the dark for 1–2 days before sowing Arabidopsis seeds (*see* **Note 4**).

3.3 Plant Preparation

1. Place a filter paper into a petri dish and add a small amount of water to make it wet.
2. Apply sterile Arabidopsis seeds onto the filter paper and cold stratify for 2 days at 4 °C in the dark before sowing.
3. Plant seeds using toothpicks on the pots with soil (*see* **Note 5**).
4. Grow plants under long day light conditions (21 °C, 16 h light–8 h dark) until the main stem produces 10 flowers (*see* **Note 6**) (Fig. 1a).

3.4 Sample Preparation

1. Put a drop of room temperature liquid 5MS medium (100 µl) on the microscopic slide ruler.
2. Cut the inflorescence using tweezers from the main stem of the plant.
3. For male meiosis, select buds 0.45 mm wide, for female meiosis, select buds 0.80 mm wide (*see* **Note 7**) (Fig. 1b).
4. To prepare samples for visualizing male meiosis, remove all sepals and petals from the appropriate-sized bud to expose anthers (*see* **Note 8**) (Fig. 1c, d) and keep it the in liquid 5MS.
5. To prepare samples for female meiosis, remove all sepals, petals, and anthers, cut off the stigma and gently remove valves of the pistil to expose ovules and separate carpels from each other (Fig. 1e, f). Keep the carpel that is still attached to the stem in liquid 5MS.

3.5 Sample Mounting

1. Melt an aliquot of solid 5MS media in an Eppendorf tube in the thermoblock at 80 °C.
2. Allow the thermoblock to cool to 37 °C and keep the media there until use.
3. Choose the appropriate piston (*see* **Note 9**).
4. Insert the piston into the glass capillary (Fig. 2f).
5. Suck the melted and cooled solid 5MS medium into the capillary using the piston like a syringe. Suck the 5MS medium approximately 2 cm into the capillary.
6. Using tweezers, gently take the prepared sample and mount it into the medium by gently pushing it inside the capillary (*see* **Note 10**).

Fig. 1 Plant material and sample preparation. (**a**) A healthy *A. thaliana* ecotype Columbia-0 in an optimal stage of development for the most consistent imaging results. Arrowhead indicates inflorescence used for the bud preparation. (**b**) A detached bud with sepals and petals. (**c**) Floral bud with sepals removed. (**d**) Floral bud without sepals embedded in solid 5MS media. (**e**) Pistil after removal of sepals and anthers prior to carpel separation. (**f**) Dissected pistil with removed carpels embedded in solid 5MS medium

7. Try to place the sample into the central axis of the capillary so that it remains centered when the capillary rotates during imaging from multiple angles (*see* **Note 11**). Wait 2 min for the 5MS media to completely solidify.
8. Suck additional liquid 5MS into the capillary and let the sample solidify for an additional 20 min at room temperature (*see* **Note 12**).
9. Before placing the sample into the microscope, push out any remaining liquid 5MS from the capillary so that only solid agarose and the mounted sample remain (*see* **Note 13**).

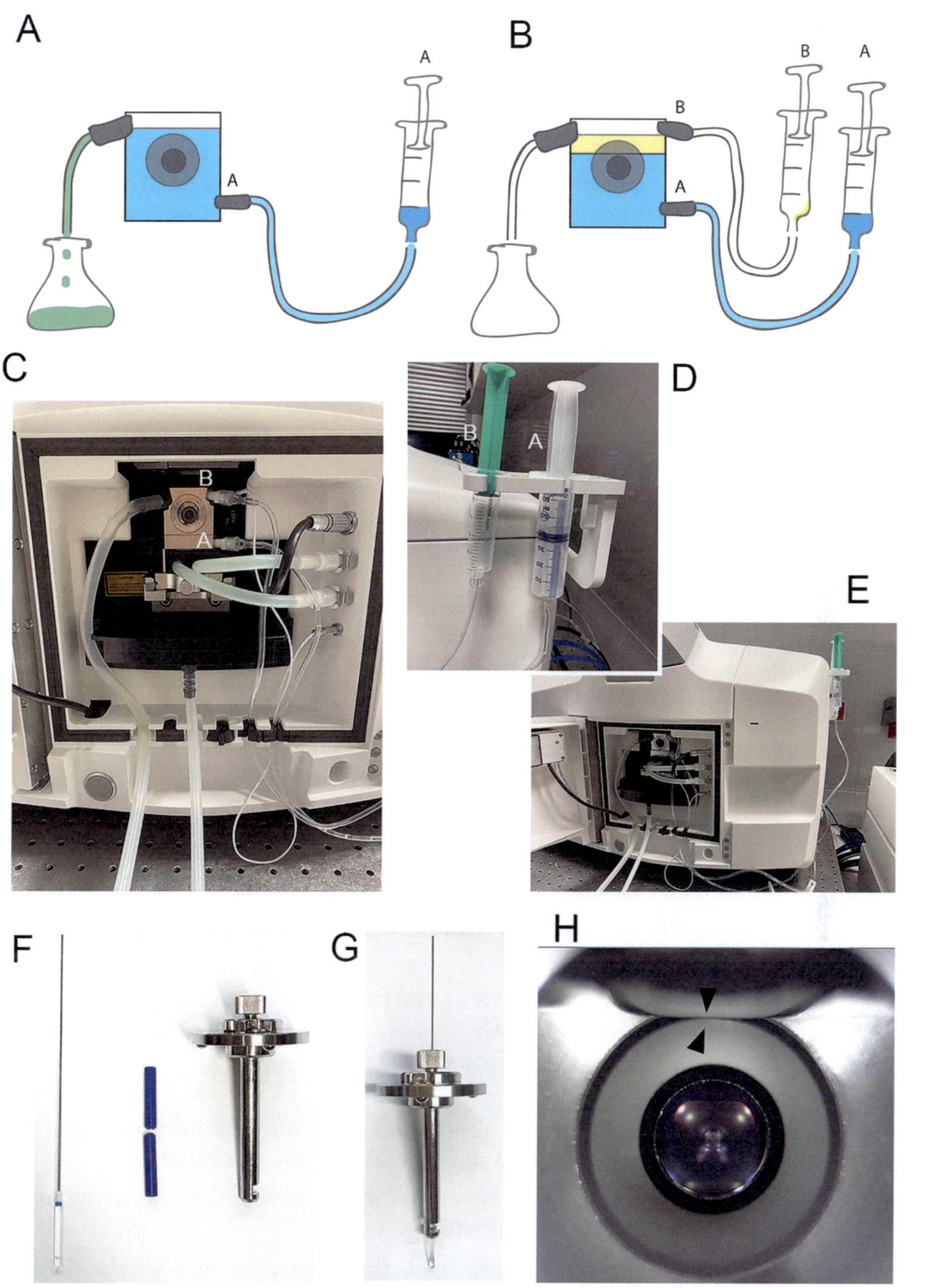

Fig. 2 Setup of the microscope. (**a**) A schematic of the buffer exchange setup. Syringe A is used to fill the chamber with liquid 5MS medium through the A opening. Excess buffer can flow through the exit opening into the waste. Media can be gently drained from the chamber by detaching the tubing from the syringe, allowing the media to flow out by gravity. After reattachment, syringe A is used to deliver inhibitor into the media. (**b**) A schematic of the buffer addition setup. No waste is generated by this method. Syringe A is used to deliver 5MS medium through the A opening for the initial imaging. Syringe B is used to deliver the inhibitor at the desired timepoint through the B opening. (**c–e**) Images of the setup for buffer addition depicting the tubing and syringes. (**f**) Capillary with solid 5MS medium, holder inserts and sample holder prior to assembly. (**g**) Assembled sample in the sample holder (**h**) Level of 5MS medium added during the top application procedure prior to inhibitor application. This amount of media is sufficient to image the sample

3.6 Imaging

1. Mount the syringe tubing to the bottom right screw of the microscope chamber (Fig. 2a, c).
2. Fill syringe A with liquid 5MS media. For the buffer addition method this volume must be exactly 23 ml (16.5 ml inside the chamber, 6.5 ml inside the tubing, and 1 ml inside syringe A for accurate application), for the buffer exchange we use usually 50 ml (Fig. 2a, d; *see* **Note 14**).
3. Press the piston to fill the microscope chamber with media, but leave the tubing and syringe filled with media (Fig. 2a, h).
4. Place the capillary with the sample into the sample holder (Fig. 2f, g).
5. Insert the sample holder into the microscope chamber.
6. Close the microscope.
7. Using ZEN software for light sheet, locate the sample.
8. Set laser parameters.

   Laser: 488-30, 561;

   2 Tracks, Switch track every: frame fast,

   Light path: LBF 405/488/561/640, SBS LP560, BP 505-454 (Camera Track2) BP 575-615 (Camera Track 1).

   Bit Depth 16 Bit.

   Laser intensity Track 1: 488 = 0, 561 = 10; Track 2: 488 = 10, 561 = 0.

   Exposure 50–70 ms.

   Incubation: 21 °C
9. Switch on pivot scan to decrease shadowing effect.
10. Set zoom to the maximum value where the sample still fits the field of view.
11. Set the best views and timepoints in Multiview and Time series (*see* **Note 15**).
12. Start imagining (*see* **Note 16**).

3.7 Application of an Inhibitor

3.7.1 Inhibitor Application from Bottom (Buffer Exchange Method)

1. Prepare 50 ml of liquid 5MS with inhibitor at a final concentration of 33 μM in a falcon tube (*see* **Note 17**).
2. Track the progress of meiosis in your sample during imagining until the desired stage occurs.
3. Between scans, open the front door and let the 5MS gently flow out of the chamber by detaching the tube from the syringe and putting the tube end below the chamber level into a waste container (*see* **Note 18**).
4. Take the detached syringe and remove the piston.
5. Reattach the tube to the syringe.
6. Fill the syringe with the prepared media with inhibitor.

7. Attach the piston and push gently (*see* **Note 19**) to transfer the media containing the inhibitor into the chamber with the sample.
8. Very carefully, close the microscope door (*see* **Note 20**).

3.7.2 Inhibitor Application from Top (Buffer Addition Method)

1. Prepare 2 ml of liquid 5MS with oryzalin in working concentration.
2. Track the progress of meiosis in your sample during imagining until the desired stage occurs.
3. Fill the smaller syringe B (volume 20 ml) with 2 ml of inhibitor in working concentration and as much air as possible.
4. Gently apply the full volume into the chamber through the B opening by pressing the air left inside of the syringe (Fig. 2b–d).
5. Very carefully, close the microscope door (*see* **Note 21**).

Results obtained with the buffer exchange and buffer addition methods are shown in Fig. 3.

3.8 Image Processing

1. Check the quality of data in ZEN lightsheet.
2. Subset data in ZEN to speed up the downstream processing (*see* **Note 22**).
3. Optional: Deconvolution in ZEN (Regularized inverse filter strength 2) (*see* **Note 23**).
4. Open data in FIJI.
5. Visualize the cells of interest by choosing single *Z* plane or create maximum intensity projection of a few layers for deeper structures. In case of drift in the *Z* axis, manually select *Z* planes for each timepoint or set of timepoints. Function Duplicate (Ctrl+Shift+D) allows you to choose timepoint and layer range. To create a time-lapse, concatenate selected images using Concatenate (Image Stacks Tools Concatenate) (*see* **Note 24**).
6. If the sample also drifted in the *XY* planes, use automatic drift correction (Plugins Registration Correct 3D drift) [16] (*see* **Note 25**).
7. If automatic drift correction does not completely correct the drift, you may select a region of interest (ROI) and add it to ROI manager (Analyze Tools ROI Manager). Manually duplicate ROI from each timepoint, or set of timepoints, and concatenate back to the time series. You may now try to correct XY automatically again.
8. Export in preferred format,.avi for movies, and png or jpg for figures.

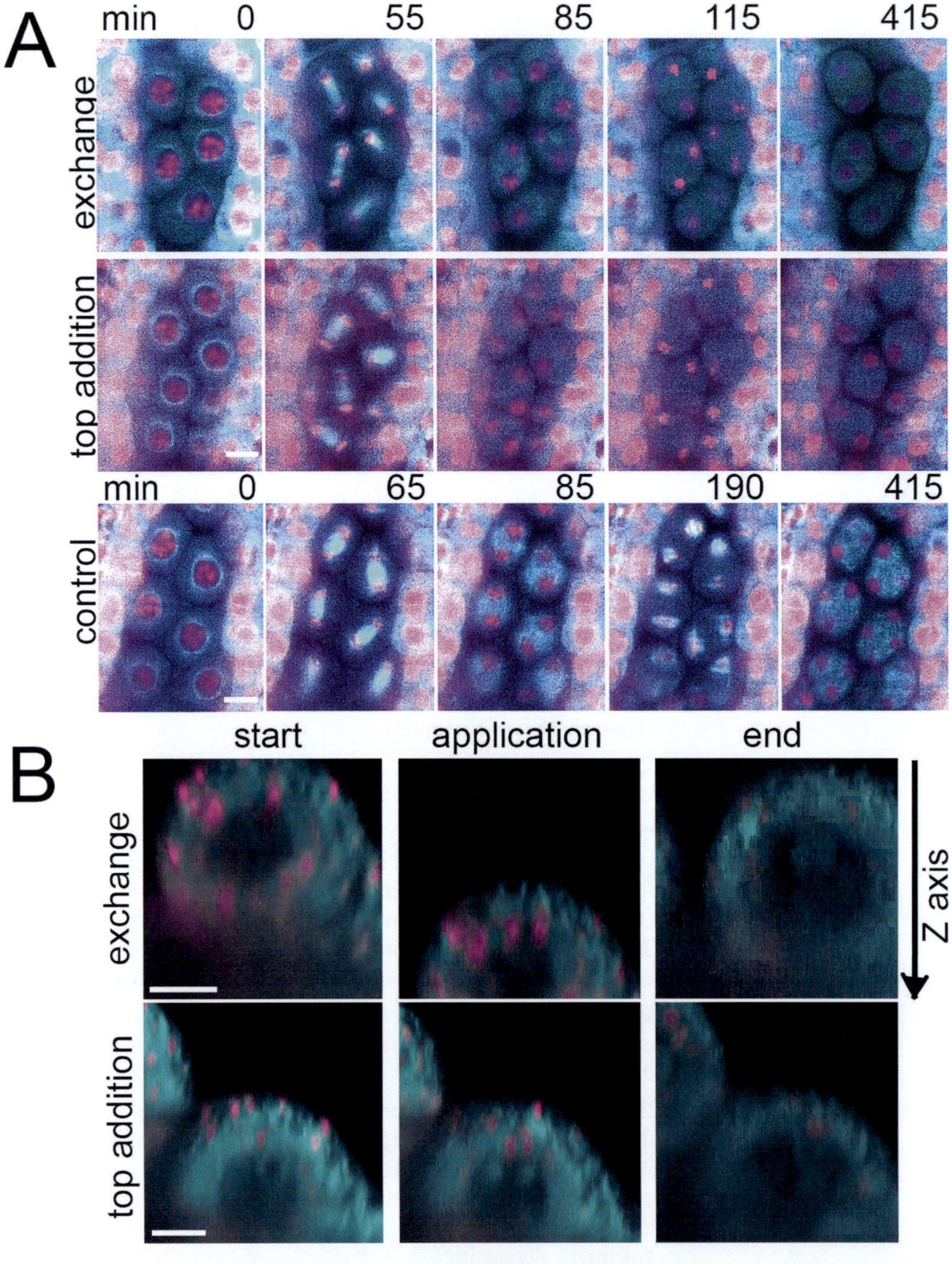

Fig. 3 Comparison of the buffer exchange and buffer addition method. (**a**) Oryzalin or 5MS medium without oryzalin (control) were added during anaphase I of male meiosis. Chromatin is visualized using the HTA10:RFP construct (violet) and spindle using the pRSP5A::G3GFP:TUA5 (blue). Both methods resulted in efficient delivery of oryzalin to PMCs as no spindle was formed in meiosis II, while meiosis completed in the control sample. (**b**) Temporary drift observed during the application of oryzalin. The buffer exchange protocol causes sample drift in the *Z* axis, but reversed over the course of the experiment. No apparent drift was observed using the buffer addition method. The view was rotated using Imaris to show shift along the *Z* axis

4 Notes

1. Monitor the optical properties of the medium. If images become unclear, consider preparing new medium.
2. This method is optimized for the Light Sheet Z.1 (Zeiss) microscope. Using other light sheet microscopes will require customized solution specifically tailored for given microscope.
3. Be very careful but quick, it is a fast chemical reaction that releases toxic chlorine gas.
4. We use nematode treatment to prevent black flies. Infestation alters plant growth and leads to poor results both biologically and optically.
5. Use yellow sticky labels as an indicator of black fly infestation and also to help control them. Black flies are attracted by yellow color.
6. The most consistent results were obtained with the 10th–25th flowers from the main stem. Optimal light, water, humidity, and a pest-controlled environment are crucial for consistency of results and image quality. Therefore, phytotron cultivation provides better results than greenhouse cultivation. Male meiosis is generally more consistent than female. Old and stressed plants give poor images.
7. Buds of this size usually contain anthers at the end of prophase I before the start of meiotic divisions.
8. Work directly in the drop of liquid 5MS medium to avoid your sample drying out.
9. The piston should not move easily in the capillary. If the piston is reused, the sealing ability is usually not sufficient and results in sample drift. To combat this, slightly peel the white part of the piston with your finger nails to enlarge the contact zone.
10. Leave approximately 1–2 mm below the sample to obtain an optimal optical path.
11. Proximity to the wall of the capillary may cause light aberration due to curvature of the media.
12. The sample needs some time to properly settle inside the medium. Proper solidification helps reduce sample drift during microscopy. Liquid 5MS medium prevents the agarose block from drying.
13. The sample stays alive for approximately 10–18 h. A continuous flow of fresh media did improve survival in some cases, but overall image quality decreased rapidly due to biofilm formation on glass slides of the chamber.

14. To fill the whole chamber, we prefer to use 50 ml of liquid 5MS. That accounts for the chamber volume, tube volume, and leaves extra medium for application error. Keeping medium inside the tubing and syringe helps prevent media flow from the chamber. However, 25 ml would be sufficient to fill the chamber.
15. We usually set as many views as useful, single illumination is usually sufficient for loculi and ovules. It is necessary to add extra layers in case of sample drift. Time increments depend on the studied process.
16. Make sure there is enough storage space on your PC and that the name of the file is not too long as this could prevent from saving your files.
17. Room temperature media are crucial to avoid cold stress to the sample, chamber volume is 25 ml, however we use extra volume to keep the tubing full and avoid drift.
18. Do not suck out the media, it worsens the sample drift.
19. This is a very crucial step, be as gentle as possible to avoid drift.
20. You must complete the procedure before the next scan occurs, the media and sample must also settle in the new environment. Write down the timepoint before and after media addition. Video remains uninterrupted. Despite the best efforts, this method generates some sample drift.
21. While diffusion lasts longer in this type of application, buffer addition minimizes sample drift and reduces image processing.
22. ZEN proved to be faster than FIJI in subsetting in our hands.
23. In case of dual side illumination, image fusion occurs. If you do not wish to fuse, subset properly in **step 2**.
24. You can duplicate more than one layer and do maximum intensity projection afterward to visualize elements present in different layers, it also helps to deal with the plane inaccuracy present in drifted samples.
25. Correct 3D drift could also be used to correct *Z* drift. However, this never worked in our hands, probably due to the poor sampling in *Z*.

Acknowledgments

Our work is supported by the Ministry of Education, Youth and Sports of the Czech Republic from the European Regional Development Fund—Project “REMAP” (No. CZ.02.1.01/0.0/0.0/15_003/0000479).

References

1. Gooh K, Ueda M, Aruga K et al (2015) Live-cell imaging and optical manipulation of Arabidopsis early embryogenesis. Dev Cell 34(2): 242–251
2. Hamamura Y, Saito C, Awai C et al (2011) Live-cell imaging reveals the dynamics of two sperm cells during double fertilization in Arabidopsis thaliana. Curr Biol 21(6):497–502
3. Sheehan MJ, Pawlowski WP (2009) Live imaging of rapid chromosome movements in meiotic prophase I in maize. Proc Natl Acad Sci U S A 106(49):20989–20994
4. Prunet N, Jack TP, Meyerowitz EM (2016) Live confocal imaging of Arabidopsis flower buds. Dev Biol 419(1):114–120
5. Prusicki MA, Keizer EM, van Rosmalen RP et al (2019) Live cell imaging of meiosis in Arabidopsis thaliana. Elife 8:e42834
6. Keller PJ (2013) Imaging morphogenesis: technological advances and biological insights. Science 340(6137):1234168
7. Ovecka M, von Wangenheim D, Tomancak P et al (2018) Multiscale imaging of plant development by light-sheet fluorescence microscopy. Nat Plants 4(9):639–650
8. Maizel A, von Wangenheim D, Federici F et al (2011) High-resolution live imaging of plant growth in near physiological bright conditions using light sheet fluorescence microscopy. Plant J 68(2):377–385
9. Ovecka M, Vaskebova L, Komis G et al (2015) Preparation of plants for developmental and cellular imaging by light-sheet microscopy. Nat Protoc 10(8):1234–1247
10. Valuchova S, Mikulkova P, Pecinkova J et al (2020) Imaging plant germline differentiation within Arabidopsis flowers by light sheet microscopy. Elife 9:e52546
11. Dejonghe W, Russinova E (2017) Plant chemical genetics: from phenotype-based screens to synthetic biology. Plant Physiol 174(1):5–20
12. Solc P, Kitajima TS, Yoshida S et al (2015) Multiple requirements of PLK1 during mouse oocyte maturation. PLoS One 10(2): e0116783
13. Yi K, Unruh JR, Deng M et al (2011) Dynamic maintenance of asymmetric meiotic spindle position through Arp2/3-complex-driven cytoplasmic streaming in mouse oocytes. Nat Cell Biol 13(10):1252–1258
14. Morejohn LC, Bureau TE, Mole-Bajer J et al (1987) Oryzalin, a dinitroaniline herbicide, binds to plant tubulin and inhibits microtubule polymerization in vitro. Planta 172(2): 252–264
15. Schindelin J, Arganda-Carreras I, Frise E et al (2012) Fiji: an open-source platform for biological-image analysis. Nat Methods 9(7): 676–682
16. Parslow A, Cardona A, Bryson-Richardson RJ (2014) Sample drift correction following 4D confocal time-lapse imaging. J Vis Exp 86: 51086

Chapter 9

Bimolecular Fluorescence Complementation to Test for Protein–Protein Interactions and to Uncover Regulatory Mechanisms During Gametogenesis

Ramakrishna Yadala, Maryia Ratnikava, and Inna Lermontova

Abstract

Bimolecular fluorescence complementation (BiFC) assay is one of the sensitive techniques that allows to investigate direct protein–protein interactions (PPI) in vivo and visualize the subcellular localization of interacting proteins. It is based on splitting of a fluorescent protein into two nonfluorescent parts accordingly fused to two putative interacting partners. If interaction between studied proteins is possible, nonfluorescent parts come to close proximity resulting in reconstitution of the functional fluorescent protein and giving fluorescence under certain wavelength. BiFC analysis implies transient or stable expression of the proteins of interest and can be used as a method to test or validate the direct PPI in various biological pathways, including the regulation of gametogenesis, which is the main focus of this book. In our protocol we give detailed information for beginners about three main steps of BiFC analysis of centromeric protein interactions. These steps include (1) generation of appropriate expression clones with the help of Gateway cloning technology, (2) infiltration of *Nicotiana benthamiana* plants by *Agrobacteria* containing generated constructs, and (3) microscopic analysis of plants under fluorescence microscope. Also, we discuss appropriate negative controls that can be used for evaluation as well as recommendable vector systems, possible artifacts and measures to avoid artifactual interactions for BiFC assay.

Key words Protein–protein interactions, BiFC, Agroinfiltration, Subcellular localization, Centromeric proteins

1 Introduction

Cellular processes are controlled by many specific proteins which are involved in the regulation of different functions in living organisms via protein–protein interactions (PPI). Understanding of these PPI can give an idea about protein function and in turn, this information helps to reveal regulating cellular networks in which these proteins are involved. In the postgenomic era it is possible to

Ramakrishna Yadala and Maryia Ratnikava contributed equally to this work.

Christophe Lambing (ed.), *Plant Gametogenesis: Methods and Protocols*, Methods in Molecular Biology, vol. 2484, https://doi.org/10.1007/978-1-0716-2253-7_9,

create a lot of protein interaction networks with the help of bioinformatical tools and these networks open the gates to functional proteomic research. Nevertheless, all of the bioinformatically predicted PPI have to be confirmed in laboratory. Nowadays, several methods for detecting and verifying of PPI exist. Bimolecular fluorescence complementation (BiFC) assay is one of such techniques that allows to study direct PPI in vivo and visualize the subcellular localization of identified proteins. The BiFC method is applicable to the analysis of PPI from all biological pathways and may be carried out in both transiently and stably transformed cells.

From the time of its development, BiFC assay gained and maintained its popularity among both the plant and animal biologists because of its principal advantages. One of the main advantages is that BiFC experiments are conducted in homologous systems. This allows to investigate PPI in more native conditions which promote true folding of analyzed proteins and give post-translational modifications important for protein function and localization. Compared to other popular in vivo techniques (yeast two-hybrid system, protein-fragment complementation assays, etc.) BiFC assay has high sensitivity with minimum background [1] and gives direct visualization of interacting proteins within the cells. No need of exogenous treatment or cell fixation and easy adaptability to different cell types transformed this technique into a high-throughput screening method with lower perturbation. Slow maturation of fluorophore [2], irreversible complex formation [3], effects of growth conditions on fluorophore [4], false positive signals limit this technique in real-time detection and monitoring dynamic interactions. Despite BiFC can produce both false positive and false negative results, it can be overcome by using appropriate experimental controls [5].

The research from Ghosh et al. [6] made first report of BiFC assay and it was adapted to animals in 2002 by Hu et al. [7] and to plants in 2004 by Walter et al. [8]. Since then several advancements in the technique [9–15] and in vector systems [16–21] have been reported. Gateway cloning technology made cloning strategies more easy, efficient, and high-throughput. Along with these advancements identification of new fluorophores [6, 7, 22–26] extended BiFC tool box for studying PPI in a wide range of applications. BiFC assay, which is based on transient overexpression in plants can be divided into three main stages: (1) Generation of BiFC vectors; (2) Agroinfiltration (3) Microscopic analysis. The major part of the experiment in terms of time is the generation of BiFC vectors (expression clones). Once the expression clone is ready, within 4–5 days of time the results can be acquired. Generation of a BiFC vector includes cloning of DNA (coding for one of the testing proteins) to the vector with a nonfluorescent tag generated by splitting a fluorescent protein (EYFP, GFP, RFP, mCherry, mVENUS, etc.) in one reading frame (*see* Fig. 1c). The

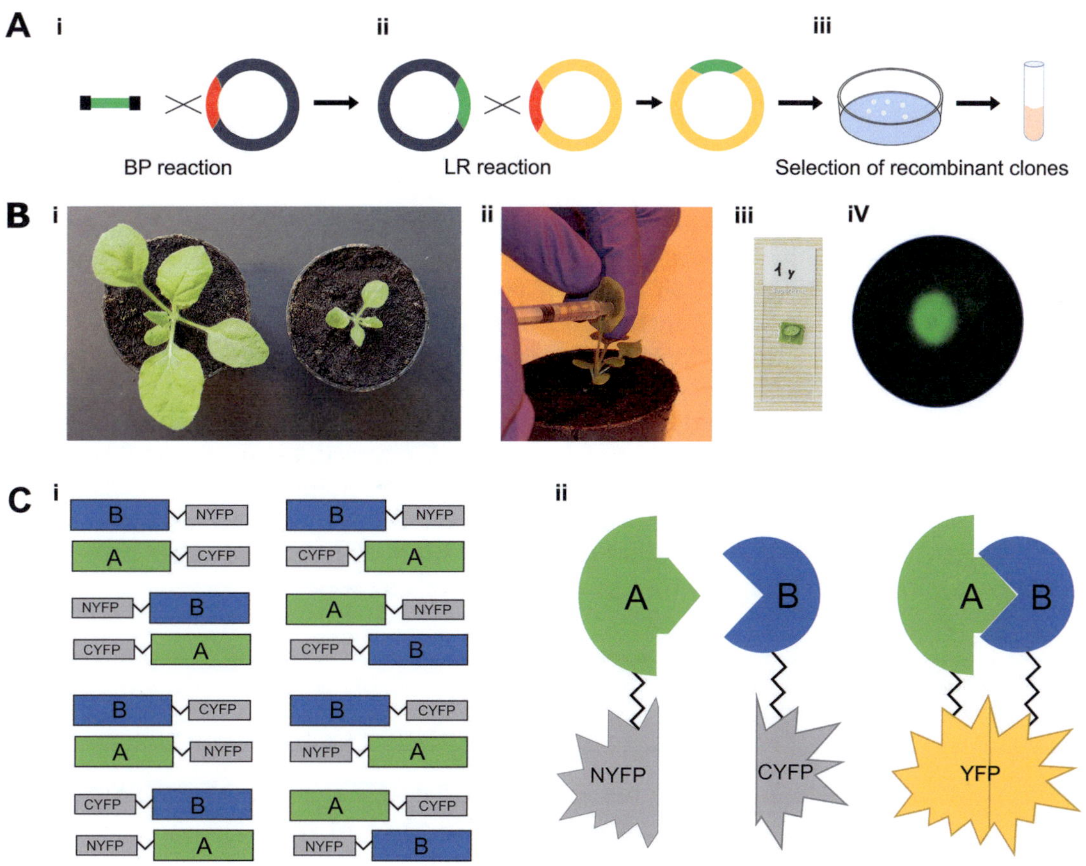

Fig. 1 Design for performing in vivo BiFC by *Agrobacterium* transient expression. (**a**) (*i*) BP reaction with plasmid containing gene of interest or PCR product with *attB*-sites; (*ii*) LR reaction with entry clone and destination vector containing split fluorescent tag; (*iii*) Selection of clones containing gene of interest with specific antibiotics. (**b**) (*i*) Two age groups of *N. benthamiana* plants for infiltration; (*ii*) Coinfiltration showing infiltrating abaxial airspace of flat tobacco leaf; (*iii*) Slide preparation for microscopy; (*iv*) Fluorescence in nucleus due to interaction of proteins through microscopic observation. (**c**) (*i*) Schemata showing different combinations of protein A and protein B fused to N-tag and C-tag of split fluorescent protein; (*ii*) Schemata showing working principle of the BiFC assay. When two studied proteins fused to nonfluorescent tags (NYFP & CYFP) generated by splitting fluorescent protein interact, both tags result in reconstruction of a functional fluorescent protein (YFP)

second vector contains DNA sequence of the probable interacting partner in the fusion with the other part of the fluorescent protein. If the studied proteins interact, the nonfluorescent tags of split fluorophore protein come to close proximity resulting in fluorescence complementation (*see* Fig. 1c(ii)) and corresponding fluorescence emission at specific wavelength that can be detected by microscopic analysis.

BiFC assay can be performed either by infiltration of plants (*Nicotiana benthamiana*) by *Agrobacterium tumefaciens (Rhizobium radiobacter)* or through protoplast transformation. The expression system can be chosen based on localization of the testing proteins. If the expressing interacting partners localize in large cellular organelles or space like plasma membrane, endoplasmic reticulum, nucleoplasm, or chloroplast, fluorescence resulting from interactions can be visualized directly in leaves. Otherwise, protoplast can be used for such refined structures as mitochondrial membrane or chloroplast envelope. Other than this, both systems have their advantages but also disadvantages such as cytosolic aggregation of target proteins due to their high expression level or short maturation time in protoplast (15–18 h) [27]. In this chapter we give a protocol of bimolecular fluorescence complementation assay in detailed description for the beginners. We have described possible artifacts, recommendable vector systems and measures to avoid artifactual interactions to study direct PPI (in our case between centromeric proteins) using *N. benthamiana* as a transient expression system. Some modifications in this protocol are possible to apply to study PPI between other proteins of interest with a proper choice of vectors.

2 Materials

2.1 Generation of Entry and Expression Clones

2.1.1 Entry Clone Generation

1. BP Clonase enzyme mix.
2. 1× TE buffer:10 mM Tris–HCl, 0.1 mM EDTA pH 8.0 (adjust pH with 1 M HCl and 5 M NaOH)
3. Donor vector (*see* **Note 1**).
4. *attB*-PCR product,
5. Proteinase K (2 μg/μl).

2.1.2 Expression Clone Generation

1. LR Clonase enzyme mix.
2. 1× TE buffer: 10 mM Tris–HCl, 0.1 mM EDTA, pH 8.0 (adjust pH with 1 M HCl and 5 M NaOH)
3. Entry clone.
4. Destination vector (*see* **Notes 2–4**).
5. Proteinase K (2 μg/μl).

2.1.3 E. coli *Transformation and Plasmid DNA Isolation*

1. Electro/chemical-competent cells: TOP10 or DH5α strains of *E. coli*.
2. 0.025 μm dialysis membranes
3. Cuvettes.
4. 1.5 ml tubes
5. Petri dishes.
6. Tips (10 μl, 200 μl, 1 ml).

7. SOC medium: 0.5% (w/v) tryptone, 0.125% (w/v) yeast extract, 0.0125% (w/v) NaCl, 2.5% 250 mM KCl, 2.5% 1 M $MgCl_2$, 5% 1 M glucose, distilled H_2O.
8. LB medium: 1% (w/v) tryptone, 0.5% (w/v) yeast extract, 1% (w/v) NaCl, distilled H_2O.

   Add 1.5% (w/v) agar-agar for LB agar
9. Kanamycin (50 μg/ml), Spectinomycin (100 μg/ml) (*see* **Note 5**).
10. GeneJET Plasmid Miniprep Kit.
11. Toothpicks.
12. 14 ml bacterial culture tubes
13. Kanamycin (50 μg/ml), Spectinomycin (100 μg/ml) (*see* **Note 5**).
14. Primers for sequencing and PCR confirmations:

(a) attB1-	GGGGACAAGTTTGTACAAAAAAGCAGGCT
(b) attB2-	GGGGACCACTTTGTACAAGAAAGCTGGGT
(c) M13-21F-	TGTAAAACGACGGCCAGT
(d) M13-27R-	CCAGGAAACAGCTATGAC
(e) 35S-SeQ-	GTGGATTGATGTGATATCTCCACTGACGT
(f) OCS-SeQ-	GCACAACAGAATTGAAAGCAAATATCA

2.2 Transformation of Agrobacteria and Agroinfiltration of N. benthamiana *Plants*

2.2.1 Transformation of Agrobacteria

1. GV3101 strain of *A. tumefaciens.*
2. The same material as described in Subheading 2.1.3, **items 2–7** is required.
3. YEB medium: 0.5% (w/v) tryptone, 0.5% (w/v) beef extract, 0.5% (w/v) sucrose, 0.1% (w/v) yeast extract, 0.4% 1 M $MgSO_4$, distilled H2O.

   Add 1.5% (w/v) agar-agar for YEB agar.
4. Spectinomycin (100 μg/ml), Rifampicin (50 μg/ml) (*see* **Note 5**).

2.2.2 Agroinfiltration of N. benthamiana *Plants*

1. 2–8-weeks-old *N. benthamiana* plants (*see* **Note 6**)
2. *A. tumefaciens* GV3101 strain, expressing clones of interest.
3. *A. tumefaciens* GV3101 strain, expressing silencing repressor protein p19 (*see* **Note 7**).
4. A. tumefaciens GV3101, expressing clones for the negative and positive controls (*see* **Notes 8** and **9**).
5. 5 ml tubes
6. 15 ml falcons

7. Tips (200 μl, 1 ml).
8. 1 ml plastic syringes without needle
9. Plastic cuvettes.
10. Paper towels.
11. Labels for plants.
12. Agroinfiltration medium (AIM), pH 5.6: 10 mM MES buffer (pH 5.3–5.5-adjust pH with 1 M HCl and 5 M NaOH), 10 mM $MgCl_2$, distilled H_2O; 0.15 mM Acetosyringone. Adjust to pH 5.6 with 1 M HCl and 5 M NaOH.
13. Acetosyringone (*see* **Note 10**).

2.3 Fluorescence Microscopy Analysis

1. Fluorescence microscope fitted with camera.
2. Super frost microscope slides.
3. 24 × 24 mm coverslips.
4. Scissors.
5. Forceps.
6. Distilled H_2O.
7. Paper towels.
8. ImageJ software (can be downloaded from https://imagej.nih.gov/ij/download.html).

3 Method

3.1 Generation of Entry and Expression Clones

3.1.1 Entry Clone Generation

BP reaction (Fig. 1a(i)) according to the manufacturer's protocol.

1. Add 1–7 μl of *attB*-PCR product (15–150 ng), 1 μl of a donor vector (150 ng/μl) in 1.5 ml tube and adjust the volume to 8 μl with 1× TE buffer (pH 8.0) (*see* **Note 11**).
2. Add 2 μl of BP Clonase enzyme, mix and spin down (*see* **Note 12**).
3. Incubate at 25 °C from 4 h to overnight.
4. Add 1 μl of 2 μg/μl of Proteinase K, mix, and spin down.
5. Incubate at 37 °C for 10 min to terminate the reaction.

3.1.2 Expression Clone Generation

LR reaction (Fig. 1a(ii)) according to the manufacturer's protocol.

1. Add 1–7 μl of entry clone (50–150 ng), 1 μl of a destination vector (150 ng/μl) in 1.5 ml tube and makeup to 8 μl with 1× TE buffer (pH 8.0) (*see* **Note 13**).
2. Add 2 μl of LR Clonase enzyme, mix, and spin down (*see* **Note 12**).
3. Incubate at 25 °C from 4 h to overnight.

4. Add 1 μl of 2 μg/μl Proteinase K, mix, and spin down.
5. Incubate at 37 °C for 10 min to terminate the reaction.

3.1.3 *E. coli* Transformation and Plasmid DNA Isolation

Dialyze the recombination mixture if more than 2 μl of reaction mixture is taken for the electroporation (*see* **Notes 14** and **15**).

1. Add 50 μl of electrocompetent *E. coli* cells with 2 μl of BP or LR recombination mixture to a cuvette and perform electroporation by giving a pulse of 1.8 kV for 0.1 cm cuvette or 2.5 kV pulse for 0.2 cm cuvette (*see* **Notes 16** and **17**).
2. Transfer the transformation mixture from the cuvette to a 1.5 ml tube containing 1 ml of SOC medium and incubate it at 37 °C with aeration (250–300 rpm) for 1–2 h for cell recovery and growth.
3. Take 25–150 μl of bacterial culture and spread it on LB agar plates with appropriate antibiotics (*see* **Notes 18** and **19**).
4. Incubate the plates at 37 °C for overnight.
5. Take 4–6 single colonies using tooth picks and grow them overnight in 5 ml of LB liquid medium with appropriate antibiotics at 37 °C, 240 rpm (*see* Fig. 1a(iii)).
6. Isolate plasmid DNA using Miniprep Kit and run 2–3 μl of isolated DNA on 0.8% agarose gel to check its quality and quantity.
7. The generated entry clone can be confirmed by colony PCR using *attB*-primers (*see* **Note 20**).
8. Confirm the correctness of the cloned gene's reading frame by Sanger sequencing and primers flanking the insert region (e.g., for pDONR221, M13-21F and M13-27R primers can be used; for pBAR, 35S-SeQ and OCS-SeQ) (*see* **Notes 21** and **22**).

3.2 Transformation of Agrobacteria and Agroinfiltration of *N. benthamiana* Plants

3.2.1 Transformation (Electroporation) of *A. tumefaciens*

1. Mix 50 μl of electrocompetent *A. tumefaciens* cells and 1 μl of expression clone confirmed by sequencing and add the resulting mixture to a cuvette (*see* **Note 16**).
2. Transfer the transformation mixture from the cuvette to a 1.5 ml tube containing 1 ml of SOC medium and incubate it at 28 °C with aeration (250–300 rpm) for 1–2 h for cell recovery and growth.
3. Take 10–50 μl of bacterial suspension and spread it on YEB agar plates with appropriate antibiotics (50 μg/ml of rifampicin for *Agrobacterium* selection and appropriate antibiotic for plasmid selection) (*see* **Notes 18** and **19**).
4. Incubate the plates at 28 °C for 2 days.
5. Pick up 4–6 single colonies and perform colony PCR using *attB*-primers (*see* **Note 20**), grow them overnight in 5 ml of

YEB liquid medium with appropriate antibiotics at 28 °C, 180 rpm (*see* **Note 5**).

3.2.2 Agroinfiltration of *N. benthamiana* Plants

1. Grow *A. tumefaciens* strains transformed by expression clones (e.g., interaction partner 1 and interaction partner 2 fused either to the N-or C-terminal parts of the selected fluorescent protein) and RNA silencing suppressor p19 in 5 ml of YEB medium with appropriate antibiotics at 28–30 °C, 180 rpm overnight (*see* **Notes 7**, **23**, and **24**).
2. Harvest 4 ml of the night cultures by centrifugation at 5000 rpm for 10 min and discard the supernatant.
3. Spin the tubes down and remove the residues of growing media with a pipette.
4. Resuspend the pellets in 4 ml of AIM (with Acetosyringone) in 5 ml tubes (*see* **Note 10**).
5. Determine the OD_{600} for each suspension and adjust the OD_{600} of samples in 15 ml falcons to 0.5–0.8 by diluting with an appropriate amount of AIM buffer (*see* **Note 25**).
6. Prepare mixtures for coinfiltrating of tobacco plants according to the ratios: interaction partner 1 + N-tag: interaction partner 2 + C-tag: p19 = 1:1:0.4
7. Incubate the coinfiltration mixtures at room temperature for 1–2 h in overhead shaker at appropriate speed rate (approximately 20 rpm).
8. Infiltrate the flat area of abaxial side of tobacco leaves by coinfiltration mixtures using a 1 ml plastic syringe by pressing a finger on the opposite side as shown on the figure (Fig. 1b(ii)). Two to three leaves of two different age plants should be infiltrated (*see* **Notes 26** and **27**).
9. Incubate infiltrated plants under normal conditions in the greenhouse for 2–4 days (*see* **Note 28**).

3.3 Fluorescence Microscopy Analysis

1. After 2–4 days of incubation the fluorophore is mature and it is possible to observe fluorescence under fluorescence microscope.
2. Prepare the slides for microscopic evaluation. Take the infiltrated leaf and cut the edges as shown on the figure (Fig. 1b (iii)).
3. Place the leaf tissue on a glass slide with the ventral side pointing upside.
4. Add distilled water on the leaf tissue and cover it with a coverslip. Adding water prevents the drying of the leaf material (*see* **Note 29**).

5. Add one drop of immersion oil on the coverslip and place on the microscope stage.
6. Use 40× objective lens and observe the fluorescence under green light filter. Capture the pictures with 8-bit RGB/grayscale format for further analysis (Fig. 1b(iv)) (*see* **Notes 30** and **31**).
7. Fluorescence intensity can be measured by using ImageJ software.

4 Notes

1. Either pENTR or pDONR vectors containing *attP*-sites and *attB*-PCR product are used for entry clone generation. There are several commercially available pDONR and pENTR vectors. PCR product with *attB*-sites coding for the studied protein can be obtained from cDNA, mRNA or genomic DNA, as a template. For generation of entry clone, we use pDONR221 Gateway vector possessing kanamycin resistance (50 μg/ml), PCR product amplified with *attB*-sites from cDNA template and BP Clonase enzyme mix.
2. Destination vector can be selected based on cloning strategy, choice of promoter, fluorophore and the subsequent application. Several Gateway cloning vectors with different fluorescent proteins (EYFP, GFP, RFP, mCherry, mVENUS, etc.) are described and available commercially. Based on the requirement, a destination vector can be chosen with the proper promoter, tag and antibiotic selection marker. There are several possible combinations in fusing fluorescence tags with proteins. They can be localized either on 5′ or 3′ end of the cloned gene. Generation of expression clones implies creating two vectors for two proteins to be checked for the direct interaction (e.g., gene 1 + N-tag on its 3′-end and gene 2 + C-tag on its 3′-end).
3. For generation of expression clones, we used 35S-VYNE [c]-GW-pBar and 35S-VYCE[c]-GW-pBar vectors possessing spectinomycin resistance (100 μg/ml) with CAMV 35S promoter and N and C-mVENUS tags fused to the 3′ end of the cloned gene (*see* **Note 4**). The LR cloning was performed with LR Clonase enzyme mix. Clones for negative control: mutations within putative or confirmed protein–protein interaction sites can be generated through Site-Directed Mutagenesis Kit.
4. Splitting the fluorescence protein in proper position and choosing proper negative control by introducing point mutations or using truncated proteins in fusion will reduce false-positive signals by preventing self-assembly [5, 28].

5. The selection of antibiotic and its concentration depends on the backbones used for generating the desired constructs.
6. The age selection of *N. benthamiana* plants depends on the type of proteins to be studied for interactions. To study proteins that are highly expressed in actively growing tissues and are involved in the regulation of mitotic, meiotic cell cycles or developmental processes, 2–4-week-old plants are recommended.
7. p19 protein [29] is an RNA silencing repressor that does not let coinjected constructs to be inactivated by plant's protection factors. The GV3101_p19 strain possesses kanamycin (50 μg/ml) and rifampicin (50 μg/ml) resistance. It is recommended to test interactions with and without the addition of p19, as increased expression of some selected proteins may lead to false-positive interactions. For our experiments, BiFC assay was tested with and without p19 silencing repressor. No difference was observed for our proteins of interest.
8. It is advised to check whether the chosen control proteins and the protein of interest localize in the same subcellular space. Appropriate controls are required to validate PPI between studied proteins while performing BiFC analysis. The best negative control for BiFC assay is protein created by introducing mutations in interacting domains of one of the interacting partners, which leads to no interaction. For positive control, already known interacting partners with the same localization and orientation of the fused tags can be used. If such combinations are not possible, several other possible appropriate controls were reported in plants and other cell types [5, 28, 30].
9. Our BiFC experiments were conducted using centromeric proteins which play key role in mitosis/meiotic cell divisions. Different combinations of truncated proteins of interest were studied for direct PPI and dimer formation. Positive control was selected from the interacting partners of the protein of interest whereas for negative control, truncated protein combinations generated by Phusion Site-Directed Mutagenesis Kit were used for all of the experiments.
10. Acetosyringone can be optionally added to AIM buffer after autoclaving. In this case AIM buffer should be stored at 4 °C. Acetosyringone (100 μM) can also be added freshly to the buffer.
11. Different variants of vector constructs containing point mutations or deleted regions of the gene of interest can be generated with the help of Phusion Site-Directed Mutagenesis Kit on the step of entry clone generation. It is recommended to use plasmids below 5 kb for this procedure. Created entry clones can be used either for construction of negative control for

validation or to identify interacting domains of the proteins of interest.

12. Volume of the Clonase enzyme can be reduced to 1 μl per reaction if the incubation time is overnight.
13. The amount of entry clone for LR reaction can be calculated using the following formula:

$$\text{Entry vector concentraion(ng/reaction)} = \frac{\text{Destination vector concentration (ng/μl)} \times \text{entry vector size (bp)}}{\text{Destination vector size (bp)}} \times 2$$

14. It is important to dialyze the reaction mixture before electroporation if you take more than 2 μl. High concentrations of salts in the transformation mixture cause arcing leading to the absence of bacterial transformation.
15. For dialysis, place 0.025 μm dialysis membrane on water and add the recombination mixture carefully on the membrane. Allow it to be dialyzed for 1 h and then transfer the mixture to a new tube.
16. While performing electroporation, consider a negative control (competent cells without addition of recombination mixture) and perform all transformation steps alongside to the samples. In a negative control, no colonies should be detected with antibiotics. If colonies are detected, either the competent cells are contaminated or the antibiotics are not working or contamination occurred.
17. If chemically competent cells are used for transformation, follow the chemical transformation protocol.
18. The volume of bacterial suspension to be spread on plates depends on the concentration of competent cells, quantity of recombination mixture taken for the transformation, incubation time after electroporation for cell recovery and copy number of the vector.
19. In initial steps it is recommended to prepare two plates with different volumes of bacterial culture (e.g., 50 and 150 μl) to obtain optimal number of colonies. In case no colonies is observed, concentrate the culture by centrifugation and resuspend the cells in 100 μl to spread them.
20. Screening for the desired constructs can be checked before plasmid isolation by performing colony PCR with *attB*-primers. For this, prepare PCR mixture with attB-primers and resuspend a single colony from the transformation plates using tooth pick. Same inoculum containing the clones should be grown on a new LB agar plate or in LB broth with antibiotics.

After confirmation of colony PCR with expected band size, the clones can be processed for plasmid DNA isolation.

21. Always confirm the plasmids by sequencing to avoid mutated or unwanted plasmid transformation. Plasmid DNA before sequencing can be quantified using a Nanodrop.
22. Prepare glycerol stocks (two parts of night culture: one part of 50% glycerol) for the future use and store them at −80 °C with proper labeling.
23. Replication time of *A. tumefaciens* is longer than that of *E. coli* (2.5–4 h). To prepare night cultures *of A. tumefaciens* for tobacco infiltration, it is important to incubate them for a minimum of 18 h prior to cell harvesting.
24. It is recommended to check the interaction in both directions (N-tag–Protein A + C-tag–Protein B and N-tag–Protein B + C-tag–Protein A) (*see* Fig. 1c(i)) and obtained results should coincide. If not, try to use the tag fused to the other end of the proteins or use linker peptides.
25. Adjusting OD_{600} between samples is important for obtaining good fluorescence picture. The difference between OD_{600} values should be not more than 0.1. Almost all dilutions end in less OD_{600} than it is expected during calculations (e.g., expected mean is 0.6, real one is 0.53). It is recommended to calculate the final dilution to 0.7–0.8 and the actual concentrations of suspensions should be higher than 0.6. $OD_{600} \geq 0.6$ provides better fluorescence.
26. Two plants of different age groups (*see* Fig. 1b(i)) should be considered for infiltration by one coinfiltration mixture. Infiltrate 2–3 leaves as technical replicates to avoid the failing of the experiment in case one plant is damaged in the process. Choose young and flat leaves and avoid central vein for infiltration. It allows to avoid the difficulty of injection movement and reduces excessive autofluorescence.
27. It is recommended to water the plants before infiltration to promote stomatal opening for better infiltration process. Use proper pressure to ensure that the coinfiltration mixture enters the leaves and do not push the syringe too firmly to avoid piercing of the leaf (*see* Fig. 1b(ii)). Remove the excess solution with tissue paper. Change the gloves and syringe after each infiltration mixture to avoid cross-contamination.
28. After infiltration keep the plants with less direct light and temperatures under 25 °C. Otherwise, injected bacteria stimulate oxidative processes in plants resulting in artifactual results and necrosis of leaves. In this case, there is a high risk to observe autofluorescence caused by different metabolites that appear during necrosis. Few fluorophores (EYFP, ECFP,

mRFP1-Q66T) are temperature sensitive for maturation. To avoid obtaining false negative results choose proper tags and conditions.

29. Limit the number of slides prepared at one time (no more than 3–5) to prevent the drying of samples.
30. Choose the proper filter based on the tag excitation and emission wavelength. To avoid false interpretation of signals, the fluorescence can be checked under different filters (red, green, blue, etc.) If the fluorescence signal is detected with other filters, the signal is likely an autofluorescence signal.
31. Absence of fluorescence even when PPI exists is possible. In such case, it is advised to check for the subcellular localization and the expression level of the interacting partners.

References

1. Kerppola TK (2008) Bimolecular fluorescence complementation (BiFC) analysis as a probe of protein interactions in living cells. Annu Rev Biophys 37:465–487
2. Tsien RY (1998) The green fluorescent protein. Annu Rev Biochem 67:509–544
3. Shyu YJ, Hu C-D (2008) Fluorescence complementation: an emerging tool for biological research. Trends Biotechnol 26:622–630
4. Miller KE, Kim Y, Huh W-K et al (2015) Bimolecular fluorescence complementation (BiFC) analysis: advances and recent applications for genome-wide interaction studies. J Mol Biol 427:2039–2055
5. Kudla J, Bock R (2016) Lighting the way to protein-protein interactions: recommendations on best practices for bimolecular fluorescence complementation analyses. Plant Cell 28: 1002–1008
6. Ghosh I, Hamilton AD, Regan L (2000) Antiparallel leucine zipper-directed protein reassembly: application to the green fluorescent protein. J Am Chem Soc 122:5658–5659
7. Hu C-D, Chinenov Y, Kerppola TK (2002) Visualization of interactions among bZIP and Rel family proteins in living cells using bimolecular fluorescence complementation. Mol Cell 9:789–798
8. Walter M, Chaban C, Schütze K et al (2004) Visualization of protein interactions in living plant cells using bimolecular fluorescence complementation. Plant J Cell Mol Biol 40:428–438
9. Kodama Y, Hu C-D (2010) An improved bimolecular fluorescence complementation assay with a high signal-to-noise ratio. BioTechniques 49:793–805
10. Seo JS, Chua N-H (2019) Trimolecular fluorescence complementation (TriFC) assay for visualization of RNA-protein interaction in plants. Methods Mol Biol 1933:297–303
11. Wang S, Ding M, Chen X et al (2017) Development of bimolecular fluorescence complementation using rsEGFP2 for detection and super-resolution imaging of protein-protein interactions in live cells. Biomed Opt Express 8:3119–3131
12. Chen M, Li W, Zhang Z et al (2015) Novel near-infrared BiFC systems from a bacterial phytochrome for imaging protein interactions and drug evaluation under physiological conditions. Biomaterials 48:97–107
13. Berendzen KW, Böhmer M, Wallmeroth N et al (2012) Screening for in planta protein-protein interactions combining bimolecular fluorescence complementation with flow cytometry. Plant Methods 8:25
14. Kwaaitaal M, Keinath NF, Pajonk S et al (2010) Combined bimolecular fluorescence complementation and Förster resonance energy transfer reveals ternary SNARE complex formation in living plant cells. Plant Physiol 152:1135–1147
15. Nickerson A, Huang T, Lin L-J et al (2014) Photoactivated localization microscopy with bimolecular fluorescence complementation (BiFC-PALM) for nanoscale imaging of protein-protein interactions in cells. PLoS One 9:e100589
16. Tanaka Y, Kimura T, Hikino K, Goto S, Nishimura M, Mano S, et al (2012) Gateway vectors for plant genetic engineering: Overview of plant vectors, application for bimolecular fluorescence complementation (BiFC) and

multigene construction. Croatia: InTech; 2012. pp.35–68

17. Nishimura K, Ishikawa S, Matsunami E et al (2015) New gateway-compatible vectors for a high-throughput protein–protein interaction analysis by a bimolecular fluorescence complementation (BiFC) assay in plants and their application to a plant clathrin structure analysis. Biosci Biotechnol Biochem 79:1995–2006
18. Kamigaki A, Nito K, Hikino K et al (2016) Gateway vectors for simultaneous detection of multiple protein–protein interactions in plant cells using bimolecular fluorescence complementation. PLoS One 11:e0160717
19. Lepur A, Kovačević L, Belužić R et al (2016) Combining unique multiplex gateway cloning and bimolecular fluorescence complementation (BiFC) for high-throughput screening of protein–protein interactions. J Biomol Screen 21:1100–1111
20. Mehlhorn DG, Wallmeroth N, Berendzen KW et al (2018) 2in1 vectors improve in planta BiFC and FRET analyses. In: Hawes C, Kriechbaumer V (eds) The plant endoplasmic reticulum : methods and protocols. Springer, New York, pp 139–158
21. Luo Y, Qiu Y, Na R et al (2018) A golden gate and gateway double-compatible vector system for high throughput functional analysis of genes. Plant Sci 271:117–126
22. Jach G, Pesch M, Richter K et al (2006) An improved mRFP1 adds red to bimolecular fluorescence complementation. Nat Methods 3: 597–600
23. Shyu YJ, Liu H, Deng X et al (2006) Identification of new fluorescent protein fragments for bimolecular fluorescence complementation analysis under physiological conditions. BioTechniques 40:61–66
24. Fan J-Y, Cui Z-Q, Wei H-P et al (2008) Split mCherry as a new red bimolecular fluorescence complementation system for visualizing protein-protein interactions in living cells. Biochem Biophys Res Commun 367:47–53
25. Chu J, Zhang Z, Zheng Y et al (2009) A novel far-red bimolecular fluorescence complementation system that allows for efficient visualization of protein interactions under physiological conditions. Biosens Bioelectron 25:234–239
26. Fujii Y, Yoshimura A, Kodama Y (2018) A novel orange-colored bimolecular fluorescence complementation (BiFC) assay using monomeric Kusabira-Orange protein. BioTechniques 64:153–161
27. Schweiger R, Schwenkert S (2014) Protein-protein interactions visualized by bimolecular fluorescence complementation in tobacco protoplasts and leaves. J Vis Exp (85):51327
28. Horstman A, Tonaco IAN, Boutilier K et al (2014) A cautionary note on the use of Split-YFP/BiFC in plant protein-protein interaction studies. Int J Mol Sci 15:9628–9643
29. Lakatos L, Szittya G, Silhavy D et al (2004) Molecular mechanism of RNA silencing suppression mediated by p19 protein of tombusviruses. EMBO J 23:876–884
30. Kodama Y, Hu C-D (2012) Bimolecular fluorescence complementation (BiFC): a 5-year update and future perspectives. BioTechniques 53:285–298

Chapter 10

Quantifying Meiotic Crossover Recombination in Arabidopsis Lines Expressing Fluorescent Reporters in Seeds Using *SeedScoring* Pipeline for CellProfiler

Nadia Kbiri, Julia Dluzewska, Ian R. Henderson, and Piotr A. Ziolkowski

Abstract

The number of crossovers during meiosis is relatively low, so multiple meioses need to be analyzed to accurately measure crossover frequency. In Arabidopsis, systems based on the segregation of fluorescent T-DNA reporters that are expressed in seeds (fluorescent-tagged lines, FTLs) allow for an accurate measurement of crossover frequency in specific chromosome regions. A major advantage of FTL-based experiments is the ability to analyze thousands of seeds for each biological replicate, which requires the use of automatic seed scoring. Here, we describe a protocol to computationally count the proportion of seeds that experienced a crossover event within the tested FTL interval and so measure the recombination frequency within that interval. We describe *SeedScoring*, a CellProfiler pipeline where the total time needed to measure crossover frequency in a single FTL line is approximately 5 min using a series of three images taken under a fluorescent stereomicroscope (3 min) and passing these images through the *SeedScoring* pipeline described in this protocol (2 min).

Key words Meiosis, Crossover, Recombination, Fluorescent-Tagged Lines (FTLs), Traffic Lines, *SeedScoring*, CellProfiler, Arabidopsis

1 Introduction

Crossover is a reciprocal DNA exchange between homologous chromosomes, which occur during meiosis. It is essential for proper chromosome segregation in the first meiotic division, therefore each pair of chromosomes must experience at least one crossover event. Crossovers also play a key role in creating genetic variation in a population as they enable the formation of new haplotypes [1]. Despite this important role, the number of meiotic crossovers is usually small and does not typically exceed two or three events per chromosome pair [1]. As a consequence, studying crossover frequency along the chromosome is challenging and requires the analysis of many meiotic events. In plants, crossover numbers are

Christophe Lambing (ed.), *Plant Gametogenesis: Methods and Protocols*, Methods in Molecular Biology, vol. 2484, https://doi.org/10.1007/978-1-0716-2253-7_10,

commonly assessed by cytological analyses, for instance by counting chiasmata on metaphase I chromosome spreads, or by counting foci of crossover-associated proteins, for example HEI10 or MLH1 [2–6]. However, those techniques require specialized expertise and can be challenging for plants with small chromosomes such as *Arabidopsis*. Genotyping-by-sequencing (GBS) is another method that is increasingly used, in which genome-wide analysis of crossover frequency and distribution at high resolution is possible [7, 8]. GBS, however, is relatively costly and laborious, and therefore not suitable for preliminary studies, initial mutant characterization and forward genetic screening.

Coinheritance of linked, heterozygous genetic markers through meiosis allows the measurement of crossover rate for a chromosomal region. Crossover detection occurs when the linkage phase of markers changes [9–12]. Both natural DNA sequence polymorphisms and T-DNA insertions can be used as genetic markers. In Arabidopsis, a convenient type of markers widely used to measure crossover recombination in selected chromosome regions, are T-DNA expressing fluorescent reporter proteins in seed or pollen [11–15]. Calculating crossover frequency using seed-based system is less time-consuming and more cost-effective than standard genetic markers [11, 15]. This assay requires the use of fluorescent-tagged lines (FTLs) [11]. Recently, a set of several dozen new FTLs in the Col and L*er* genetic backgrounds, called *Traffic Lines*, were released, which define intervals located throughout the entire genome, enabling analysis of crossover frequency in diverse chromosome regions [16]. In FTLs/*Traffic Lines*, transgenes expressing green (eGFP) and red (dsRed) fluorescent proteins define specific regions of the genome, including centromeric, interstitial, or subtelomeric regions, and are expressed under seed-specific napin promoter [11, 16]. FTLs with a pair of genetic markers defining a chromosome interval of interest can be crossed with a nonfluorescent Arabidopsis line. The segregation of fluorescent proteins in seeds obtained from F_1 plants allows for the assessment of crossover rate in a tested region (Fig. 1a). For analysis of lines carrying a recessive trait, for example mutant lines, it is necessary to first backcross the line to the selected FTL.

Our crossover recombination frequency (RF) measurement protocol, which is a detailed version of the method described in [15], is based on images of seeds collected in three channels: brightfield, UV through RED filter and UV through GFP filter (Fig. 1b). The micrographs are then processed by the CellProfiler image analysis software [17]. Our CellProfiler-based pipeline called *SeedScoring* identifies seed boundaries in images and allows to adjust a threshold for dsRed and eGFP fluorescence intensity of recognized objects (i.e., seeds). The pipeline serves to count fluorescent vs. nonfluorescent seeds for crossover measurement. Using this method, it is possible to score approximately 2000

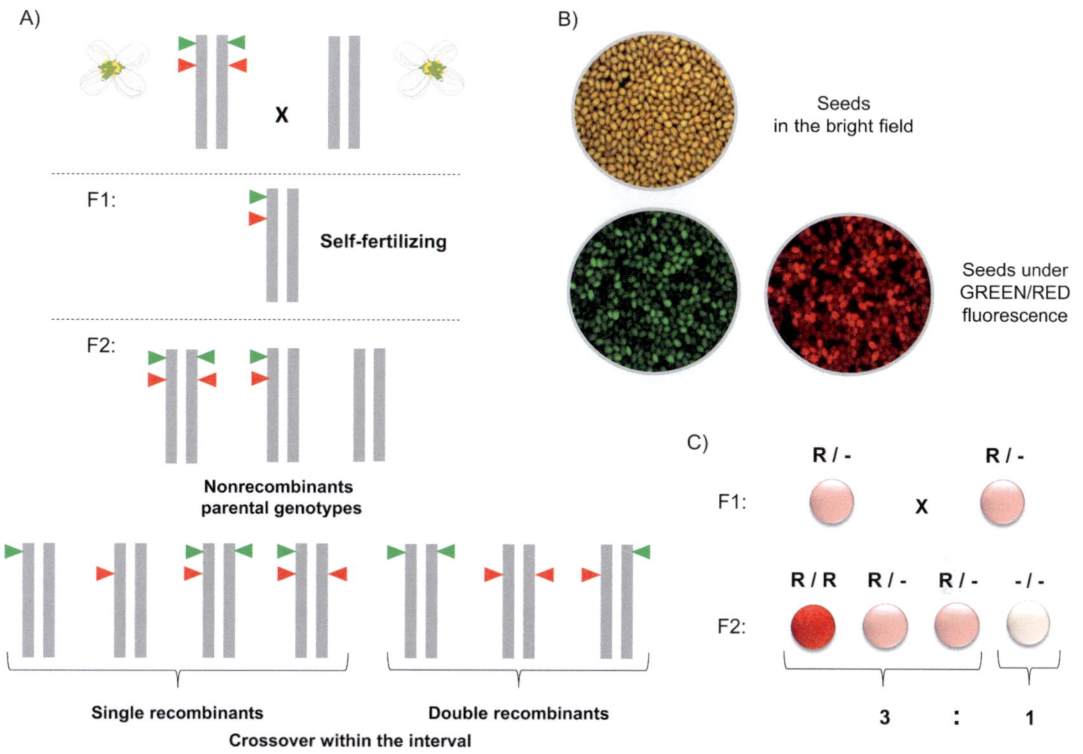

Fig. 1 Seed-based system used to quantify crossover frequency in *Arabidopsis thaliana*. (**a**) Crossing scheme with FTLs in order to obtain F_2 seeds that show segregation of fluorescent reporters. Gray bars represent chromosomes. A noncolor line of interest is crossed with an FTL line containing fluorescent transgenes marking a specific chromosomal region. The F_1 plant is self-fertilized to produce F_2 offspring seeds, in which one can observe presence or absence of seed-expressed fluorescent proteins (*see* **Note 14**). As F_2 seeds are diploid and generated through selfing, each seed is a product of both male and female meioses. Possible genotypes of F_2 seeds are shown. (**b**) Dosage-dependent expression of dsRed/eGFP genes under the control of seed-specific napin promoter. Images of Arabidopsis seeds in the bright field, under green and red fluorescence are shown. Strong green or red fluorescence indicate the presence of the fluorescent transgene copy on both chromosomes (FTL homozygous), whereas medium fluorescence values indicates hemizygote transgenes. Nonfluorescent seed does not inherit FTL transgenes. (**c**) Green/red fluorescent transgenes should show Mendelian inheritance, with the ratio of color to noncolor seeds close to 3:1

seeds per self-fertilized individual within 5 min. This method gives statistically identical results to manual scoring; however, it is 20–25 times faster [15]. We have previously successfully used our *SeedScoring* pipeline for the calculation of RF within many genomic intervals in different genetic backgrounds [15, 18–23].

2 Materials

1. Arabidopsis F_2 seeds obtained from a cross between an FTL individual, expressing fluorescent proteins, with a nonfluorescent plant. The seeds should be cleaned of plant tissue debris.

2. Stainless steel fine mesh sieve.
3. Glassine paper bags.
4. Epifluorescence stereomicroscope equipped with a CCD camera, dsRed filter and GFP filter.
5. An exposed X-ray film.
6. Toothpicks.
7. Glass microscope slides.
8. CellProfiler software (version 2.1.1) (*see* **Note 1**).
9. Microsoft Excel software, or equivalent spreadsheet software (*see* **Note 2**).

3 Methods

3.1 Image Acquisition

1. Collect dry seeds and clean them of any plant debris using a stainless-steel fine mesh sieve. The seeds should be kept in glassine paper bags or similar and stored in a dark and dry environment (*see* **Notes 3 and 4**).
2. Place the seeds under the epifluorescence stereomicroscope on the exposed X-ray film and manually remove any remaining plant debris using a humidified toothpick.
3. Place a clean microscope slide on the seeds and press gently to spread them evenly creating a single layer. This step is important to facilitate identification of single seed objects during image processing.
4. Gently shift the film with the seeds under the microscope in such a way to locate them in the center of the field of view. The seeds should ideally cover the entire field of view. If this is not possible, ensure that the seeds are well centered in the field of view (*see* **Note 5**).
5. Set the microscope on the lowest magnification. The described pipeline is based on imaging with ×6.4 magnification.
6. For each sample of seeds, acquire a set of three pictures using bright field, green fluorescence and red fluorescence (Fig. 1b). The single layer field must be identical for the three pictures. This is important for the alignment of the channels during image processing. An image resolution of 912 × 736 pixels is used (*see* **Note 6**).
7. We recommend saving the images in series of three with a file naming system (e.g., 1-1, 1-2, 1-3; 2-1, 2-2, 2-3; 3-1, 3-2, 3-3; …).

3.2 CellProfiler: SeedScoring *Pipeline*

The *SeedScoring* pipeline utilizes the set of images that were chosen (bright field, red and green; Fig. 1b). The images are colored to a gray scale and the intensity is corrected so that it is homogenous across the whole picture. The boundaries of objects (seeds) are identified using the bright field image. The red and green images are overlaid onto the bright field picture so that the fluorescence can be measured per individual object. The fluorescence intensity threshold is set to distinguish between noncolor and color objects, for both green and red separately. The objects are then classified as green/nongreen and red/nonred, accordingly. Such information for all objects and for both colors is saved in a text file, which is further used to compute the number of seeds with recombinant and parental fluorescence patterns, and crossover frequency calculation.

1. Create a ".csv" file that provides the instructions for *CellProfiler*, indicating which sets of pictures are to be used. This file should have three columns and two rows; in the first row enter the labels given to each picture, for bright field, red, green. These names are used as an input for the *SeedScoring* pipeline in *CellProfiler*. The second row contains the file name of each picture, including the format extension which should be ". jpg". This row changes for each plant/sample to be scored. It is essential that the picture names in the second row correspond to the correct color in the first row. Save the file in ".csv" format. For an example file, *see* Table 1.
2. Install *CellProfiler*, an open access software available at https://cellprofiler.org (Fig. 2). The *SeedScoring* pipeline was built using the 2.1.1 version (*see* **Note 1**).
3. Create the *SeedScoring* pipeline by selecting "New Project" in *File* menu (*see* **Note** 7). Add the following modules by selecting them from the menu:
   (a) Use *LoadData* to set the input folder where the set of three images and the ".csv" file to call for them are stored. Select "Yes" for "Load images based on this data?" and "Rescale intensities."
   (b) Use *ColorToGray* (bright field) to convert the bright field picture from RBG to a gray scale. The image is called using the name given in the ".csv" file, for example, *OrigBright*. Relative weights of all three channels (red, green and blue) should be set for 1. A name is given to the output image for the following steps.

Table 1
An example ".csv" input file for *SeedScoring* pipeline

Image_FileName_OrigBright	Image_FileName_OrigRed	Image_FileName_OrigGreen
1-1.jpg	1-2.jpg	1-3.jpg

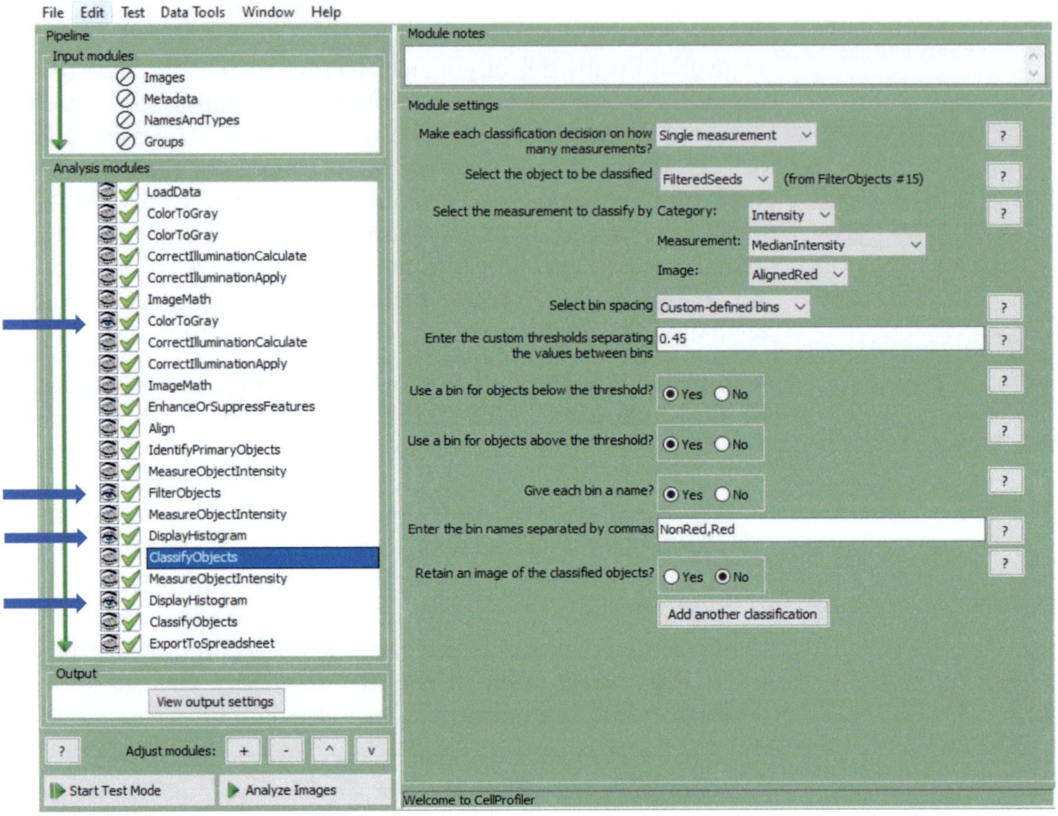

Fig. 2 Overview of the CellProfiler (2.1.1) software with the *SeedScoring* pipeline. In the "Analysis modules" window one can find two *ClassifyObjects* modules, for red and green fluorescent seeds, where the user specifies the threshold value separating fluorescent vs. nonfluorescent seeds for each color separately; the threshold values can be read out of histograms. Blue arrows indicate modules for which "eye" icons should be opened

(c) Use *ColorToGray* (for red channel). The red picture is called (e.g., *OrigRed*) and undergoes the same process as for the bright field picture, only the red channel is given a weight of 1. The weights of green and blue are given 0. Give a name to the output image to use it in the following steps.

(d) Illuminating the seeds with fluorescent light through the objective may cause the seeds located in the center of the field of view to exhibit a higher fluorescence than those located on the sides. Use **CorrectIlluminationCalculate** module to overcome this issue. The input image for this step is the output image of ColorToGray for the red picture. Set the calculation of illumination function on "Regular" and the rescaling of the illumination function on "yes." Choose that the function is calculated separately for each image, and set the smoothing method on "Convex Hull." A name is given to the output image.

(e) **CorrectIlluminationApply** module applies the image correction calculated in the previous step to correct the red image colored-to-gray (output from ColorToGray). Set the image illumination function application on "Divide." A new name is given to the output image.

(f) Use **ImageMath** to merge the output image of red colored-to-gray and the corrected red image. This is required since the CorrectIlluminationApply may modify the figure too extensively, leading to false results. Choose "Add" in the operation box. The type of files used is "Image." In our hands the best results are obtained when the color-to-gray image is multiplied by 0.5 and the corrected image (output from ImageMath) by 1. The result is multiplied by 1 and raised by 1. If values are less than 0, they are to be increased to equal 0 and when greater than 1 they are to be decreased to equal 1. Give a new name to the output image.

(g) Generate **steps "c" to "f"** for the green image in a similar way.

(h) Use **EnhanceOrSuppressFeatures** to suppress small debris that could be mistaken by the software as objects. It is applied to the bright field picture for a feature size equal to 10. Give a new name to the output picture.

(i) **Align** is the module, during which the three images (bright field, red and green) are aligned. The output images from modules EnhanceOrSuppressFeatures for bright field, and ImageMath for red and green, are used as input. Set the alignment method to "Mutual information" and the crop mode to "Crop to aligned region." The alignment is applied Similarly for all the images. All four images are given new output names. In addition, use the output of ColorToGray for red color and also choose "Similarly" as a method to apply the alignments. Name the output as "Filter," which will be further used for filtering objects.

(j) Apply **IdentifyPrimaryObjects** to outline each seed to use it later as an object (Fig. 3). The input is the output image for the bright field from step Align. The primary objects to be identified need to be named, for instance "seeds." If you collect seed images using a magnification of around ×0.6–0.7, then set the expected diameter of the objects between 10 and 60 pixels. Objects outside this diameter range, or those at the border of the image

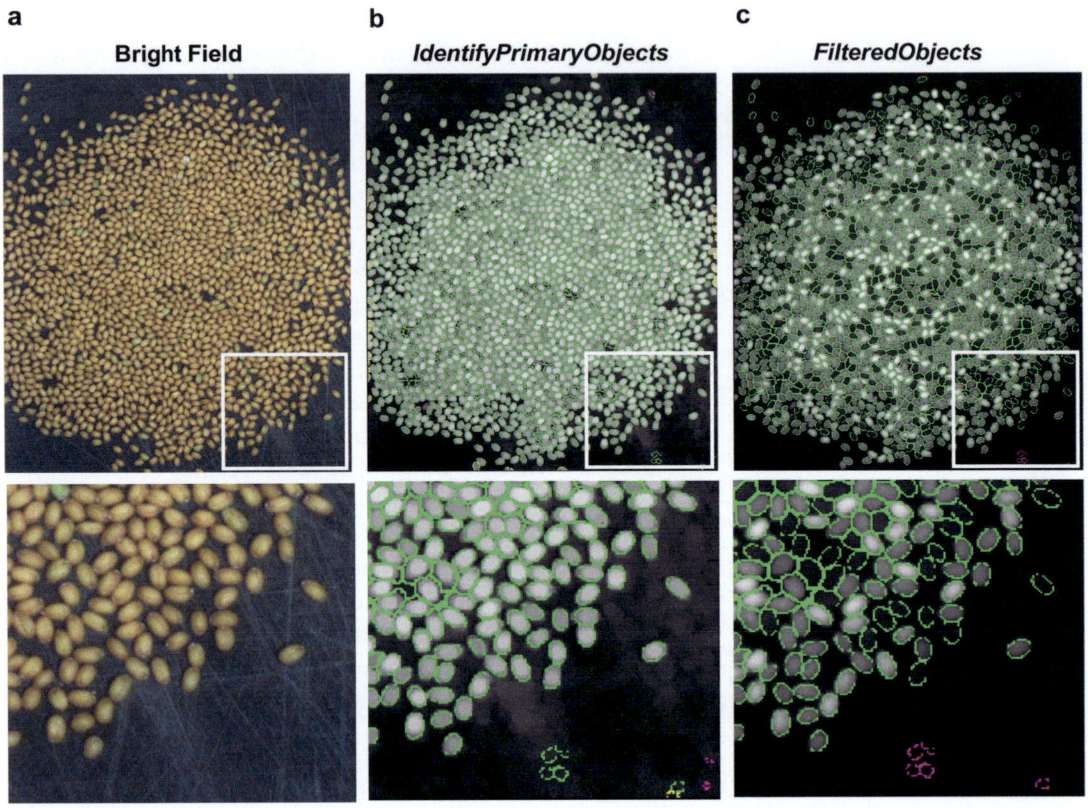

Fig. 3 Operation of modules *IdentifyPrimaryObjects* and *FilteredObjects*. (**a**) Example of a seed image taken in bright field. (**b**) The *IdentifyPrimaryObjects* module serves to outline in green each seed to use it later as an object. (**c**) In *FilteredSeeds,* objects in magenta are identified as nonseeds and are discarded from further calculations. The bottom row pictures show zoom-in of areas in white squares

are discarded. For thresholding, select “Adaptive” strategy and use the “Otsu” method for two classes. Select the “Minimize the weighted variance” option and “Automatic” smoothing method for thresholding. When running the pipeline, you can adjust the threshold correction factor, though the starting value should be set for 1. Also, lower and upper bounds on threshold should be set on 0 and 1, respectively. Select “Image size” for calculation of the adaptive window size parameter. Select “Intensity” for both the method of distinguishing clumped objects, and for drawing dividing lines between clumped objects. Turn on automatic calculation of size of smoothing filter for declumping, automatic calculation of minimum allowed distance between local maxima and the speeding up option, which uses a lower-resolution image to find local maxima. The holes in the identified objects are filled after both thresholding and declumping, while the handling is to be continued even if the number of objects identified is excessive.

(k) Following seed identification, the next step is to measure the fluorescence intensity for each seed. Use **MeasureObjectIntensity** module for this task. The input for this step is the output image of Align for the red image. Set the objects measured as "seeds."

(l) Seed images may contain residual fragments of petals and other tissue debris. To filter them out so they are not treated as seeds, use **FilterObjects** (Fig. 3). The identified objects (seeds) are filtered using "Measurements" as mode and "Limits" as method. The measurement used for filtering is categorized as "Intensity" using the "UpperQuartileIntensity" on the output image of Align step for the red image named "Filter." We set the minimum value for measurement filter between 0.19 and 0.24. Use a new name to the output objects.

(m) Use **MeasureObjectIntensity** with the output image of Align for red.

(n) Use **DisplayHistogram** to plot the filtered objects with fluorescence intensity on the X axis and the number of objects on the Y axis. "Median intensity" is the displayed measurement. A name can be given to this plot, so it will be easier to identify it later, as this histogram will be used to set the fluorescence intensity threshold between the red/nonred seeds (Fig. 4a).

(o) **ClassifyObjects** is to classify each seed object as red or nonred. The decision is made based on a "single measurement" of each filtered seed (obtained in FilterObjects step). The measurements are classified by "Intensity" using the "Median intensity" of the aligned red image (obtained as an output of Align). The bin spacing is set on "Custom-defined bins" and corresponds to the intensity value of the lowest point between the two peaks (the fluorescence intensity with the lowest seed count), which can be read from the histogram generated in step Display Histogram. The bin is used for both groups of objects, below and above that threshold. Name the two groups, for example "NonRed" and "Red" (the objects below the threshold are nonred and the objects above it are red). You can find the location of the threshold in Fig. 4a.

(p) Repeat **steps** from **"m" to "p"** with the green aligned image.

(q) **ExportToSpreadSheet**, is the module which allow to export the intensities and the classification of each seed object as red or nonred and green or nongreen in the form of a ".txt" file to the output folder. Use Tab as the column delimiter.

a

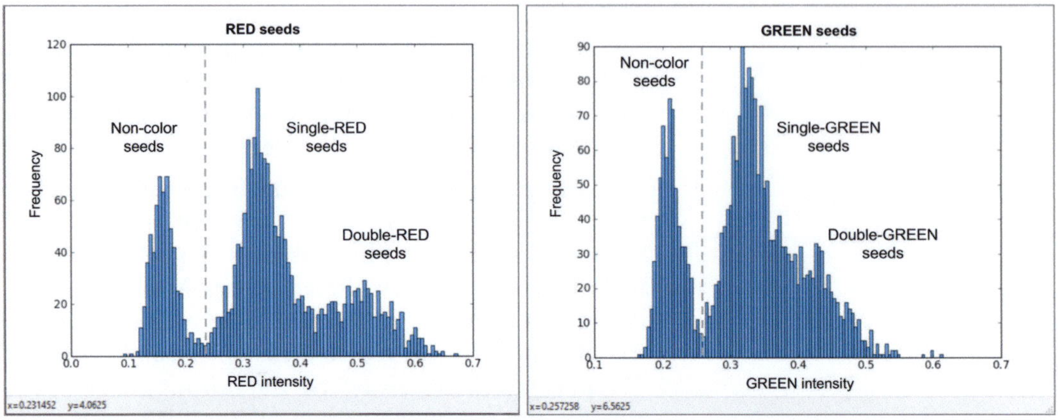

b

Green only	Red only	Both	None	Total	G / nonG	R / nonR	Green / Red	RF (cM)
211	229	1652	400	2492	2.96	3.08	0.92	19.57

c $cM = 100 \times (1 - [1 - 2(N_G + N_R) / N_T]^{1/2})$

N_G – green-only fluorescent seeds
N_R – red-only fluorescent seeds
N_T – total seed number

Fig. 4 Calculating crossover frequency using the *SeedScoring* pipeline. (**a**) Histograms of seed object fluorescence intensities from FTL segregating seed. Three peaks can be differentiated: noncolor seeds (carrying no fluorescent transgene—homozygous negative), single-red or single-green seeds (containing a fluorescent transgene on only one chromosome—hemizygote positive) and red and green fluorescent seeds (containing fluorescent transgene copies on both chromosomes—homozygous positive). For crossover measurement the threshold value of fluorescence separating noncolor and fluorescent seeds is the lowest value between the two first peaks on the *x*-axis of the histograms shown here by gray dashed line. The threshold must be specified separately for green and red fluorescent seeds. (**b**) An example of crossover calculation based on three pictures of seeds. The table shows numbers of green-only, red-only, both colors, no-color, and total seeds, which are used to calculate crossover frequency (centiMorgan, cM). The green to nongreen (G/nonG) and red to nonred seed (R/nonR) ratios should fall within the range of 2.7–3.3 and are used to evaluate Mendelian segregation of markers. The green to red seed ratio (G/R) should also equal ~1. (**c**) The formula used for cM calculation based on data from a table shown in (**b**)

4. Once the pipeline is ready, run it using a set of seed pictures taken as described in **Image acquisition** section. Open "eye" icons on the left-hand side of IdentifyPrimaryObjects, FilterObjects and DisplayHistograms for red and green (Figs. 2 and 4a). Test Mode can be used to verify whether all the steps worked correctly.

5. Save the project if the CellProfiler does not indicate a pipeline error.

6. Run the pipeline by clicking on "Analyze Images" button. Check whether seed identification and filtering work properly—most of the seeds and their boundaries on figures created by the IdentifyPrimaryObjects module should be well identified and wrongly recognized seeds should be filtered out by FilterObjects (for an example, *see* Fig. 3), *see* **Notes 8 and 9**.

7. Find the threshold value for red and green seeds by placing the cursor on the lowest value of the valley in both histograms (Fig. 4a; *see* **Notes 10 and 11**). Write down both thresholds (fluorescence intensities) and then enter them in the "Enter the custom thresholds separating the values between bins" box in ClassifyObjects (separately for red and green). Rerun the pipeline ("Analyze images" button).
8. The CellProfiler output file contains six columns: Image number (column A), Object number (Column B), and binary classifications of seeds (0 or 1) as green (Column C), nongreen (Column D), nonred (Column E), and red (Column F), for each object. Each row corresponds to a single object (a seed), therefore the number of rows is equal to the number of filtered and classified objects. Create an "output" Excel file and import, copy-and-paste or create a hyperlink to the output CellProfiler file in cell A1. The Excel file is a template to calculate RF from the row data file produced by CellProfiler.
9. Create an additional column next to the CellProfiler output (Column G) using the function shown in Table 2 for *Operation 1*. This will define the color of each seed.
10. To calculate RF, the number of individuals for each fluorescence profile is counted using a "COUNTIF" Excel function. Use the functions from Table 2 for *Operations 1–6* (*see* also Fig. 4b). The last formula needs to be adjusted to include the proper cell addresses. The pipeline does not distinguish between homozygous (G/G) or hemizygous (G/-) color seeds. To correct this missing information, the sum of the seeds that are only red or only green is divided by the total number of identified objects (Fig. 4c; *see* **Notes 12** and **13**).

Table 2
Formulas used to calculate RF from the *SeedScoring* output file

Operation	Calculation	Function
1	Calculation of the fluorescence of each object	IF(C3=1,IF(F3=1,"Both","Green"),IF (F3=1,"Red","None"))
2	Counting of the number of objects with green fluorescence	COUNTIF(G:G,"Green")
3	Counting of the number of objects with red fluorescence	COUNTIF(G:G,"Red")
4	Counting of the number of objects with both fluorescences	COUNTIF(G:G,"Both")
5	Counting the total number of objects	SUM(A:A)
6	Calculation of the recombination frequency	(1-SQRT(1-2*(no. of none color objects/total no. of objects))*100

11. To investigate segregation of the two colors, use three quality ratios: green/nongreen, red/nonred, and green/red. For Mendelian segregation, one expects green/nongreen and red/nonred to be in the range 2.7–3.3, and color/total ratios around 1 (Figs. 1c and 4b). This is important as deviation from the ratios may indicate silencing of FTL T-DNA genes (ratio lower than 2.5), aneuploidy (ratio above 5 for both colors), or segregation distortion within the FTL interval (both ratios biased). In such cases we suggest to rerun the protocol and verify whether the threshold values are correct (*see* **Notes 10–12**).

4 Notes

1. Other CellProfiler versions may require modification of the pipeline.
2. The protocol uses Microsoft Excel for the final RF calculation, but this procedure can be easily adapted to any spreadsheet with logical functions.
3. It is essential to keep seeds dry, as humidity causes green fluorescence to deteriorate rapidly. For this reason, we do not use containers but rather tie plants to sticks and let the siliques dry naturally.
4. The seeds should be as clean as possible; significant amounts of flower debris, especially petals, can interfere with subsequent seed identification.
5. Seeds obtained from a single plant usually cover the entire field of view, so that the images do not contain seedless areas. Figure 3 shows a situation where only a part of the field of view is covered with seeds to better illustrate the operation of the CellProfiler.
6. The image resolution is not very important, but the lower resolution images are analyzed faster by CellProfiler. Usually we use ".jpg" format and the images are approx. 500–600 kB in size.
7. For *SeedScoring* pipeline instructions, if not specified, the option to choose is "*No*" or whatever is more convenient for the user.
8. Identification of seeds (*IdentifyPrimaryObjects* module) depends on the stereomicroscope magnification, image size and resolution. If required, adjust the values in *Typical diameter of objects* box and/or *Suppress local maxima* box.
9. Filtering is based on red fluorescence, as even noncolor seeds show some level of autofluorescence, which allows them to be distinguished from debris. In the case that some seeds are

filtered out, or if impurities are recognized as seeds, decrease or increase the *Minimum value* for the *Filter using a minimum measurement value* option accordingly.

10. The histograms should resemble those shown in Fig. 4a, with a distinct valley between noncolor and color seeds. The causes of incorrect histograms are most often a shift between bright field and fluorescence images, or poor fluorescence quality. In order to distinguish between these two possibilities, it is best to manually check the matching of images and their qualities.
11. If the images do not have a homogenous fluorescence, or if fluorescence is weak, cropping the central part of the images can help analysis. The different intensities of fluorescence can affect the accuracy of the binning (object classification based on color). The cropping must be applied in the same way for all three images (bright field, red and green). This can be done by incorporating the *CropImages* module into the *SeedScoring* pipeline, after *ColorToGray* steps.
12. Check if the fluorescence profile is calculated for all the identified objects (Column G in the "output" Excel file). If the profile is not computed for some seeds, usually at the end of the list, these seeds will not be included in the scoring, which will affect RF calculation.
13. A crossover frequency above 50% indicates that the GFP and dsRed reporter cassettes are on different homologous chromosomes (in *trans*). By subtracting RF from 100, one can estimate RF, but this is less accurate than measurement in *cis*, due to low number of noncolor seeds. We suggest not to use measurements made with the FTL T-DNAs in *trans*.
14. In many cases, especially in mutant analysis, it is very useful to measure RF in filial generations. To do this, preselect single red and single green seeds (RG/--) from F_2 seeds under the stereomicroscope. Toothpicks are very convenient to move and pick Arabidopsis seeds under a microscope.

Acknowledgments

This work was supported by Polish National Science Centre grants 2019/35/N/NZ2/02933 to N.K., 2020/37/N/NZ2/01226 to J.D., 2016/22/E/NZ2/00455 to P.A.Z., a Foundation for Polish Science grant POIR.04.04.00-00-5C0F/17-00 to P.A.Z, and European Research Council Consolidator Award ERC-2015-CoG-681987 "SynthHotSpot" to I.R.H. J.D. is the holder of Foundation for Polish Science START stipend (START 10.2021) and AMU Foundation scholarship for Ph.D students.

References

1. Mercier R, Mézard C, Jenczewski E et al (2015) The molecular biology of meiosis in plants. Annu Rev Plant Biol 66:297–327
2. Sanchez-Moran E, Armstrong SJ, Santos JL et al (2002) Variation in chiasma frequency among eight accessions of Arabidopsis thaliana. Genetics 162:1415–1422
3. López E, Pradillo M, Oliver C et al (2012) Looking for natural variation in chiasma frequency in Arabidopsis thaliana. J Exp Bot 63: 887–894
4. Armstrong SJ, Sanchez-Moran E, Franklin FCH (2009) Cytological analysis of Arabidopsis thaliana meiotic chromosomes. Methods Mol Biol 558:131–145
5. Chelysheva L, Grandont L, Vrielynck N et al (2010) An easy protocol for studying chromatin and recombination protein dynamics during Arabidopsisthaliana meiosis: Immunodetection of cohesins, histones and MLH1. Cytogenet Genome Res 129:143–153
6. Chelysheva L, Vezon D, Chambon A et al (2012) The Arabidopsis HEI10 is a new ZMM protein related to Zip3. PLoS Genet 8: e1002799
7. Rowan BA, Patel V, Weigel D, Schneeberger K (2015) Rapid and inexpensive whole-genome genotyping-by-sequencing for crossover localization and fine-scale genetic mapping. G3 Genes Genomes Genet 5:385–398
8. Sun H, Rowan BA, Flood PJ et al (2019) Linked-read sequencing of gametes allows efficient genome-wide analysis of meiotic recombination. Nat Commun 10:1–9
9. Copenhaver GP, Browne WE, Preuss D (1998) Assaying genome-wide recombination and centromere functions with Arabidopsis tetrads. Proc Natl Acad Sci U S A 95:247–252
10. Drouaud J, Camilleri C, Bourguignon P et al (2006) Variation in crossing-over rates across chromosome 4 of Arabidopsis thaliana reveals the presence of meiotic recombination "hot spots". Genome Res 16:106–114
11. Melamed-Bessudo C, Yehuda E, Stuitje AR, Levy AA (2005) A new seed-based assay for meiotic recombination in Arabidopsis thaliana. Plant J 43:458–466
12. Francis KE, Lam SY, Harrison BD et al (2007) Pollen tetrad-based visual assay for meiotic recombination in Arabidopsis. Proc Natl Acad Sci U S A 104:3913–3918
13. Berchowitz LE, Copenhaver GP (2008) Fluorescent Arabidopsis tetrads: a visual assay for quickly developing large crossover and crossover interference data sets. Nat Protoc 3:41–50
14. Yelina NE, Ziolkowski PA, Miller N et al (2013) High-throughput analysis of meiotic crossover frequency and interference via flow cytometry of fluorescent pollen in Arabidopsis thaliana. Nat Protoc 8:2119–2134
15. Ziolkowski PA, Berchowitz LE, Lambing C et al (2015) Juxtaposition of heterozygous and homozygous regions causes reciprocal crossover remodelling via interference during Arabidopsis meiosis. Elife 4:1–29
16. Wu G, Rossidivito G, Hu T et al (2015) Traffic lines: new tools for genetic analysis in Arabidopsis thaliana. Genetics 200:35–45
17. Carpenter AE, Jones TR, Lamprecht MR et al (2006) CellProfiler: image analysis software for identifying and quantifying cell phenotypes. Genome Biol 7:R100
18. Blackwell AR, Dluzewska J, Szymanska-Lejman M et al (2020) MSH 2 shapes the meiotic crossover landscape in relation to interhomolog polymorphism in Arabidopsis. EMBO J 39:e104858
19. Ziolkowski PA, Underwood CJ, Lambing C et al (2017) Natural variation and dosage of the HEI10 meiotic E3 ligase control Arabidopsis crossover recombination. Genes Dev 31: 306–317
20. Lawrence EJ, Gao H, Tock AJ et al (2019) Natural variation in TBP-ASSOCIATED FACTOR 4b controls meiotic crossover and germline transcription in Arabidopsis. Curr Biol 16: 2676–2686
21. Serra H, Lambing C, Griffin CH et al (2018) Massive crossover elevation via combination of HEI10 and recq4a recq4b during Arabidopsis meiosis. Proc Natl Acad Sci 115:2437–2442
22. Zhu L, Fernandez-Jimenez N, Szymanska-Lejman M et al (2021) Natural variation identifies SNI1, the SMC5/6 component, as a modifier of meiotic crossover in Arabidopsis. Proc Natl Acad Sci U S A 118:e2021970118
23. Nageswaran DC, Kim J, Lambing C et al (2021) HIGH CROSSOVER RATE1 encodes PROTEIN PHOSPHATASE X1 and restricts meiotic crossovers in Arabidopsis. Nat Plants 7: 452–467

Part III

Plant Manipulation and Genome Editing

Chapter 11

Generation of Mutants by Combined Treatment of Physical and Chemical Mutagens in Rice

Yuanyuan Tan and Shan Li

Abstract

Mutants can be generated by physical or chemical mutagens. The procedure includes mutagenesis treatment, post-treatment, planting of M_1, harvest of M_1, planting of M_2, phenotypic observation or mutation scanning in M_2 plants. Here we describe the combined treatment of the rice seeds from the *indica* variety Tenuo with ^{137}Cs gamma rays at 100 Gy and a chemical mutagen—ethyl methanesulfonate (EMS). Theoretically, the method can be used to study various plant biological processes which include plant gametogenesis in addition to the generation of mutants by straightforward physiological screens. As an example of the procedure, we screened for mutants resistant to imazethapyr, an herbicide which can kill grass weeds.

Key words Mutants, Gamma rays, EMS, Combined treatment, Rice

1 Introduction

Since Staedler (1928) first demonstrated the irradiation could cause mutations in maize and barley, induced mutations were rapidly adopted in plant breeding. In addition, mutants and TILLING populations are considered as important genetic resources for gene discovery and functional analysis. Until now, various physical and chemical mutagens proved to be able to produce mutations. The Joint FAO/IAEA (Food and Agriculture Organization/International Atomic Energy Agency) Division edited the manual on mutation breeding and constantly update the content based on the research progress. The procedures of irradiation by physical and chemical mutagens, management of M_1, M_2, M_3 populations and the application in plant breeding have been elaborately described in the third edition of the manual [1]. However, it is complex for the breeders or researchers to generate mutants or TILLING

Christophe Lambing (ed.), *Plant Gametogenesis: Methods and Protocols*, Methods in Molecular Biology, vol. 2484, https://doi.org/10.1007/978-1-0716-2253-7_11,

population. To simplify the procedures of irradiation and make it more effective, we treated rice seeds with chemical and/or physical mutagens.

It is commonly recommended that the irradiation dose of indica and japonica is 350 and 250 Gy according to the range of GR 50 (Growth Reduction by 50%) [2], but Li et al. (2016) found that similar mutation rates were produced in the whole genome by varied irradiation dose. This suggests that a low dose could generate as many generation mutations as the high dose, but the survival rate of M_1 could be increased [3]. Combined treatment of chemical mutagens resulted in increased mutation frequency in barley and this approach was used to create a TILLING population in barley and rice [4, 5]. The mechanism of damaging the DNA by physical mutagen is different from that of chemical mutagen. A combined treatment with physical and chemical mutagen may broaden the spectrum of induced mutations. Ethyl methanesulfonate (EMS) mutagenesis was performed according to Konzak and Mikaelsen, Jankowicz-Cieslak and Till with some modifications [6, 7].

Here we describe the combined treatment with ^{137}Cs gamma rays at 100 Gy and the chemical mutagen-EMS on rice seeds from the *indica* variety Tenuo. To save labor and cost, direct seedling of M_1 and M_2 together with bulk harvest of M_1 seeds were adopted. Mutants resistant to herbicide could be identified (Fig. 1.) with the ratio of about 3.7×10^{-6}. In addition to breeding new varieties, new genes or mechanisms on herbicide resistance may be discovered by using the mutants. Therefore, the method can also be used to study more complicated plant biological processes including plant gametogenesis.

2 Materials

Prepare all solutions using distilled water and chemical grade reagents except for EMS which is an analytical grade reagent. Prepare and store all reagents at room temperature except for EMS which is kept at 4 °C in a refrigerator.

1. Seeds of the commercial *indica* variety Tenuo was chosen for the protocol (*see* **Note 1**).
2. Phosphate buffered solution (PBS): 0.2 M Na_2HPO_4, 0.2 M NaH_2PO_4. Weigh 2.84 g Na_2HPO_4 and add water to the total volume of 100 mL. Separately, weigh 2.40 g NaH_2PO_4 and add water to the total volume of 100 mL. Mix two solutions above mentioned to prepare PBS (pH 7.0). The pH reaches 7.0 by mixing the two solutions together. Store at room temperature.

Fig. 1 Selection of mutants with herbicide resistance from M_2 seedlings

3. 0.5% EMS incubation solution: 5 ml EMS concentration, 20 ml DMSO, 975 ml PBS (*see* **Note 2**).
4. 1 M sodium thiosulfate: Weight 150 g $Na_2S_2O_3$ and add water to a volume of 1 l.
5. imazethapyr solution: Dissolve the commercialized imazethapyr in water and make the final concentration at 0.3–0.4 g/l.

3 Methods

Carry out all procedures at room temperature.

3.1 Physical Mutagenesis Treatment (Gamma Rays)

Treat 2.5 kg rice seeds with ^{137}Cs gamma rays at 100 Gy (1 Gy/min) (*see* **Notes 3–7**).

3.2 Chemical Mutagenesis Treatment (EMS)

3.2.1 Presoaking

1. Put the seeds in a net bag of appropriate size.
2. Add water and make sure every single seed is immersed in the water.
3. Soak the seeds overnight (16–24 h).
4. Take the seeds out and drain the water.

3.2.2 EMS Incubation

1. Add 3 l of 0.5% EMS incubation solution to the seeds (*see* **Notes 6–8**).
2. Incubate the seeds for 2 h at room temperature. Move the bag of seeds several times to make sure every single seed is in contact with EMS (*see* **Note 9**).

3.2.3 Post Treatment

1. Add an equal amount of freshly prepared 1 M sodium thiosulfate to the mutagenized seeds and incubate for 15 min to deactivate the EMS. During the incubation, turn over the bag several times. Repeat the incubation of 1 M sodium thiosulfate twice.
2. Rinse the seeds with tap water and incubate for 15 min. During the incubation, turn over the bag in the solution several times. Repeat the rinsing and incubation twice.
3. Collect and put all the liquid into the toxic waste bottle/ bucket. Add an equal volume of 200 mM sodium thiosulfate into the bottle/bucket to inactive the EMS. Clean and screw the lid tightly.
4. Wash the seeds using flowing tap water for 4–6 h.
5. Keep the seeds in the moist condition for gemination. The seeds can be planted in the paddy field for transplanting or directly sowed. Otherwise, the seeds should be dried for storage as soon as possible.
6. Dry the seeds in the sun for several days, until the water content of the seeds approximately reaches 13%.
7. Store the seeds at room temperature or at lower temperature (*see* **Note 10**).

3.3 Planting and Analysis of Mutant Plants

1. Germinate and sow the irradiated seeds. The plants grow into M_1 population.
2. Bulk harvest seeds of all M_1 plants.
3. Grow M_2 population by single plant (*see* **Note 11**).
4. The M_2 population is used for mutant selection through phenotypic observation and analysis or mutation scanning as a TILLING population.
5. Treat 10–20 day-old plants with a solution of imazethapyr (*see* **Note 12**).

6. Observe the phenotype and screen the mutants (Fig. 1).
7. Transplant the mutants and assemble the plants in the same plot.
8. Harvest the seeds from single M_2 plant (*see* **Note 13**).

4 Notes

1. Select dry healthy, uniformly sized seeds with moisture content of ca. 14%.
2. DMSO can be replaced with ethanol to improve the solubility of EMS in water.
3. Commonly 0.25–1 kg of seeds need to be treated for generation of mutants or development of TILLING population. Increasing the quantity of M_1 plants may improve the frequency of mutants. Also, analyzing more M_2 plants can help to identify more possible mutants.
4. *Indica* rice is more tolerant to irradiation than *Japonica* rice. The *indica* rice seeds should be subjected to gamma ray irradiation at 150 Gy when the single mutagen is used.
5. Physical mutagenesis treatment should be performed prior to chemical mutagenesis when the combined treatment is applied.
6. Here, the combined treatment of the *indica* variety Tenuo is gamma rays at 100 Gy and 0.5% EMS for 2 h. When applying the combined treatment to a *japonica* variety, the irradiation dose or EMS concentration or duration of EMS treatment can be decreased.
7. Different species showed varied tolerance to mutagens. The optimum irradiation dose or EMS concentration can be recommended based on the previous literatures. For the combined treatment, different treatment combinations should be applied to confirm the optimum combination. Germination rate and seedling emergence can be used to determine the optimal dose of physical and chemical mutagenesis treatment.
8. Practically, 1 kg of rice seeds can be immersed in 1 l of water. Therefore, in consideration of the cost, mutation efficiency and environmental protection, the volume of EMS incubation solution should be 0.5 l more.
9. Lowering the EMS concentration could be compensated by increasing duration of EMS treatment.
10. The storage temperature of the mutated seeds depends on the planting time. Lower temperature can keep the seeds in a good germination capability for a longer time.

11. Keep a portion of M_2 seeds in case of the extraordinary circumstances that no or fewer plants can survive.
12. The application of herbicide is used to screen for a mutant that is herbicide resistant but the step should not be performed otherwise.
13. Harvest the seeds by single plant from the mutants and confirm the phenotype in M_3 plants or seeds.

Acknowledgments

This work was supported by the grant from the National Key Research and Development Program of China (2016YFD0102103). Sincere thanks to Mr. Fuli Yuan and Mr. Yuexiao Zhang for their help in planting, herbicide spraying and phenotypic observation.

References

1. Spencer Lopes MM, Forster BP, Jankuloski L (2018) Manual on mutation breeding, 3rd edn. Sustria, Vienna
2. Brunner H (1995) Radiation induced mutations for plant selection. Appl Radiat Isotopes 46(6–7):589–594
3. Li S, Zheng YC, Cui HR et al (2016) Frequency and type of inheritable mutations induced by γ rays in rice as revealed by whole genome sequencing. J Zhejiang Univ Sci B 17(12): 905–915
4. Jankowicz-Cieslak J, Mba C, Till BJ (2017) Mutagenesis for crop breeding and functional genomics. In: Biotechnologies for plant mutation breeding. Springer, New York, pp 3–18
5. Till BJ, Cooper J, Tai TH et al (2007) Discovery of chemically induced mutations in rice by TILLING. BMC Plant Biol 7(1):19
6. Konzak C, Mikaelsen K (1977) Induced mutation techniques in breeding the seed propagated species. In: Manual on mutation breeding. Technical Report Series 119, pp 125–138
7. Jankowicz-Cieslak J, Till BJ (2016) Chemical mutagenesis of seed and vegetatively propagated plants using EMS. Curr Protoc Plant Biol 1(4): 617–635

Chapter 12

Generation of Mutant Plants by Gamma Ray Exposure and Development of Low-Cost TILLING Population in Chickpea (*Cicer arietinum* L.)

Wahiba Amri-Tiliouine, Joanna Jankowicz-Cieslak, Bradley J. Till, and Meriem Laouar

Abstract

Induced mutations have been used to facilitate plant breeding for more than 80 years. Success requires the development of a mutant population and methods to evaluate that population. In this protocol we provide methods for the development of a chickpea mutant population using gamma irradiation, and low-cost methods for the molecular characterization of the mutant population. Specifically, this chapter provides detailed methods for (1) mutation induction by gamma rays and determination of LD_{50} and RD_{50}, (2) phenotypic assessment of the M_2 generation, (3) low-cost extraction of genomic DNA, and (4) identification of induced mutations using low-cost agarose-gel based TILLING. The methods are low-cost and designed to be applicable in most research settings.

Key words Legumes, Mutation induction, Mutation discovery, Radiosensitivity, M_2 generation, Phenotypic screening, Enzymatic mismatch cleavage, Low-cost TILLING

1 Introduction

Mutagenesis, which induces novel variation directly at the level of DNA sequence, is a rapid instrument, which can be effective for plant breeding and functional genomics [1–3]. It can provide a species or a variety with an additional exploitable genetic variability [4]. Several thousand varieties are derived from induced mutations, especially as a result of irradiation [2, 5, 6]. Gamma rays were the most common technique used to modify genes in 64% of mutant varieties officially released and registered within the Mutant Variety Database (MVD) (http://mvd.iaea.org) [5, 6] In general, ionizing radiations such as X-rays and gamma rays are used more frequently due to their easy application, good penetration, reproducibility, and high mutation frequency. Different mutagens increase the frequency of mutations that can be subjected to selection, but no

Christophe Lambing (ed.), *Plant Gametogenesis: Methods and Protocols*, Methods in Molecular Biology, vol. 2484, https://doi.org/10.1007/978-1-0716-2253-7_12,

random mutagen is specific to one type of mutation in terms of genomic location and effect on gene function. However, ionizing radiation mainly produces chromosome breaks, leading to structural changes such as deletions, insertions, inversions, and translocations [7]. The applied dose of a mutagen is an important factor to consider in any mutation-breeding program [8]. There is a significant genotypic effect of susceptibility to mutagenic treatments in plants. It is therefore recommended to carry out preliminary tests to determine the appropriate dose. Determining the dose which would give the greatest amount of desirable mutation per unit while producing the least density of adverse effects is known as the "radiation sensitivity test"; but more commonly called "radiosensitivity test." Determination of the mutagenic dose that results in a 50% reduction in seed germination or seedling viability, known as the LD_{50}, (lethal dose 50) was used to accurately predict the most effective mutagen dose. The determination of the mutagenic dose which results in the reduction in the average growth of the seedlings by 50% is also commonly used and is abbreviated as RD_{50} (Reduction Dose 50) or GR_{50} (Growth Reduction 50) [9, 10]. Factors influencing the effects of mutagens can have biological (nuclear volume, chromosome volume, and DNA content), environmental (oxygen, water state and temperature) or chemical nature [11]. The moisture content of the seeds is a critical factor affecting the frequency of mutation. Even small differences in the water content can have a very pronounced influence on the final biological effect. A difference of only 0.2–0.3% moisture content could dramatically change the radiosensitivity of some species. Seeds containing 12–14% moisture usually give a higher mutation frequency by irradiation [8, 10].

The process of mutation breeding involves the following steps: mutation induction, detection of mutations, fixation of mutations, and development of mutant lines and release of new mutant cultivars [2, 12]. Plant material, prior to the mutagenic treatment, is being called M_0 and after the treatment, M_1. Plants that are produced directly from mutagenized seeds (or gametes) are M_1 plants, and the seeds are called M_2, followed by next generations called M_3, M_4, etc. Following the process of induction of mutations, the plants are evaluated for heritable traits that are different from the nonmutagenized parental line. However, when using physical or chemical mutagenesis, the first generation (M_1) is a genetic mosaic because each mutagen-treated cell is able to accumulate different mutations. This chimeric generation is therefore not suitable for phenotypic or genotypic evaluation because different tissues can have different genotypes and thus express different characters. Ideally, evaluation of a mutant population is carried out starting in the first nonchimeric generation (M_2) to reduce the loss of desirable combinations of favorable alleles [13–15]. Once the potentially interesting traits have been recovered, work is undertaken to fix the new allele(s) in the population and ensure the stability and heritability of the desired traits [2]. Kharkwal and Shu [16] indicated that mutant populations have also been used as a resource for genomics and molecular enhancement [17]. Mutagenesis has received great attention as a method for revealing gene function (reverse genetics). Targeting Induced Local Lesions IN Genomes (TILLING) is an efficient early-screening approach and usually applied in combination with chemical mutagenesis, but it has been shown that it can be also used in physical mutagenesis, such as gamma rays and fast neutron irradiation [18]. The mutation detection system used in traditional TILLING is based on screening libraries of PCR fragments of targeted regions that are amplified from genomic DNA samples that are typically pooled prior to PCR to increase screening throughput. The mutant lines are then identified using the bioinformatic platform appropriate for the screening technology. A low-cost TILLING process was developed for chickpea in collaboration with the Plant Breeding and Genetics Laboratory (PBGL, Seibersdorf, Austria) [19]. The low-cost TILLING does not require specialized equipment and does not rely on toxic chemicals. It includes approaches for genomic DNA extraction and extraction of single-strand–specific nucleases and agarose-gel based enzymatic mismatch cleavage assays for mutation discovery [20, 21]. The methods are appropriate for the recovery of single nucleotide substitution and small insertion/deletion mutations [22]. Whole genome resequencing of gamma-irradiated plants has established that mutagenic treatments result in the accumulation of a broad spectrum of mutation types including a high proportion of small mutations discoverable by the low-cost methods described in this protocol [23].

The approach used to reach the goal of chickpea improvement breeding, is based on two ways: forward (agronomic assessment) and reverse genetic (TILLING) (Fig. 1). In this chapter, we describe methods for the development of a gamma-irradiated mutant chickpea population, and low-cost molecular methods to screen this population including DNA extraction and agarose-gel based enzymatic mismatch cleavage.

2 Materials

2.1 Mutation Induction

1. Seeds of Kabuli chickpea variety Ghab4. Ghab 4, is a selection from FLIP 93-93C developed by the International Center for Agricultural Research in the Dry Areas (ICARDA) (*see* **Notes 1** and **2**).
2. Paper bags and stapler.
3. Desiccator, silica gel or 60% glycerol: distilled water mixture (v/v).

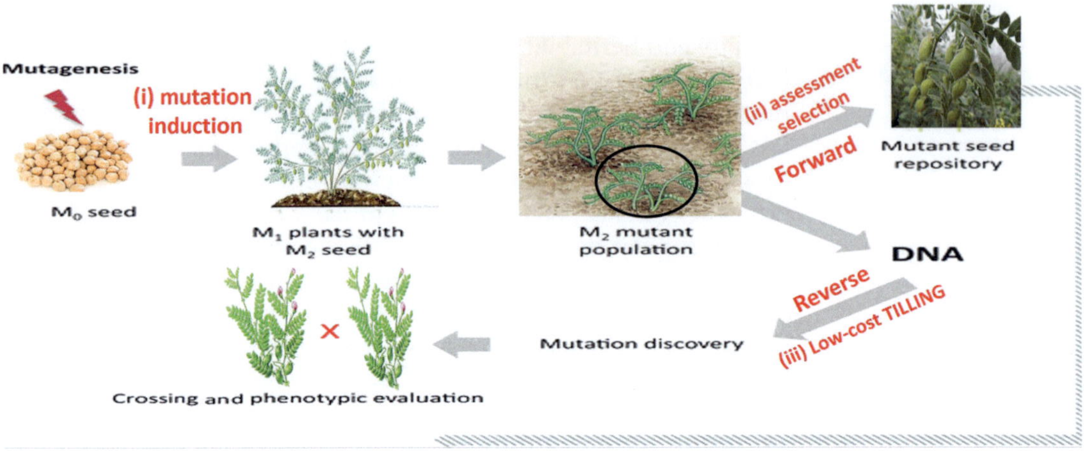

Fig. 1 Relation between forward (agronomic assessment) and reverse genetic (TILLING). (Modified from [3])

4. Source of radiation (gamma rays): Cobalt 60 (^{60}Co).
5. Fertile pots, sterilized sand or potting mix, and plastic containers.
6. To evaluate the response to irradiation the following tools were used: a ruler for seedling height measurement, a scale for weight and a laboratory dryer to obtain the dry weight.

2.2 Phenotypic Evaluation

1. Chickpea seeds of Ghab 4 variety mutagenized with LD_{50}/RD_{50} dose (for M_1) and healthy and dry seeds of M_1 plants for M_2 population.
2. Control: nonirradiated chickpea seeds of Ghab 4 variety.
3. 1 or 2% Acetocarmine stain.

2.3 Low-Cost TILLING

1. Chickpea M_2 putative mutants (81 samples chosen for optimization of the low-cost TILLING assays).
2. Control samples: nonmutated Ghab4 variety and three different Desi chickpea genotypes.
3. silica gel (with humidity indicator).
4. Celite 545 silica powder.
5. RNase (10 μg/ml).
6. Lysis Buffer: 0.5% SDS, 10 mM EDTA, 2.5 μg/ml RNase A.
7. Silica slurry (*see* **Note 3**).
8. Silica binding solution (SBS): 700 μl of 6 M KI, 50 μl of silica slurry. Prepared the solution fresh each time.
9. Wash Buffer: 50 mM of NaCl, 95% of Ethanol (*see* **Note 4**).
10. 3 M sodium acetate (pH 5.2). Adjust pH with glacial acetic acid.

2.4 Genomic DNA Quantification and Normalization

1. Agarose gel electrophoresis (horizontal electrophoresis), NanoDrop 1000 spectrophotometer.
2. Ethidium bromide.
3. Lambda DNA ladder.
4. 1×TBE buffer: 89 mM Tris base, 89 mM boric acid, 2 mM EDTA. Adjust pH to 8.0 with NaOH.
5. Tris–EDTA (TE) buffer (10×): 100 mM Tris–HCl, pH 8.0, 10 mM EDTA.

2.5 Primer Testing and Enzymatic Digestion

1. Primer designed using the Primer3program http://bioinfo.ut.ee/primer3-0.4.0/) to produce amplicons between 600 and 1200 bps.
2. Taq polymerases, dNTPs and forward and reverse primers for the marker genes.
3. Crude enzyme extract containing enzymatic mismatch cleavage activity [20]: 14 μl of H_2O, 4 μl of 10× CELI buffer and 2 μl of CJE nuclease.
4. 0.25 M EDTA (pH = 8.0). Adjust pH with NaOH.

3 Methods

3.1 Determination of LD_{50} and RD_{50} (See Note 5)

1. For radiosensitivity test use homogenous and dry chickpea seeds (12–14% moisture content) of Ghab 4 (672 Seeds) (*see* **Notes 6** and 7).
2. Use different doses of Cobalt 60 (^{60}Co): six radiations (150, 200, 250, 300, 350, and 400 Gy) (*see* **Note 8**).
3. Use untreated seeds as control.
4. Count out the number of seeds required per dose and pack them in paper bags (96 seeds per dose).
5. Label the paper bags with information on species, variety, and the treatment dose.
6. Pack the seeds in sealed plastic bags to maintain moisture content and ship them to the irradiation facility.
7. After having received treated seeds, record information on the dose rate and the exposure time (*see* **Note 9**).
8. Carry out the experimental trial in a greenhouse.
9. Use a complete random block experiment design (RCBD) with four blocks, 24 seeds per dose and per block (Fig. 2).
10. Sow seeds in fertile pots filled with approximately 500 g of sand. Keep the sowing depth uniform with 2–3 times the size of the seed.

Fig. 2 Chickpea radiosensitivity test in a greenhouse

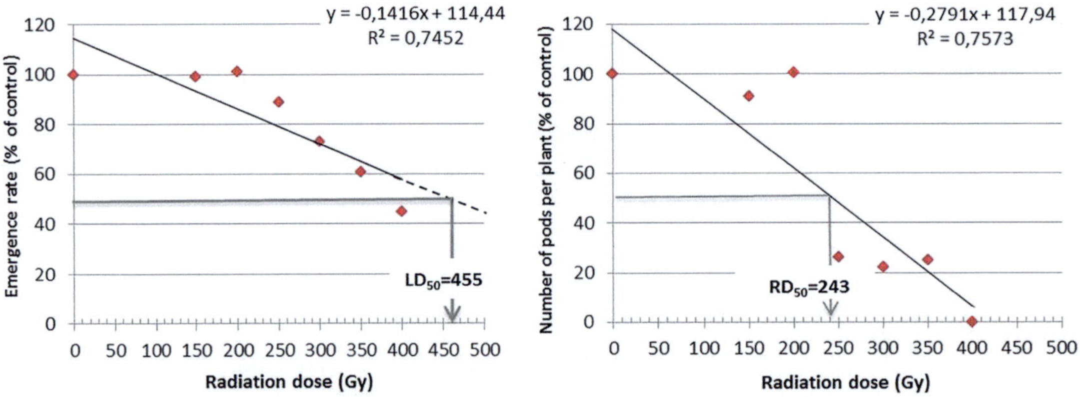

Fig. 3 Average of emergence rate and number of pods per plant (% of control) and regression equation of chickpea variety (Ghab4) exposed to different doses of gamma radiation

11. Place the fertile pots in plastic containers in the greenhouse and irrigate according to the needs of the plant (*see* **Note 10**).
12. Data collection: 22 days after sowing, randomly pull out ten seedlings per dose and per block and record the emergence rate (Fig. 3), seedling height (Fig. 4), the dry weight of shoots and roots.
13. Transplant the remaining seedlings 45 days after sowing into plastic pots containing 5 kg of equivalent amount of soil, sand, and potting soil. Observe phenotype of these plants up to the maturity stage.

Fig. 4 Seedling height decreasing gradually with increasing gamma radiation dose

14. At maturity, record the number of pods per plant (Fig. 3) and the survival rate.
15. Use linear regression analysis to estimate LD_{50} and RD_{50}. Express all measured parameters as a percentage of the control.

3.2 Raising M_1 and M_2 Generations

All important stages are illustrated in Fig. 5.

3.2.1 Raising M_1 Generation

1. To raise M_1 generation, prepare healthy and dry seeds (12–14% moisture content), and ship them to the irradiation facility (for the purpose of this study, 9000 seeds of Ghab 4 variety were used). For Ghab 4 variety, use the irradiation dose of 280 Gy (optimal dose found through the radiosensitivity test).
2. After having received treated seeds, record information on the dose rate and the exposure time.
3. Carry out the experimental trial in the field.
4. Use a total randomization experimental design.
5. Ensure good soil and seedbed tillage and optimum humidity conditions (*see* **Note 11**).

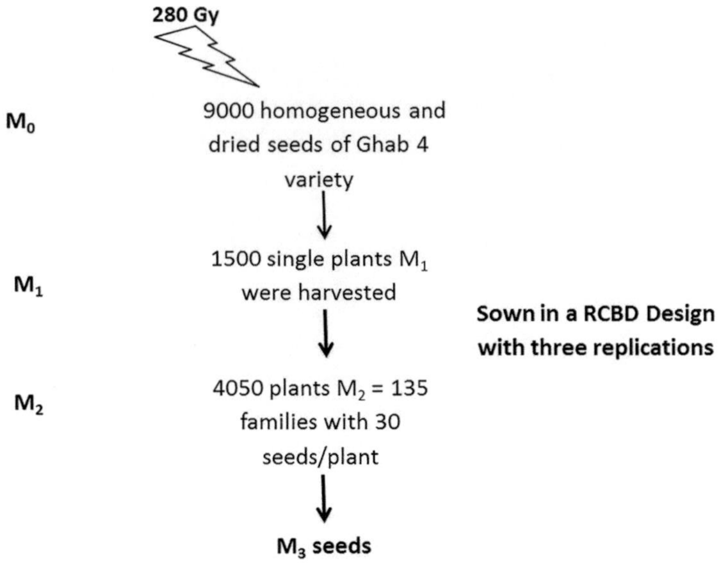

Fig. 5 Schematic representation of M_2 population of chickpea

6. Sow the seeds in the field manually or with a precision seeder with following sowing distances: 30 cm × 20 cm.
7. Record phenological developmental stages, emergence, seedling survival, chimera induction in M_1, survival to maturity (survival relative to the control population) and sterility in M_1.
8. Harvest all the M_1 plants (1500 plants) separately at complete physiological maturity (*see* **Note 12**).

3.2.2 Raising M_2 Generation

1. To raise M_2 generation prepare healthy and dry seeds harvested from M_1 plants (Example: seeds from 135 M_1 plants) (*see* **Note 13**).
2. Carry out the experimental trial in the field.
3. Use a complete random block experiment design (RCBD) with three blocks.
4. Sow 10 seeds per row, one by one, per family (seeds coming from the same M_1 plant) per block.
5. Include in each block 135 M_2 families with a sowing distance of 50 cm × 35 cm (Fig. 6).
6. Sow rows (10 seeds/row) of nonirradiated variety in each block. Include a control row for every ten family's rows (in total 14 control rows per block).
7. Assure optimum cultivation practices, including supplemental irrigation if necessary, weed control by herbicide or by mechanical means and prevention of severe disease levels.

Fig. 6 Photo illustrating field trial of M_2 chickpea generation (total area is 884 m^2). (Photo reproduced from [24] under the license CC BY 4.0)

8. Label the families and all plants at the seedling stage.
9. Harvest all the M_2 plants separately at complete physiological maturity.

3.3 Screening of M_2 Generation

Two assessments can be conducted: the determination of the irradiation efficiency and the agronomic interest; for this the variability between families for many traits can be studied.

3.3.1 Efficiency Assessment

1. Evaluation of chlorophyll mutation: Albina, Xantha, Chlorina, Viridis, and others (Method of evaluation; Modified classification of [25, 26].
2. Evaluation of pollen viability (1 or 2% acetocarmine stain test).

3.3.2 Agronomic Assessment

1. Phenological evaluation: days to emergence, days to flowering, days to pod formation, and days to physiological maturity.
2. Evaluation of the vegetative part: height of first pod (cm), plant height (cm), number of primary branches, growth habit (erect, semierect, semispreading, spreading, and prostrate).
3. Evaluation of the yield and yield components: number of seeds/pod, total number of pods/plant, total number of seeds/plant, seed yield (g), 100 seed weight.

3.4 Genomic Low-Cost DNA Extraction

1. Collect leaf samples from chickpea seedlings grown in the greenhouse using a scissor disinfected with ethanol, and place them quickly in labeled paper envelopes.
2. Immediately after collection, put these envelopes containing the leaf samples in a container containing silica gel. Seal the container. Dry for at least 24 h. When completely dried, the tissue should be brittle.
3. Place the dried tissue in 2 ml tubes each containing three metal beads and label them with the sample number.
4. For grinding the tissue, use the vortex until a fine powder is obtained.
5. Add 800 μl of lysis buffer and 4 μl of RNase A (10 μg per μl) to each tube.
6. Mix well for about 2 min with the vortex until the powder is fully hydrated and mixed with the buffer.
7. Incubate 10 min at room temperature.
8. Add 200 μl of 3 M sodium acetate (pH 5.2). Mix by inversion of the tubes and incubate on ice for 5 min.
9. Centrifuge at 15,000 × *g* for 5 min at room temperature.
10. Label the tubes with aliquots of silica binding solution (SBS), (SBS, 700 μl).
11. Transfer the liquid to the properly labeled tubes containing SBS. Completely suspend the silica powder by vortexing and inverting the tubes for 20 s.
12. Incubate for 15 min at room temperature on a shaker at 14 × *g* and or by inverting the tubes from time to time.
13. Centrifuge at 15,000 × *g* for 5 min at room temperature.
14. Remove the supernatant with a pipette. The DNA is bound to the silica.
15. Add 500 μl of freshly prepared Wash Buffer to each tube.
16. Completely resuspend the silica powder by vortexing and inverting the tubes for approximately 20 s.
17. Centrifuge at 15,000 × *g* for 3 min at room temperature to pellet the silica. Remove the supernatant and keep the pellet.
18. Repeat **steps 15–17**.
19. Centrifuge the pellet for 30 s and remove any ethanol residue with a micropipette.
20. Open the lid of the tubes and place them in a fume hood for 30 min (This can be done for a longer period on the bench).
21. Add 200 μl of TE buffer to each tube to elute the DNA.
22. Completely resuspend the silica powder in the vortex and invert the tubes for approximately 20 s.

23. Incubate at room temperature for 5 min.
24. Centrifuge at 15,000 × *g* for 5 min at room temperature to pellet the silica.
25. Label new 1.5 ml tubes with sample numbers.
26. Collect the liquid containing genomic DNA and place it in the new 1.5 ml tubes.
27. DNA can be stored at 4 °C before checking for quality and quantity.

3.5 Genomic DNA Quantification

Quantify the extracted genomic DNA using a 1% agarose gel with lambda DNA standards and a NanoDrop 1000 spectrophotometer.

3.5.1 Agarose Gel

1. Estimate the amount of DNA using gel electrophoresis where the extracted DNA and a dilution series of a standard DNA can be visualized on a 1% agarose gel in 0.5× TBE buffer.
2. Staining with 0.15 μg/ml ethidium bromide.
3. Expose the gel with a UV light.
4. Unknown DNA concentrations can be estimated by comparing the intensity of the bands of the extracted DNA with the bands of the standard DNAs (*see* **Note 14**).

3.5.2 Nanodrop Spectrophotometer

1. Measurements of DNA concentration can be obtained directly by measuring absorbance with the use of a Nanodrop spectrophotometer.
2. Quantification is based on the 1 mm optical path with the use of 1 μl of DNA sample without prior dilution (*see* **Note 15**).

3.5.3 Concentration of Genomic DNA

1. Normalize the genomic DNA samples to a common concentration using the results obtained from 1% agarose gel electrophoresis with lambda DNA standards and by reading on a NanoDrop 1000 spectrophotometer.
2. DNA normalization is necessary for pooling samples for TILLING, so that each sample in a pool is equally represented in the downstream PCR amplification step (*see* **Note 16**).

3.6 Database Search for Candidate Genes

1. Sequences of candidate genes are obtained from a public database such as NCBI (https://www.ncbi.nlm.nih.gov/assembly/GCF_000331145.1/). BLAST searches are carried out to identify putative orthologs in chickpea.
2. A minimum of five primer pairs producing nonoverlapping amplicons are suggested to control for variations in PCR performance (*see* **Note 17**).

3.7 Primer Testing

1. Use the primers developed in Subheading 3.6 for the polymerase chain reaction (PCR) test.

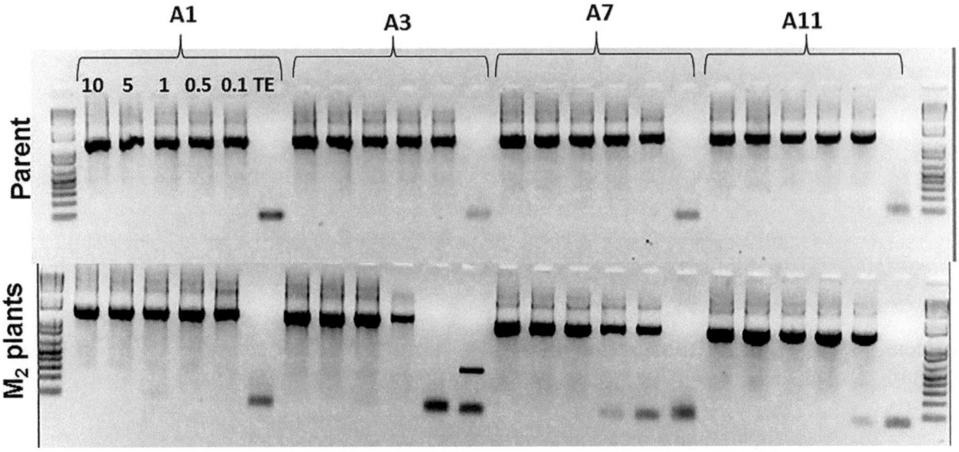

Fig. 7 Agarose gel evaluation of PCR primers. Test of 4 primers with 5 different concentrations of genomic DNA (10, 5, 1, 0.5, 0.1 ng/μl and TE control) of parent Ghab 4 and M_2 plants [24]. (Image taken from [24] under the license CC BY 4.0)

2. Test different template DNA concentrations (0.1, 0.5, 1, 5, and 10 ng/μl). For subsequent assays, choose the concentration which has a good amplification (Example: 1 ng/μl).
3. Use the following PCR mix: 82.5 μl H_2O, 15 μl 10× Ex Taq buffer, 12 μl dNTP mix (2.5 mM), 1.5 μl of L primer (10 μM), 1.5 μl of R primer (10 μM), 0.38 μl Ex Taq (5 U/μl).
4. Use a volume of 7.5 μl DNA and 22.5 μl of PCR mix.
5. The thermocycling conditions are: 95 °C for 2 min; loop1 for 8 cycles (94 °C for 20 s, 73 °C for 30 s, reduce the temperature by 1 °C per cycle, ramp to 72 °C at 0.5 °C/s, 72 °C for 1 min); loop 2 for 45 cycles (94 °C for 20 s, 65 °C for 30 s, ramp to 72 °C at 0.5 °C/s, 72 °C for 1 min); 72 °C for 5 min; 99 °C for 10 min; loop 3 for 70 cycles (70 °C for 20 s, reduce temperature 0.3 °C per cycle; hold at 8 °C).
6. To evaluate performance of different primer combinations, use 5 μl of the PCR products and separate them on a 1.5% agarose gel electrophoresis (Fig. 7).

3.8 Enzymatic Digestion (Enzymatic Mismatch Cleavage)

1. Use 10 μl of the PCR products (approximately 100 ng) for enzymatic mismatch cleavage using crude celery juice extract (CJE) containing the nuclease CELI.
2. Test different enzyme concentrations. The following digestion reaction conditions can be used for subsequent assays: 10 μl of PCR product, 14 μl of H_2O, 4 μl of buffer (CJE) and mix 2 μl of CJE nuclease. Incubate for 15 min at 45 °C. Stop the digestions by cooling the reaction to 8 °C and add 10 μl of 0.25 M EDTA (ethylene diamine tetra acetic acid) (pH = 8.0).

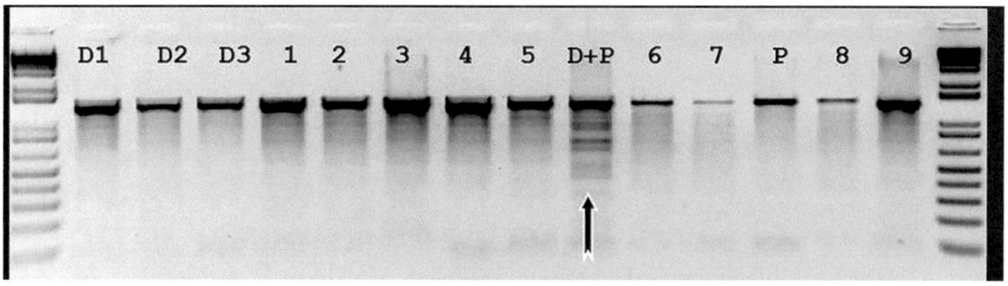

Fig. 8 Agarose gel evaluation of enzymatic digestion using A3 (ca_erlp_3) primers with genomic DNA M_2 plants. M_2 plants are labeled with 1, 2, 3, 4, 5, 6, 7, 8 or 9. *D1* Desi genotype n°1, *D2* Desi genotype n°2, *D3* Desi genotype n°3 and *D + P* mixture of four DNAs, three Desi genotypes and Ghab4 parent [24]. (Image taken from [24] under the license CC BY 4.0)

3. Separate 10 μl of the final reaction by gel electrophoresis on a 1.5% agarose gel in 0.5×TBE containing 0.2%g/ml ethidium bromide (Fig. 8).
4. Visual evaluation of gel images is carried out. The presence of a putative mutation is identified by the presence of a pair of bands of lower molecular weight than the full-length PCR product.

3.9 Sequence Validation of Low-Cost TILLING Results

Perform Sanger sequencing reactions to evaluate the results obtained in the enzymatic mismatch cleavage assays. In addition to confirming true mutations, this assay provides information on the exact position and type of DNA mutation. PCR products are prepared for commercial sequencing. The main sequencing steps are described below.

1. PCR amplification is carried out using the same parameters used for low-cost TILLING.
2. Verification of DNA concentration on a 1% agarose gel with the use of a size marker (Low DNA mass ladder).
3. Dilution of samples according to the specifications for Sanger sequencing provided by the sequencing provider.
4. Purification of the DNA product for sequencing. Excess primers and dNTPs need to be removed before sequencing. Check with the sequencing provider if this is included in the service. If not, incubate PCR product with exonuclease I to remove excess primer and shrimp alkaline phosphatase to remove phosphates from dNTPs, (incubation at 37 °C for 15 min, Inactivation of enzymes by incubation at 80 °C for 15 min and cooling to 8 °C).
5. Submit samples and primers to the sequencing provider according to their instructions.
6. When sequencing is complete, the sequencing provider should deliver chromatograms. Chromatograms are evaluated for the presence of sequence variations using a software such as ApE (A plasmid Editor, https://jorgensen.biology.utah.edu/wayned/ape/).

4 Notes

1. Ghab 4 has been described in this protocol but other varieties of chickpea can be used.
2. Ghab4 was developed from a three-way cross between FLIP 85-122C and FLIP 82-150C and FLIP 86-77C by the International Center for Agricultural Research in Arid Zones (ICARDA) in Aleppo, Syria [27]. This variety is characterized by high yield, Ascochyta blight tolerance and is suitable for mechanical harvesting.
3. Prepare Silica slurry by suspending Celite 545 in an equal volume of deionized water, centrifuge to pellet, pout off water and repeat this three times. Suspend the final pellet in an equal volume of clean deionized water.
4. Dissolve NaCl in water before adding to Ethanol. Some precipitate may form and will not affect the assay.
5. Radiosensitivity test is considered as a first step in any mutation-breeding program. LD_{50}: The mutagenic dose that results in a 50% reduction in seed germination or seedling viability. It is used to accurately predict the most effective and efficient mutagenic dose. RD_{50}: the mutagenic dose, which leads to the reduction in the average growth of the seedlings by 50%. RD_{50} is also useful in predicting the mutagenic dose. Chickpea is known to be a radiation sensitive species; it can germinate but wilt at an advanced stage of development or, in other cases, do not produce seeds. This is why the study of RD_{50} is complementary to the LD_{50} in chickpea. The objective is to choose the optimal irradiation dose (LD_{50}/RD_{50}) which causes the least loss of seeds while ensuring their germination to produce M_2 plants. The success of this approach relies on the application of an irradiation dose, which induces sufficient genetic variation while maintaining fertility.
6. Homogeneous seeds are used to avoid confusion between genetic and phenotypic variability resulting from natural allelic heterogeneity versus newly induced mutations.
7. To balance the humidity at 12–14%, the seed samples are placed in paper bags and are stored for at least 7 days in a desiccator containing silica gel or 60% glycerol, before irradiation. Remove the seeds from the desiccator just before radiation treatment. Dry seeds are suitable material for mutagenesis.
8. The choice of this range of doses was based on an extensive literature search on several radiomutagenesis studies carried out in chickpeas as well as in some other large-seeded legumes.

9. Plant the seeds as soon as possible after irradiation to avoid any damage that may result from prolonged storage after irradiation.
10. In order to avoid any water stress constraining the normal development of the plants, the times of intervention and the corresponding irrigation frequencies were determined according to the needs of the crop.
11. M_1 generation should be grown in favorable conditions.
12. For example, in our own experiment, only 1500 M_1 plants out of 9000 subjected to irradiation had given seeds. The number of seeds per plant varied from 3 to 150.
13. From the M_1 (1500) plants, only the plants with sufficient harvested seeds (30 seeds) are sown which corresponded to 135 M_2 families in our own experiment (each family with 30 plants).
14. An important step in the development of the low-cost TILLING is the production of genomic DNA of high molecular weight and of sufficient quantity for the intended assays.
15. Low-cost TILLING assays use a crude enzyme mix that recognizes and cuts heteroduplex DNA formed by denaturing and hybridizing PCR products that contain sequence differences. As such, the test will not detect homozygous mutations if the screening of M_2 plants is not combined with that of the control. Therefore, an equal amount of parental (unmutated) genomic DNA is mixed with each sample prior to PCR amplification. In this way, homozygous and heterozygous mutations can be detected.
16. For the quantification of genomic DNA on a 1% agarose gel, extreme care must be taken to put the same amount of DNA and lambda DNA standards in the wells. Without it, we cannot compare our samples accurately.
17. Parameters for primer design: Primer size is between 600 and 1200 bp and primer T_m is between 67 and 73 °C.

    Example of primers for candidate genes involved in the early flowering of chickpea;

    Gene/primer **ca_efg_**(1306-bp PCR product).

Primer name	Primer sequence
ca_efg_1F	GAGTATCCGCATCCACCAAGGCAAC
ca_efg_1R	TGTCGCTCCCAAGTCCTAACATCCTG

References

1. Fossati A (1972) Mutagenèse et amélioration du blé tendre. Rev Suisse agro 4:168–172
2. Jankowicz-Cieslak J, Mba C, Till BJ (2017) Mutagenesis for crop breeding and functional genomics. In: Jankowicz-Cieslak J, Tai T, Kumlehn J, Till B (eds) Biotechnologies for plant mutation breeding. Springer, Cham, pp 3–18
3. Till BJ, Datta S, Jankowicz-Cieslak J (2018) TILLING: the next generation. In: Scheper TH, Belkin S, Bley TH, Bohlmann J, Gu MB, Hu WS, Mattiasson B, Nielsen J, Seitz H, Ulber R, Zeng AP, Zhong JJ, Zhou W (eds) Advances in biochemical engineering/biotechnology. Springer, Berlin, pp 1–22
4. Deshayes A, Cornu A (1980) La mutagénèse expérimentale chez les végétaux supérieurs : devenir de la cellule mutée et nature des événements génétiques induits. In: Chaussat R, Bigot C (eds) La multiplication végétative chez plantes superieures. Gauthier-Villars, pp 233–258
5. Maluszynski KN, Zanten LV, Ahluwalia BS (2000) Officially released mutant varieties. The FAO/IAEA Database. Mutat Breed Rev 12:1–12
6. Ahloowalia BS, Maluszynski M, Nichterlein K (2004) Global impact of mutation derived varieties. Euphytica 135:187–204
7. Bouharmont J (1995) Création variétale et amélioration des plantes. In: Agronomie modern- Bases physiologiques et agronomiques de la production végétales. Hatier, Paris, pp 313–337
8. Solanki RK, Gill RK, Verma P et al (2011) Mutation breeding in pulses: an overview. In: Khan S, Kozgar MI (eds) Breeding of pulse crops. Kalyani Publishers, Ludhiana, pp 85–103
9. Kodym A, Afza R, Forster RBP et al (2012) Methodology for physical and chemical mutagenic treatments. In: Shu QY, Forster BP, Nakagawa H (eds) Plant mutation breeding and biotechnology, vol C14. Plant Breeding and Genetics Section. Joint FAO/IAEA Division of Nuclear Techniques in Food and Agriculture. International Atomic Energy Agency, Vienna, pp 169–180
10. Spencer-Lopes MM, Jankuloski L, Mukhtar AGA et al (2018) Physical mutagenesis. In: Spencer-Lopes MM, Forster BP, Jankuloski L (eds) FAO/IAEA manual on mutation breeding, 3rd edn. Food and Agriculture Organization of the United Nations, Rome, pp 5–49
11. Sigurbjönsson B (1983) Induced mutations. In: Wood DR (ed) Crop breeding. American Society of Agronomy and Crop Science Society of America, Madison, pp 153–176
12. Ali Ghanim AM, Forster BP (2014) Techniques for acceleration of mutation breeding in crop plants. In: Mutagenesis and high-throughput screening CORNET Projects Lehr-conference of the association of plant breeders and seed merchants, Austria, p 7. ISBN: 978-3-902849-00-7
13. Solanki IS, Sharma B (2001) Differential behaviour of polygenic characters to mutagenic treatments and selection in macrosperma lentil (*Lens culinaris* Medik). Indian J Genet Plant Breed 61:242–245
14. Solanki IS, Sharma B (2002) Induced polygenic variability in different groups of mutagenic damage in lentil (*Lens culinaris* Medik). Indian J Genet Plant Breed 62:135–139
15. Sneep J (1977) Selection for yield in early generations of self-fertilizing crops. Euphytica 26: 27–30
16. Kharkwal MC, Shu QY (2009) The role of induced mutations in World Food Security. In: Shu QY (ed) Induced plant mutations in the genomics era. Food and Agriculture Organization of the United Nations, Rome, pp 33–38
17. Gulfishan M, Bhat TA, Oves M (2015) Mutants as a genetic resource for future crop improvement. In: Al-Khayri JM et al (eds) Advances in plant breeding strategies: breeding, biotechnology and molecular tools, vol 4. Springer, Switzerland, pp 95–112
18. Shu QY (2009) A summary of the international symposium on induced mutations in plants. In: Shu QY (ed) Induced plant mutations in the genomics era. FAO/IAEA, Rome/Vienna, pp 15–18
19. Till BJ, Jankowicz-Cieslak J, Huynh OA et al (2015) Low-cost methods for molecular characterization of mutant plants, tissue desiccation, DNA extraction and mutation discovery: protocols. Springer, Cham, 35p, ISBN: 978-3-319-16258-4
20. Hofinger BJ, Huynh OA, Jankowicz-Cieslak J et al (2017) A protocol for benchtop extraction of single-Strand-specific nucleases for mutation discovery. In: Biotechnologies for plant mutation breeding. Springer, Cham, pp 241–251
21. Huynh OA, Jankowicz-Cieslak J, Saraye B et al (2017) Low-cost methods for DNA extraction and quantification. In: Jankowicz-Cieslak J, Tai

TH, Kumlehn J, Till BJ (eds) Biotechnologies for plant mutation breeding. Springer, Cham, pp 227–239

22. Sato Y, Shirasawa K, Takahashi Y et al (2006) Mutant selection of progeny of gamma-ray-irradiated rice by DNA hetero-duplex cleavage using brassica petiole extract. Breed Sci 56: 179–183
23. Li F, Shimizu A, Nishio T et al (2019) Comparison and characterization of mutations induced by gamma-ray and carbon-ion irradiation in Rice (*Oryza sativa* L.) using whole-genome resequencing. G3 Genes Genomes Genet 9(11):3743–3751. https://doi.org/10.1534/g.3.119.400555
24. Amri-Tiliouine W, Laouar M, Abdelguerfi A et al (2018) Genetic variability induced by gamma rays and preliminary results of low-cost TILLING on M2 generation of chickpea (*Cicerarietinum* L.). front. Plant Sci 9: 1568
25. Lamprecht H (1960) Über Blattfarben von Phanerogamen. Klassifikation, Terminologie und Gensymbole von chlorophyll und anderen Farbmutanten. Agric Hortic Gen 18:135–168
26. Kharkwal MC, Jain HK, Sharma B (1988) Induced mutations for improvement of chickpea, lentil, pea and cowpea. In: Improvement of grain legume production using induced mutations. IAEA, Vienna, STI/PUB/766 ISBN 92-0-111188-6, pp 89–109
27. Malhotra RS, Nassif AM, Singh KB et al (2005) Registration of 'Ghab 4' kabuli chickpea cultivar. International Center for Agricultural Research in the Dry Areas. Crop Science Society of America

Chapter 13

Efficient Generation of CRISPR/Cas9-Based Mutants Supported by Fluorescent Seed Selection in Different Arabidopsis Accessions

Tomasz Bieluszewski, Maja Szymanska-Lejman, Wojciech Dziegielewski, Longfei Zhu, and Piotr A. Ziolkowski

Abstract

Investigating the process of gamete formation in plants often requires the use of mutants of selected genes in various genetic backgrounds. For example, analysis of meiotic recombination based on sequencing or genotyping requires the generation of hybrids between two lines. Although T-DNA mutant collections of *Arabidopsis thaliana* are vast and easily accessible, they are largely confined to Col-0 background. This chapter describes how to efficiently generate knock-out mutants in different Arabidopsis accessions using CRISPR/Cas9 technology. The presented system is based on designing two single-guide RNAs (sgRNAs), which direct the Cas9 endonuclease to generate double-strand breaks at two sites, leading to genomic deletion in targeted gene. The presence of seed-expressed dsRed fluorescence cassette in the CRISPR construct facilitates preselection of genome-edited and transgene-free plants by monitoring the seed fluorescence under the epifluorescent microscope. The protocol provides the detailed information about all steps required to perform genome editing and to obtain loss-of-function mutants in different Arabidopsis accessions within merely two generations.

Key words CRISPR/Cas9, sgRNA, Mutant, Knock-out, Arabidopsis

1 Introduction

Discovery of gene function in gametogenesis is an important aspect of understanding the plant reproduction. Thus, it is indispensable to use gene mutants, especially loss-of-function alleles, to investigate their role in this process. For many years the Arabidopsis research community has relied on lines from a collection of insertional mutants [1]. Through large insertion mutagenesis projects, it is possible to obtain mutants for virtually any Arabidopsis gene [2–4]. These projects, combined with the existence of germplasm collections such as NASC and ABRC, represent extensive resources that cannot be overestimated, enabling dynamic progress of

Christophe Lambing (ed.), *Plant Gametogenesis: Methods and Protocols*, Methods in Molecular Biology, vol. 2484, https://doi.org/10.1007/978-1-0716-2253-7_13,

Arabidopsis research. However, insertional mutants were shown to be often associated with large chromosomal rearrangements that can induce partial loss of fertility due to instability of the genetic material and abnormal chromosome pairing in meiosis [5]. This is especially undesirable in the study of gametogenesis, because it can lead to wrong conclusions about the role of individual genes in this process. Moreover, the vast majority of the insertional mutant collections were constructed in the genetic background of the reference, Col-0 accession. As the understanding of the processes related to gametogenesis progresses, it is becoming common to use multiple mutant combinations, which are often difficult to obtain by crossing due to the genetic linkage. A similar problem concerns tandemly repeated genes, where only targeted mutagenesis allows obtaining the knockout alleles [6, 7].

Many experiments on gametogenesis require the use of mutants in different genetic backgrounds or hybrids. For example, molecular techniques to study genetic recombination are based on DNA sequence polymorphism that allows the identification of the exact recombination sites [8–12]. The use of such methods to study the functions of genes requires an access to mutant lines in additional accessions [7, 13–18]. Traditionally, such lines can be obtained by multiple backcrossing of the Col-0 mutant line to the selected accession. However, backcrossing is time-consuming and often results in extensive introgressions from the donor parent, especially within the region spanning the mutant allele [19]. As local reduction in heterozygosity may affect crossover distribution, this may lead to serious bias in the chromosome recombination pattern [20]. Therefore, the ability to edit the genome sequence in order to generate targeted mutations in genes has great potential in the plant gametogenesis research.

Currently, CRISPR/Cas9 is the most widely used system for genome editing in plants due to its simplicity, high efficiency and versatility [21]. It is based on designing a single-guide RNA (sgRNA), containing a 20-nt spacer sequence targeting Cas9 endonuclease activity toward matching genomic DNA fragment [22, 23]. A prerequisite for Cas9 cleavage is the presence of a protospacer-adjacent motif (PAM), which is located downstream of the target sequence. For the most frequently used type II *Streptococcus pyogenes* Cas9, PAM sequence corresponds to 5′-NGG-3′ (N represents any nucleotide; G represents, guanine), located directly downstream of the target sequence. Two nuclease domains of Cas9 cleave targeted DNA 3 bp upstream from the PAM sequence and introduce the DNA double-strand breaks (DSBs). They are subsequently repaired either by nonhomologous end joining (NHEJ) or homology-directed repair (HDR). The error-prone NHEJ can result in random insertions or deletions (indels), contributing to frameshift mutations. To increase chances of generating the loss-of-function mutants, it is possible to use two closely located sgRNAs, which result in the deletion of the sequence between two cut sites. An additional benefit of this approach is that mutants are genotyped much more easily, based on indel markers [24]. The most widely used promoters for sgRNA expression in plants come from *U3* and *U6* snRNA genes that are able to recruit RNA polymerase III [25]. The sgRNA expression cassettes have to be inserted into the binary vector expressing Cas9 endonuclease. This construct is integrated with the genomic DNA of Arabidopsis via *Agrobacterium*-mediated transformation. Generation of the desired mutation is triggered by expression of CRISPR/Cas9 components from the transgene. The selective markers enable to select transformed T_1 plants and transgene-free T_2 plants.

In this chapter, we provide a comprehensive CRISPR/Cas9 protocol for generation of new mutant alleles in different *Arabidopsis thaliana* backgrounds, which we developed and successfully used to generate knockout mutants in multiple genes [6, 17, 26]. It is based on dual sgRNA approach, in which a gene is targeted by two sgRNAs simultaneously, resulting in a deletion of the sequence between two Cas9 cutting sites (Fig. 1) [27].

Our genetic construct incorporates several elements facilitating generation of mutant lines in *A. thaliana*, as demonstrated previously by others. The *Cas9* gene is under the control of the *ICU2* promoter, which is expressed in dividing cells, so that the resulting mutations are usually inherited over the generation [28]. Thanks to the use of the dsRed fluorescent reporter expressed in seeds, it is possible not only to easily identify transformants (fluorescence-based positive selection), but also to quickly eliminate the transgene from the obtained mutants (fluorescence-based negative selection) [29]. Combination of these elements proved to be highly effective in our hands.

The protocol describes in detail all steps necessary to obtain a deletion mutant, starting with designing the sgRNAs, transformation of different Arabidopsis accessions based on previous methods [30, 31] and preselection of the transformants. We also provide simple yet effective methods for sgRNA vector mutagenesis, multi-sgRNA vector assembly, and clone screening not described elsewhere.

2 Materials

2.1 Plant Growth, Transformation, and Selection

1. *Arabidopsis thaliana* seed stocks. Many accessions are available from Nottingham Arabidopsis Stock Centre (NASC, arabidopsis.info) or Arabidopsis Biological Resource Center (ABRC, abrc.osu.edu).
2. Plastic pots with holes in the bottom (e.g., 11 cm diameter, 5.5 cm square) or plastic trays for Jiffy-7 Peat Pellets.

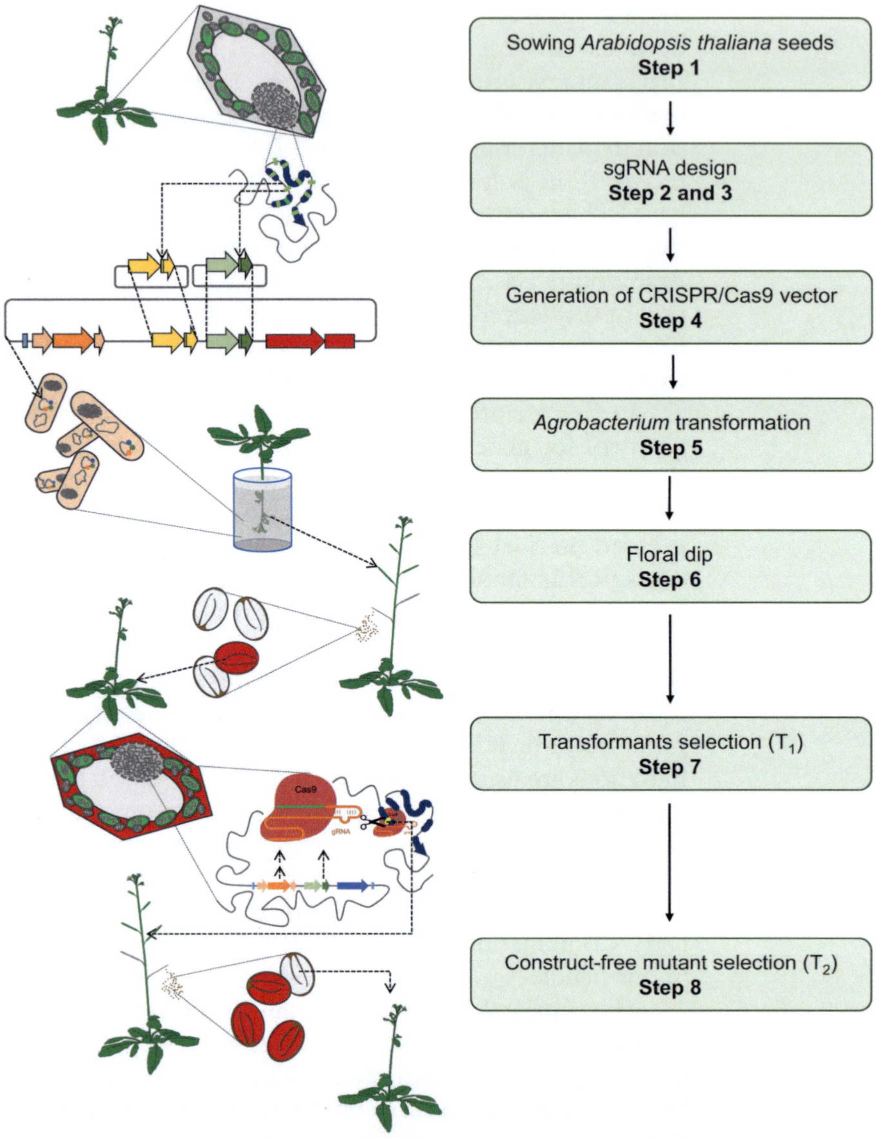

Fig. 1 Schematic workflow of the protocol to generate the mutants using CRISPR/Cas9. A pair of sgRNAs is designed to target the Cas9 cutting sites within the gene of interest. Targeted mutagenesis of the gRNA expression cassettes is performed to clone the designed sgRNAs to *pJET1.2*-based vectors. Next, the sgRNA expressing cassettes are subcloned into the binary vector carrying Cas9 expression cassette and dsRed reporter cassette as a selective marker. The binary vector is introduced first to *Agrobacterium* and then to plants via *Agrobacterium*-mediated transformation. Transformants are preselected based on seed fluorescence. Seeds showing red fluorescence correspond to T_1 carrying the transgene. Genomic deletions within coding regions of the target genes are induced by dual-sgRNA-Cas9. Seeds from T_1 plants with the expected deletion are harvested and nonfluorescent T_2 seeds are preselected as they do not carry the T-DNA construct

3. Jiffy-7 Peat Pellets or any soil suitable for Arabidopsis growth.
4. Amistar 250 SC or equivalent fungicide.
5. Glufosinate-ammonium (Basta) herbicide.
6. Arasystem or equivalent container for plant growth.
7. Sterile toothpicks.
8. Stainless-steel fine mesh sieve for cleaning seeds.
9. Glassine paper bags for seed storage.
10. Silwet L-77.
11. Sucrose.

2.2 Bacteria Strains, Media and Bacteria Transformation

1. Competent *Escherichia coli* strain DH5α or equivalent.
2. Competent *Agrobacterium tumefaciens* GV3101 strain or equivalent.
3. Liquid LB medium: 10 g tryptone, 5 g yeast extract, 10 g NaCl per liter.
4. LB solid medium: 10 g tryptone, 5 g yeast extract, 10 g NaCl, and 15 g Bacto agar per liter.
5. 14 mL polypropylene round-bottom tubes.
6. Liquid nitrogen.
7. Antibiotics (ampicillin, kanamycin, gentamicin, rifampicin).
8. Flasks (200 mL, 500 mL).
9. Disposable inoculating loop.

2.3 Molecular Cloning, PCR, and Plant Genotyping

1. 100 μM solutions of oligonucleotides shown in Table 1.
2. *pJET1.2-U3* and *pJET1.2-U6* vectors (Fig. 2a).

   pJET1.2-derrived cloning vectors containing sgRNA scaffold under *A. thaliana U3* or *U6* promoters for inserting spacer sequence. These plasmids are available at Addgene (www.addgene.org, plasmid #173156 and #173157)
3. *pFGC-I2Cas9* binary vector (Fig. 2b).

   A *pFGC-RCS*-derived binary plasmid carrying *hSpCas9* under the *ICU2* promoter, *dsRed* under the *NapA* promoter and Basta resistance cassette (for selection in plants), kanamycin resistance cassette (for selection in bacteria), containing restriction sites for inserting up to six sgRNA expression cassettes. *Cas9* gene was recloned from *psgR-Cas9-At* vector [32]. This plasmid is designed for *Agrobacterium*-mediated DNA plant transformation and is available at Addgene (www.addgene.org, plasmid #173158).
4. High-fidelity polymerase.
5. ClonExpress II One Step Cloning Kit or equivalent Gibson assembly–based cloning kit.

Table 1
Oligonucleotides necessary for this protocol

Primer name	Primer sequence (5′-3′)
Spacer_F	GTTTTAGAGCTAGAAATAGCAAG
SpacerU3_R	NNNNNNNNNNNNNNNNNNNN**T**GACCCTTGATGCTTTCTATGCA
SpacerU6_R	NNNNNNNNNNNNNNNNNNNN**C**AATCACTACTTCGACTCTAGCTGT
VRF1	ATGTTACTAGATCGGGGATCCGGATGGCTCGAGTTTTCAGC
R1VF	AGAATTCCCATGGAAGGATCCTCGAGGCTGCAGGAATTCGATATCAAGC
R2F1	CTTGGCGTAATCATGGGGATGGCTCGAGTTTTCAGC
R3F1	CAAGATTTTCAGGCTGGGATGGCTCGAGTTTTCAGC
R4F1	CTCATGAAAACTACGAGGATGGCTCGAGTTTTCAGC
R5F1	TGAGACAGGAATTTTAGGATGGCTCGAGTTTTCAGC
R6F1	TTTGACATGTTTAAAGGGATGGCTCGAGTTTTCAGC
R2	CCATGATTACGCCAAGCTCG
R3	CAGCCTGAAAATCTTGAGAGAATAAAAG
R4	TCGTAGTTTTCATGAGAGTCGATTG
R5	TAAAATTCCTGTCTCAGTTTCCTGAAG
R6	CTTTAAACATGTCAAAAGAGACGTC
construct_F	CTGTTCGAGCTGGAAAACGG
construct_R	CCGATCTAGTAACATAGATGACACC
pJET_F	CGACTCACTATAGGGAGAGCGGC
pJET_R	AAGAACATCGATTTTCCATGGCAG

6. Blunt-end cloning kit.
7. T4 DNA Ligase.
8. T4 polynucleotide kinase.
9. Restriction enzymes *Bam*HI and *Xho*I.
10. Taq DNA Polymerase, Taq Buffer, and dNTPs.
11. Phire Plant Direct PCR Master Mix or equivalent kit for direct plant genotyping.
12. PCR Purification Kit.
13. Gel Extraction Kit.

2.4 Specific Lab Equipment

1. Growth chambers or greenhouse, adjustable to long-day condition of 16 h light/8 h dark, 20 °C.
2. Thermocycler.

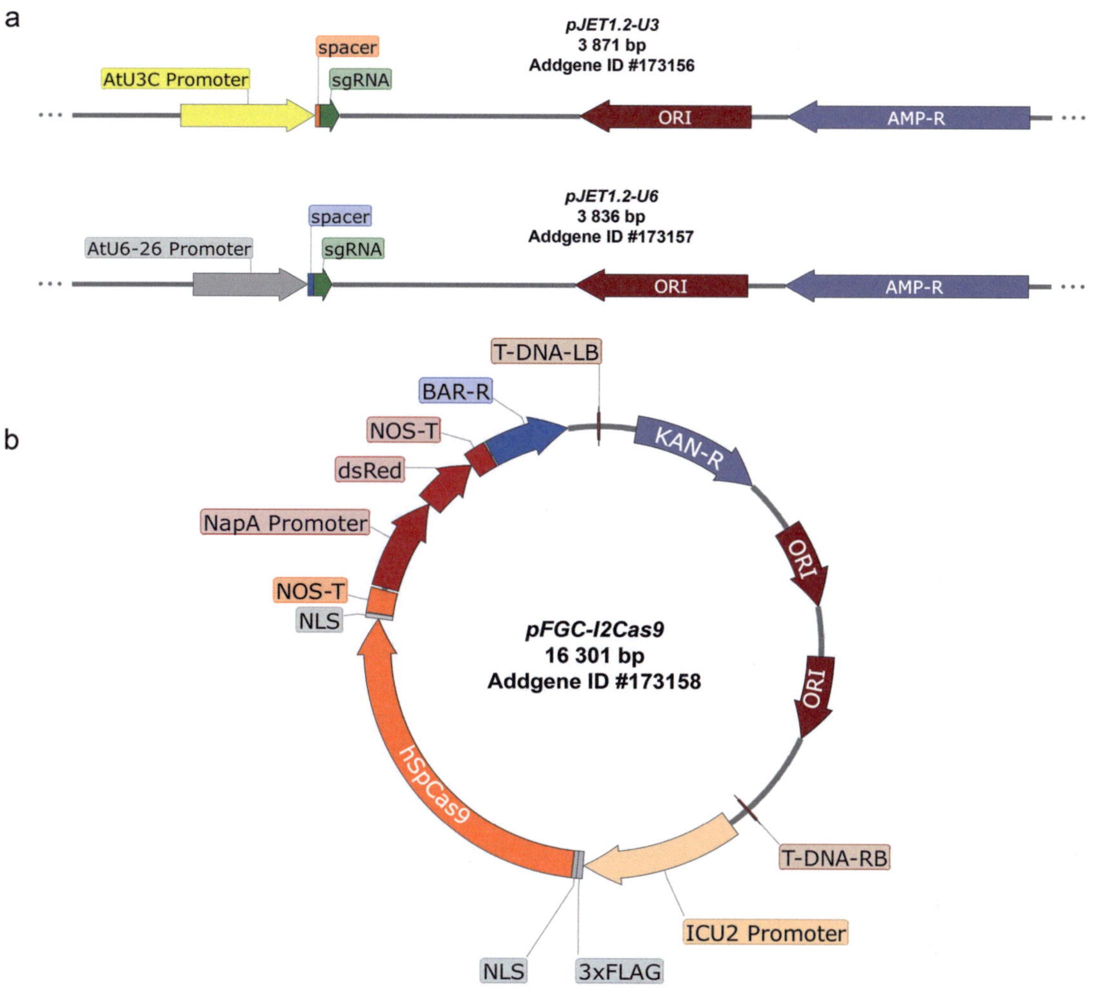

Fig. 2 Maps of vectors used in the protocol. (**a**) gRNA expression vectors with sgRNA scaffold under *AtU3C* or *AtU6-26* promoters to introduce the 20-nt spacer sequence. The plasmids confer ampicillin resistance (AMP-R). (**b**) Binary plasmid *pFGC-I2Cas9* for *Agrobacterium*-mediated expression of hSpCas9 and sgRNA cassettes. The plasmid confers kanamycin resistance (KAN-R) to bacteria and Basta resistance (BAR-R) to plants. Additionally, it has a fluorescent reporter cassette that is expressed in Arabidopsis seeds (dsRed). T-DNA-LB and T-DNA-RB indicate the T-DNA left and right borders respectively

3. Thermal mixer.
4. Centrifuges (for 1.5 mL tubes and 50 mL falcons).
5. Incubators (37 °C and 28 °C).
6. Nanodrop or Qubit Fluorometer.
7. Epifluorescent stereomicroscope equipped with dsRed filter.

3 Methods

3.1 Growing Arabidopsis Plants for Agrobacterium-Mediated Transformation

1. Sow Arabidopsis seeds onto hydrated pellets or the surface of the soil when growing in pots (*see* **Note 1**). For efficient transformation we normally sow one seed per pellet and 5 seeds per pot.
2. Spray the soil with the fungicide to prevent fungal infestation.
3. Cover the tray with pots or pellets with a plastic dome or use the plastic wrap.
4. Transfer the tray to cold room for 48 h to stratify the seeds (*see* **Note 2**).
5. Place the tray in a growth chamber with long-day (16 h of light) conditions at 20 °C. If the plants are growing on pellets, they need to be fertilized regularly.
6. Start plant transformation (**step 6** in Fig. 1) when plants initiate flowering and the first open flower buds appear.

3.2 Designing a Pair of Single-Guide RNAs

Single-guide RNA (sgRNA) contains a 20-nt spacer sequence that targets Cas9 endonuclease to a specific genomic locus. Any mismatches, especially in the "seed" region of gRNA (positions 8–13), can decrease the efficiency and specificity of Cas9 binding [33]. A knock-out mutation in the gene of interest is generated by designing a pair of closely located targeting sequences to induce a dual cleavage leading to a frameshift deletion. Different bioinformatic tools are available for spacer design, for instance CRISPOR (http://crispor.tefor.net) and CHOPCHOP (https://chopchop.cbu.uib.no) [34]. These applications generate a list of possible spacer sequences present in the input sequence predicting on-target efficiency and evaluating potential off-targets in the genome.

1. Copy the genomic sequence corresponding to the first 200–500 bp of the coding sequence of the targeted gene. The best option is to copy the first exon, but if this is too short, select another exon located close to the 5′ end of the gene or a genomic fragment containing several exons.
2. Use the sequence to query an online guide sequence design software.
3. Select *A. thaliana* as a target genome/organism to check the potential off-targets for the possible spacer sequences.
4. Select "NGG" as PAM. In most programs including CRISPOR and CHOPCHOP, "NGG" is selected by default.
5. Run the software. From the list of possible spacer sequences select a pair based on the following criteria:
   (a) Select a pair of spacer sequences that targets the coding region, preferentially the first exon. This will create a

nonsense frameshifting indel mutation at the start of the gene leading to production of severely truncated protein products, significantly reducing the chance of producing hypomorphic or gain-of-function mutants (*see* **Notes 3** and **4**).

(b) Select a pair of spacer sequences leading to generation of a deletion in which the number of deleted coding base pairs is not divisible by three (*see* **Notes 5** and **6**). This will increase a chance of producing an out-of-frame mutation. The predicted size of the deletion is equal to the distance between both Cas9 cut sites, which are located three nucleotides upstream of the PAM (5′-NGG-3′) sequence (Fig. 3a) (*see* **Note 7**).

(c) The expected deletion size should not be less than 100 bp. Deletions of 100–300 bp in size are recommended. This will facilitate subsequent genotyping of the mutants by designing indel markers.

(d) Choose the spacer sequences with the highest predicted efficiency score and the lowest number of off-targets.

(e) In case you intend to generate mutants in a genetic background other than Col or in several different accessions, make sure that the selected spacer sequence is conserved between these lines. The easiest way to do this is by comparing the gene sequence using the 1001 Arabidopsis Genome Project in the http://signal.salk.edu/atg1001/3.0/gebrowser.php browser (*see* **Notes 8** and **9**) [35].

6. Check if the spacer sequences start with an A or a G. If this is not the case, add an extra nucleotide (A or G) at the 5′ terminus (Fig. 3b) (*see* **Note 10**).
7. Convert spacer sequences into their reverse complement. You can use an online tool, e.g., Reverse Complement http://bioinformatics.org/sms/rev_comp.html.
8. Add the sequence complementary to the fragment of *AtU3C* or *AtU6-26* promoter to the 3′ end of the obtained sequence, depending on the first nucleotide at the 5′ end of the spacer (A or G, respectively). This will allow incorporation of newly designed spacer sequence into the sgRNA expression cassette (Fig. 3b).
9. Repeat this procedure for each selected spacer sequence. Order the obtained sequences from an oligonucleotide synthesis company. Additionally, order also the *Spacer_F* primer, which is universal for all sgRNAs (for the *Spacer_F* sequence, refer to Table 1).

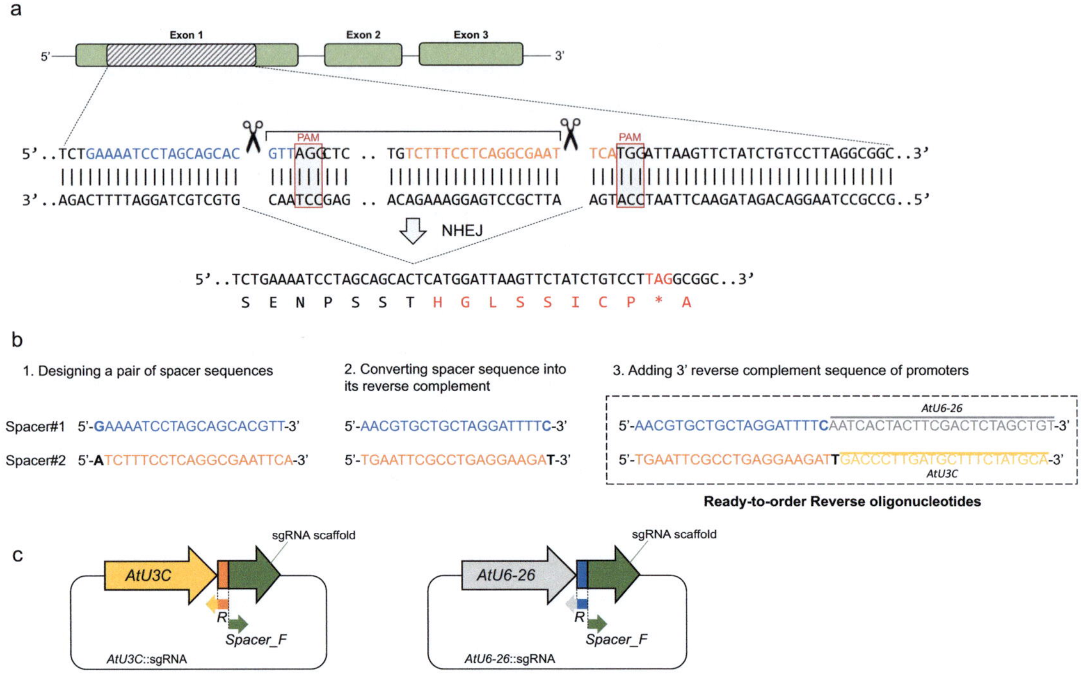

Fig. 3 Designing of a pair of spacer sequences and generation of sgRNA expression cassettes. (**a**) An example showing a spacer designed to target the first exon. 20-nt spacer sequences, marked in blue and orange, precede the PAM sequence (5′-NGG-3′) indicated by a red square. Two Cas9 domains generate DSBs 3 bp upstream of PAM sequence (shown as scissors). The repair of the DSBs is based on nonhomologous end joining (NHEJ), which result in deleting the genomic fragment between two cut sites. The distance between those cut sites should not be divisible by three, which leads to a frameshift and the appearance of the premature STOP codon (indicated by red asterisk). (**b**) Oligonucleotides design to perform mutagenesis. The PAM sequence is not included in the oligomers. As the Spacer#2 does not start with A or G, an extra A nucleotide is added at the 5′ end of the spacer sequence (bold type). First, the designed spacer sequences, marked in blue and orange, are converted into reverse complement sequences. Next, the appropriate overhang of the 3′ reverse complement promoter sequence is added. If the spacer starts with G nucleotide, the overhang for *AtU6-26* promoter is used (marked in grey), whereas the spacer starts with A nucleotide, the overhang for *AtU3C* promoter is used (marked in yellow). The resulting oligonucleotides are ready-to-order. (**c**) Scheme illustrating targeted mutagenesis of sgRNA expressing cassettes. The amplification of *AtU3C*/*AtU6-26*::sgRNA *pJET1.2*-based vectors is performed using *Spacer_F* (which corresponds to 5′ end of both sgRNAs scaffolds) and the designed Reverse oligonucleotide (*SpacerU3_R* primer or *SpacerU6_R* primer) containing the reverse complement sequence of the designed spacer sequence and appropriate promoter overhang

10. Design and order genotyping primers to confirm the generated deletion by PCR and sequencing (Subheading 3.7, **step 4** and Subheading 3.8, **step 3**). Primers should flank the gRNA sites at least 50 bp away from the Cas9 cutting sites. The expected amplicon sizes for wild-type and deletion mutant should differ by at least 20% to allow easy distinction in the gel.

3.3 Generation of sgRNA Expression Cassettes

sgRNA expression cassettes are created by a PCR reaction, in which the designed spacer sequences are incorporated into the plasmid containing *AtU3C* or *AtU6-26* promoter and sgRNA scaffold sequences (Fig. 3c). This is followed by recircularization of the plasmids. Locations and orientations of all primers required for directed mutagenesis are shown in Fig. 3c, while the primer sequences are shown in the Table 1.

1. Prepare the PCR reaction by mixing:

   0.5 μL of 10 μM *Spacer_F* primer.

   0.5 μL of 10 μM *SpacerU3_R* primer or *SpacerU6_R* primer (designed oligonucleotide from Subheading 3.2, **step 10**).

   0.5 μL of 0.5 ng/μL *pJET1.2-U3* or *pJET1.2-U6* (*see* **Note 11**).

   7.5 μL of CloneAmp HiFi PCR Premix.

   6 μL of ddH_2O.
2. Proceed to thermal cycling as follows:

   98 °C for 2 min.

   (98 °C for 20 s, 60 °C for 20 s, 72 °C for 30 s) × 30.

   72 °C for 1 min.

   4 °C.
3. Take 2 μL of the PCR product and visualize in 1% agarose gel to estimate the size of the amplicon. If the amplicon is ~3.8 kb, purify the PCR product using a PCR Purification Kit.
4. Prepare the phosphorylation reaction of the purified PCR product by combining the following reagents in a PCR tube:

   10 ng of purified PCR product.

   0.5 μL of T4 Polynucleotide Kinase 10 U/μL.

   1 μL of 10× T4 Ligase Buffer.

   ddH_2O up to 10 μL.
5. Incubate the phosphorylation reaction for 30 min at 37 °C.
6. Prepare ligation of the vector containing the new spacer sequences in sgRNA cassette by adding:

   8.5 μL of ddH_2O

   1 μL of 10× Ligase Buffer

   0.5 μL of T4 DNA Ligase 5 U/μL
7. Incubate the ligation for 30 min at room temperature.
8. Transform chemically competent DH5α cells with the ligated plasmids.

   (a) Combine 100 μL of the cells with 5 μL of the reaction mix in a 1.5 mL Eppendorf tube. Incubate for 30 min on ice.

(b) Perform a heat shock by transferring the tube to 42 °C for 60 s.

(c) Quickly move to ice and incubate for 5 min.

(d) Add 700 μL of LB medium. Incubate for 40 min at 37 °C with gentle shaking to recover the bacteria.

(e) Centrifuge the tube at 4000 rpm for 2 min and discard the supernatant leaving ~100 μL of the medium to resuspend the bacteria.

(f) Spread the bacteria on LB plates containing ampicillin (100 μg/mL) with a sterile loop or a cell spreader. Incubate at 37 °C for 16 h.

9. After overnight incubation, transfer three single colonies to separate 14 mL round-bottom tubes with 3 mL of LB containing ampicillin (100 μg/mL) for plasmid miniprep. Incubate for 16 h at 37 °C with vigorous shaking.
10. Isolate plasmids with Plasmid Miniprep Kit according to manufacturer's instruction.
11. Verify sequence accuracy of the sgRNA expression cassette by Sanger sequencing using *pJET_F* and *pJET_R* primers.

3.4 Generation of CRISPR/Cas9 Binary Vector Construct

The next step is to introduce the confirmed sgRNA expression cassettes to the binary vector which will be later used for *Agrobacterium*-mediated Arabidopsis transformation. Transfer of multiple gRNA cassettes is possible through recombination cloning or a Gibson assembly kit.

3.4.1 Preparation of pFGC-I2Cas9 Vector for sgRNA Cassettes Cloning

1. Linearize *pFGC-I2Cas9* with *Bam*HI digest by combining in a 1.5 mL Eppendorf tube:

   1 μg of *pFGC-I2Cas9* vector

   10 U of *Bam*HI restriction enzyme

   2 μL of 10× *Bam*HI Buffer

   ddH_2O up to 20 μL.
2. Mix the content and incubate for 1 h at 37 °C.
3. Run 1 μL of the digested plasmid in 1% agarose gel to assess the digestion efficiency. The length of linearized *pFGC-I2Cas9* vector is 16,301 bp.
4. Purify the reaction with PCR Purification Kit (*see* **Note 12**) and quantify the DNA concentration with NanoDrop or Qubit. Elute the digested plasmid into 20 μL elution buffer.

3.4.2 Preparation of sgRNA Expression Cassettes

1. To amplify sgRNA cassettes, combine the following reagents in two PCR tubes (*see* Table 2).
2. Proceed to thermal cycling as follows.

Table 2
Reagents to amplify sgRNA cassettes

	Amplification of sgRNA cassette 1	Amplification of sgRNA cassette 2
Verified plasmids from Subheading 3.3, **step 11** (0.2–0.4 ng)	Plasmid with sgRNA 1	Plasmid with sgRNA 2
Forward primer (0.5 μL)	*VRF1*	*R2F1*
Reverse primer (0.5 μL)	*R2*	*R1VF*
CloneAmp HiFi PCR premix	7.5 μL	7.5 μL
ddH_2O	Up to 15 μL	Up to 15 μL

98 °C for 2 min.

(98 °C for 20 s, 60 °C for 20 s, 72 °C for 30 s) × 30.

72 °C for 1 min.

4 °C

3. Purify the PCR products using PCR Purification Kit according to manufacturer's protocol and assess the size of the purified PCR product in 1% agarose gel. Correct amplicon sizes are as follows.
   - VRF1—R2 (construct with *AtU3C* promoter)—928 bp.
   - VRF1—R2 (construct with *AtU6-26* promoter)—893 bp.
   - R2F1—R1VF (construct with *AtU3C* promoter)—850 bp.
   - R2F1—R1VF (construct with *AtU6-26* promoter)—815 bp.

     For cloning of more than two gRNA cassettes, *see* **Note 13**.
4. Quantify the DNA concentration of purified amplicons with NanoDrop or Qubit.

3.4.3 Cloning sgRNA Cassettes into the Binary Vector

1. Mix linearized vector (Subheading 3.4.1) and inserts (Subheading 3.4.2) in the appropriate ratio (1:2) with ClonExpress II One Step Cloning Kit reagents according to the manufacturer's protocol.
2. Transform DH5α with the cloning reaction (without purification) as described in Subheading 3.3, **step 8**, but using kanamycin (50 μg/mL) instead of ampicillin as the selection marker.
3. After overnight incubation, transfer 4–10 single colonies to separate 14 mL round-bottom tubes with 3 mL of LB containing kanamycin (50 μg/mL) for plasmid miniprep. Incubate for 16 h at 37 °C with vigorous shaking.

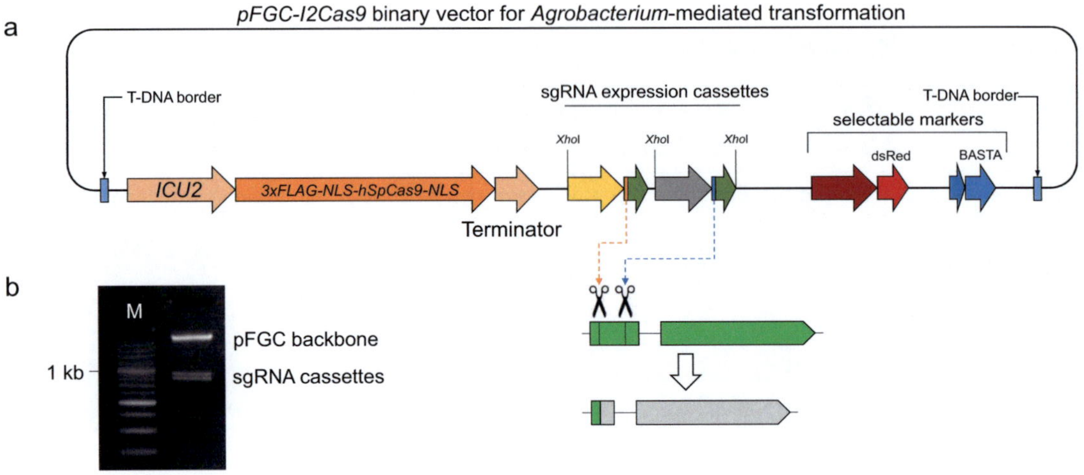

Fig. 4 Generation of CRISPR/Cas9 binary vector construct for Arabidopsis targeted mutagenesis. (**a**) The schematic representation of *pFGC-I2Cas9* binary vector ready for *Agrobacterium*-mediated transformation. The T-DNA, which is integrated with the plant genome, carries Cas9-expression cassette, two sgRNA expression cassettes targeting the Cas9 cutting sites within the same exon resulting in nonfunctional deletion. Additionally, the selective markers of Basta resistance cassette and seed-expressed dsRed reporter cassette are present to allow screening for transformants and monitoring the presence of the transgene. (**b**) The successful cloning of two gRNA expression cassettes to *pFGC-I2Cas9* binary vector can be confirmed by *XhoI* digestion. Two bands (~800 bp and ~900 bp) correspond to two gRNA cassettes, while the longer product of ~16 kb is the linearized *pFGC-I2Cas9* vector backbone

4. Isolate plasmids with a Plasmid Miniprep Kit according to the manufacturer's instruction.
5. Confirm correct cloning by *Xho*I restriction analysis. This is possible as the *Xho*I sites are included in the overhangs of the primers used for amplification of the sgRNA cassettes (*VRF1*, *R2F1*, and *R1VF*) (Fig. 4a).
   (a) Combine in a 1.5 mL Eppendorf tube:

   1 μg of *pFGC-I2Cas9* vector with cloned sgRNA cassettes

   5 U of *Xho*I

   2 μL of 10× *Xho*I Buffer

   ddH_2O up to 20 μL.

   (b) Incubate at 37 °C for 1 h.
   (c) Visualize the reaction in 2% agarose gel. The successful cloning of both cassettes should have a pattern of three bands. Two shorter fragments (~800 and ~900 bp) correspond to two sgRNA cassettes, while ~16 kb fragment is the *pFGC-I2Cas9* backbone. Exemplary result of restriction analysis is shown in Fig. 4b.

3.5 Transformation of CRISPR/Cas9 Construct to Agrobacterium tumefaciens

Once the CRISPR/Cas9 construct (*pFGC-I2Cas9* plasmid with introduced sgRNA expression cassettes) is ready, it is transformed into *A. tumefaciens* cells via standard freeze-thaw method. The bacteria strain used in this protocol carries resistance to gentamycin and rifampicin. Additionally, kanamycin is used for selection of construct-positive colonies.

1. Add 0.1–1 μg of CRISPR/Cas9 construct (obtained in Subheading 3.4.3, **step 4**) to a 1.5 mL Eppendorf tube containing 100 μL of *A. tumefaciens.* Incubate on ice for 5 min.
2. Transfer the tube to liquid nitrogen for 5 min. Next, move the tube to a thermomixer and incubate at 37 °C for 5 min.
3. Add 700 μL of antibiotic-free LB medium and incubate for 2–3 h at 28 °C with gentle shaking to recover the bacteria.
4. Centrifuge cells at 5000 rpm for 2 min and discard the supernatant.
5. Resuspend the bacteria in 100 μL of LB and spread on a solid LB medium with kanamycin, rifampicin and gentamycin (50 μg/mL each) with a sterile loop or a cell spreader.
6. Incubate the plates with transformed *A. tumefaciens* for 48 h at 28 °C.
7. In order to confirm the presence of the construct in bacterial cells it is recommended to perform colony PCR on 3–4 colonies. Touch the edge of each tested colony using a pipette tip or sterile toothpick and transfer to a 1.5 mL tube containing 30 μL of ddH_2O. Vortex the tube and use the suspension as a template in successive steps.
8. Perform colony PCR using primers specific to *pFGC-I2Cas9* vector. The sequences of the primers are shown in Table 1. Mix in a PCR tube:

   0.5 μL of 10 μM *construct_F* primer.

   0.5 μL of 10 μM *construct_R* primer

   0.5 μL of 10 mM dNTPs

   1 μL of 10× Taq Buffer

   0.2 μL Taq DNA polymerase

   0.5 μL DNA template

   ddH_2O up to10 μL.
9. Proceed to thermal cycling as follows:

   95 °C for 5 min.

   (95 °C for 20 s, 60 °C for 20 s, 72 °C for 1 min) × 30.

   72 °C for 5 min.

   4 °C.

10. Visualize the PCR products in 2% agarose gel. The correct amplicon size is 839 bp.
11. Prepare 20 mL of liquid LB supplemented with kanamycin, rifampicin and gentamycin (each 50 μg/mL) in a 100 mL flask. Inoculate with a single PCR-confirmed colony and incubate at 28 °C for 24 h with vigorous shaking.
12. Transfer 50–100 μL of the culture to 100 mL of liquid LB medium with above-mentioned antibiotics and incubate for additional 24 h at 28 °C with vigorous shaking.

3.6 Agrobacterium-Mediated Transformation of Arabidopsis thaliana via Floral Dip

1. Spin down the 100 mL *Agrobacterium* culture at 5000 × *g* for 10 min at room temperature in a 50 mL tube and resuspend the pellet in 200 mL of 5% sucrose solution.
2. Transfer the *Agrobacterium* cell suspension to a 200 mL beaker and add 10 μL of Silwet L-77 (*see* **Note 14**).
3. Transform the plants sown in Subheading 3.1. by dipping inflorescences in the *Agrobacterium* suspension. Dip the inflorescence for 30 s with gentle agitation.
4. Put the plants on a side in a tray, which should be covered with paper towel. Cover the plants with the second tray and transfer them to a dark, humid room for at least 3 h.
5. Transfer each plant to a separate Arasystem to prevent cross-fertilization and uncontrolled spread of GMO seeds.
6. Repeat the floral dip after 1 week, to increase the transformation efficiency.
7. Collect seeds from transformed T_0 plants in 4–6 weeks after the transformation.

3.7 Selection of T_1 Transformants Using Epifluorescent Stereomicroscope

Transformants can be selected either by BASTA-selection or based on fluorescent reporter expression using epifluorescent stereomicroscope (Fig. 5a, b). We recommend the later method as it allows fast screen of transformants in T_1 and easier selection of the seeds carrying the desired mutation but lacking the transgene in successive generations.

1. Harvest seeds from T_0 plants, clean them on strainers and store in separate envelopes.
2. Place the seeds under the epifluorescence stereomicroscope on the exposed X-ray film and use a toothpick to manually preselect ~30 seeds presenting red fluorescence (Fig. 5a).
3. Sow the preselected seeds on pellets (1–2 seeds per pellet), stratify the seeds for at least 2 days and transfer to growth chambers (*see* **Note 2**).
4. Screen the T_1 transformants for the desired deletion in targeted gene. For this purpose, use a kit for fast plant genotyping.

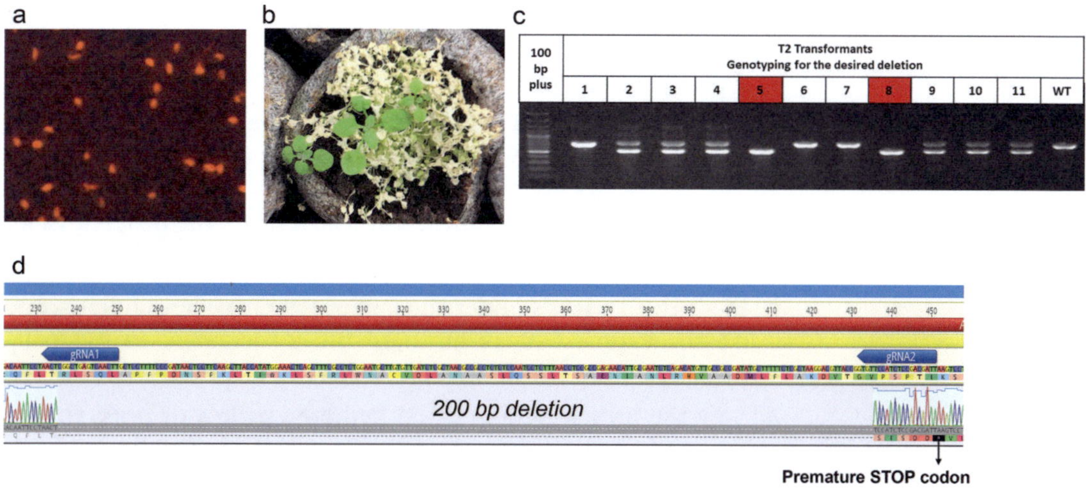

Fig. 5 Screening for transformants and verification of the deletion generated by CRISPR/Cas9. (**a**) Preselection of transformants is based on seed fluorescence as the seed-expressed dsRed reporter cassette is present in the construct. The seed fluorescence is monitored under the epifluorescent stereomicroscope; the red-fluorescent T_1 seeds (with the transgene) are preselected, while in the next generation, non-fluorescent T_2 seeds are preselected, as they correspond to transgene-free individuals. (**b**) Additionally, the T_1 transformants can be selected by spraying with Basta. After 1 week of Basta treatment, the transformed plants carrying the BASTA resistance cassette develop properly, whereas non-transformed plants turn yellow and wilt. (**c**) Checking the length of the deletion in the agarose gel using two primers flanking the designed spacer sequences. The wild-type (WT) amplicon is ~630 bp, while the shorter product corresponds to successfully mutated allele. Plants number 5 and 8 are homozygous mutants for the desired deletion (marked in red). (**d**) The PCR product obtained from homozygous mutants is subsequently verified by Sanger sequencing to confirm the frameshift deletion. The 200-bp deletion in the first exon results in the appearance of a premature stop codon (indicated by a black square)

(a) Collect material from cotyledons of 2.5–3-week-old T_1 plants and macerate the tissue in the buffer provided with the kit. Collect the material from a non-transformed wild-type plant as a negative control.

(b) Use such prepared solution as a template for a PCR with the primers flanking the expected deletion site (designed in Subheading 3.2, **step 10**). Mix in a PCR tube:

0.5 μL of 10 μM *Forward* primer.

0.5 μL of 10 μM *Reverse* primer.

0.5 μL of DNA template.

5 μL of Phire Plant Direct PCR Master Mix 2×.

ddH_2O up to10 μL.

5. Proceed to thermal cycling as follows:

98 °C for 5 min.

(98 °C for 15 s, 60 °C for 25 s, 72 °C for 30 s) × 30.

72 °C for 2 min.

4 °C.

6. Visualize the PCR products in 2% agarose gel. The plants that exhibit presence of additional amplicons in comparison to wild-type have been gene-edited by Cas9 (*see* **Note 15**).
7. Harvest seeds from several T_1 plants showing a strong band corresponding to mutated alleles.

3.8 Selection of T_2 Lines Carrying Homozygous Mutations

Selection for the optimal lines is performed using the progeny of T_1 plants, which showed the deletion within the targeted gene.

1. Place the T_2 seeds under the epifluorescence stereomicroscope on the exposed X-ray film and use a toothpick to manually preselect ~30 non-fluorescent seeds (*see* **Note 16**).
2. Sow the preselected seeds on hydrated pellets or pots, stratify the seeds and continue growth in normal conditions.
3. Verify the deletion in the targeted gene by collecting the material and performing the PCR using the same primers and conditions as described above (Subheading 3.7, **steps 4** and **5**).
4. Visualize the PCR products in 2% agarose gel. Mark the plants presenting different sizes of the amplicons when compared to wild type as they carry mutated alleles (Fig. 5c).
5. Confirm the lack of the CRISPR/Cas9 transgene by genotyping using primers for *pFGC-I2Cas9* (*construct_F* and *construct_R*). Use the PCR conditions as described above (Subheadings 3.7, **steps 4** and **5**).
6. Visualize the PCR products in 1.5% agarose gel. Identify the plants without the transgene as they do not show any product after the amplification. The appearance of the 839 bp band indicates the presence of the transgene.
7. Select the transgene-free plants with edited alleles.
8. Verify the size and the exact location of the deletion by Sanger sequencing. For this purpose, cut the bands corresponding to the mutated alleles from the gel and purify using a Gel Extraction Kit. Proceed with Sanger sequencing of the purified amplicon with primers used for the amplification (*see* **Notes 17** and **18**).
9. Analyze the sequencing data by aligning them to the genomic sequence of the gene of interest obtained for the appropriate Arabidopsis accession (Fig. 5d).
   (a) Check whether the generated deletion leads to a frame shift resulting with the appearance of a premature STOP codon after the deletion site.
   (b) Verify whether the deletion does not affect the splicing site, which may result in the generation of a splicing variant.

10. It is recommended to confirm that the generated mutant is not functional by RT-PCR followed by sequencing of the mutated transcript.

4 Notes

1. In the abovementioned conditions, it takes 3–4 weeks for early-flowering accessions (e.g., Col, *Ler*, Cvi) to flower. This time is sufficient to prepare the constructs for CRISPR/Cas9. Therefore, we usually sow plants for transformation before starting to design the constructs. This is even more important when mutants for late-flowering lines are to be generated.
2. Accessions vary greatly in terms of seed dormancy. While for lines like Col and *Ler*, seed stratification only helps to synchronize plant germination, lines like Cvi require stratification to break seed dormancy.
3. When a gene has multiple alternative splicing variants, it is recommended to design the targeting spacer sequences in the first coding exon common to all splicing variants.
4. Avoid targeting the ends of the exon, which may result in the loss of splicing site and, in consequence, create an alternative splicing variant of the gene.
5. If spacer sequences are located in two different exons, subtract intervening intron length from the predicted deletion size to make sure that the deleted coding fragment is not divisible by three.
6. Look for in-frame ATG codons downstream of the cut sites. If any potential sites of translation reinitiation are present, shift the second cut site behind them.
7. There is no difference if the targeting spacer with PAM is located on forward or reverse strand. The pair of spacers can be on the same or different strands.
8. The SALK-provided browser does not always display all polymorphisms. Therefore, change the display from *Graphic* to *Text* in the *Display* menu in the browser navigation panel (*Display/View/Text*). You may need to copy the sequences to compare them using another software.
9. We recommend designing two pairs of different spacer sequences for each gene. Some sgRNAs are inefficient making it difficult to obtain a desired knockout deletion. The use of two independent pairs of sgRNAs significantly reduces this possibility. In addition, it is generally a good practice to compare at least two mutants generated with different sgRNAs in order to exclude any off-target related unspecific effects.

10. Exceeding the recommended template amount may result in contamination of the mutagenized clone pool with the template vector.
11. There are no restrictions regarding the promoter choice for sgRNA expression cassette used in one binary vector. For example, it is possible to use two *AtU6-26* promoters, as well as two *AtU3C* promoters, or one *AtU6-26* and one *AtU3C* promoter in the same construct.
12. It is recommended to perform the purification directly from the restriction digestion reaction as it provides better DNA quality when compared to gel extraction.
13. It is possible to amplify and clone up to six different sgRNA cassettes into a single Cas9-sgRNA construct. In such a case, use the following primers for the amplification of the sgRNA cassettes: sgRNA cassette #1: VRF1, R6; sgRNA cassette #2: R6F1, R5; sgRNA cassette #3 R5F1, R4; sgRNA cassette #4: R4F1, R3; sgRNA cassette #5: R3F1, R2; sgRNA cassette #6: R2F1, R1VF. The primers are shown in Table 1. The successful cloning of six gRNA cassettes can be confirmed by *Xho*I restriction digestion: consecutive cassettes differ by ~100 bp.
14. It is recommended to add Silwet L-77 just before dipping. As the surfactant is very viscous, cut the tip before pipetting.
15. It is advised to repeat the genotyping on generative tissue samples. As the T_1 generation consists of mostly mosaic plants, the genomic edits can be different in distinct parts of the plant. Confirming the presence of mutation by re-genotyping of generative parts increases the chance that the alleles will be present in the T_2 generation.
16. By selecting non-fluorescent T_2 seeds it is possible to select plants without transgene, but with heritable mutated alleles. This is crucial, because plants carrying a transgene are expressing CRISPR/Cas9, thus the identified mutations may not be heritable, while it can create additional off-target mutations. Elimination of the transgene is possible also by genotyping, but in this case, a significant number of T_2 plants have to be sown and tested.
17. It is likely that in some plants only one of the two alleles will carry the desired deletion (heterozygous or biallelic mutant). In such a case we advise to proceed to the T_3 generation and then genotype the plants to identify individuals homozygous for the mutation.
18. If the sequencing results are ambiguous, it is recommended to purify the PCR reaction using a PCR Purification Kit. Subsequent cloning of the products using a blunt-end cloning kit will allow to sequence the generated mutation.

Acknowledgments

The authors thank Dr. Jian-Kang Zhu for the *3×Flag-NLS-hSpCas9-NLS::NosT* cassette which was used to construct *pFGC-I2Cas9* plasmid. This work was supported by Polish National Science Centre grants 2016/22/E/NZ2/00455 to P.A.Z., and a Foundation for Polish Science grant POIR.04.04.00-00-5C0F/17-00 to P.A.Z. M.Sz.-L. is the Adam Mickiewicz University Foundation scholarship holder in the academic year 2021/2022.

References

1. Malley RCO, Barragan CC, Ecker JR (2015) A user's guide to the Arabidopsis T-DNA insertion mutant collections. Methods Mol Biol 1284:323–342
2. Sessions A, Burke E, Presting G et al (2002) A high-throughput Arabidopsis reverse genetics system. Plant Cell 14:2985–2994
3. Alonso M, Stepanova AN, Leisse TJ et al (2003) Genome-wide insertional mutagenesis of Arabidopsis thaliana. Science 301:653–657
4. Rosso MG, Li Y, Strizhov N et al (2003) An Arabidopsis thaliana T-DNA mutagenized population (GABI-Kat) for flanking sequence tag-based reverse genetics. Plant Mol Biol 53: 247–259
5. Nacry P, Camilleri C, Caboche M et al (1998) Major chromosomal rearrangements induced by T-DNA transformation in Arabidopsis. Genetics 149:641–650
6. Bieluszewski T, Sura W, Dziegielewski W et al (2022) NuA4 and H2A.Z control environmental responses and autotrophic growth in Arabidopsis. Nat Commun 13:277
7. Capilla-Pérez L, Durand S, Hurel A et al (2021) The synaptonemal complex imposes crossover interference and heterochiasmy in Arabidopsis. Proc Natl Acad Sci U S A 118: e2023613118
8. Giraut L, Falque M, Drouaud J et al (2011) Genome-wide crossover distribution in Arabidopsis thaliana meiosis reveals sex-specific patterns along chromosomes. PLoS Genet 7: e1002354
9. Drouaud J, Camilleri C, Bourguignon P et al (2006) Variation in crossing-over rates across chromosome 4 of Arabidopsis thaliana reveals the presence of meiotic recombination "hot spots". Genome Res 16:106–114
10. Drouaud J, Khademian H, Giraut L et al (2013) Contrasted patterns of crossover and non-crossover at Arabidopsis thaliana meiotic recombination hotspots. PLoS Genet 9: e1003922
11. Rowan BA, Patel V, Weigel D et al (2015) Rapid and inexpensive whole-genome genotyping-by-sequencing for crossover localization and fine-scale genetic mapping. G3 (Bethesda) 5:385–398
12. Sun H, Rowan BA, Flood PJ et al (2019) Linked-read sequencing of gametes allows efficient genome-wide analysis of meiotic recombination. Nat Commun 10:4310
13. Yelina NE, Lambing C, Hardcastle TJ et al (2015) DNA methylation epigenetically silences crossover hot spots and controls chromosomal domains of meiotic recombination in Arabidopsis. Genes Dev 29:2183–2202
14. Ziolkowski PA, Underwood CJ, Lambing C et al (2017) Natural variation and dosage of the HEI10 meiotic E3 ligase control Arabidopsis crossover recombination. Genes Dev 31: 306–317
15. Serra H, Lambing C, Griffin CH et al (2018) Massive crossover elevation via combination of HEI10 and recq4a recq4b during Arabidopsis meiosis. Proc Natl Acad Sci 115:2437–2442
16. Underwood CJ, Choi K, Lambing C et al (2018) Epigenetic activation of meiotic recombination near Arabidopsis thaliana centromeres via loss of H3K9me2 and non-CG DNA methylation. Genome Res 28:519–531
17. Blackwell AR, Dluzewska J, Szymanska-Lejman M et al (2020) MSH2 shapes the meiotic crossover landscape in relation to interhomolog polymorphism in Arabidopsis. EMBO J 39:e104858
18. Nageswaran DC, Kim J, Lambing C et al (2021) HIGH CROSSOVER RATE1 encodes PROTEIN PHOSPHATASE X1 and restricts meiotic crossovers in Arabidopsis. Nat Plants 7: 452–467

19. Yadav NS, Khadka J, Grafi G (2018) Arabidopsis mutants may represent recombinant introgression lines. BMC Res Notes 11:227
20. Dluzewska J, Szymanska M, Ziolkowski PA (2018) Where to cross over? Defining crossover sites in plants. Front Genet 9:609
21. Ma X, Zhang Q, Zhu Q et al (2015) A robust CRISPR/Cas9 system for convenient, high-efficiency multiplex genome editing in monocot and dicot plants. Mol Plant 8:1274–1284
22. Jinek M, Chylinski K, Fonfara I et al (2012) A programmable dual-RNA-guided DNA endonuclease in adaptive bacterial immunity. Science 337:816–821
23. Mali P, Yang L, Esvelt KM et al (2013) RNA-guided human genome engineering via Cas9. Science 339:823–826
24. Zhao Y, Zhang C, Liu W et al (2016) An alternative strategy for targeted gene replacement in plants using a dual-sgRNA/Cas9 design. Sci Rep 6:23890
25. Ma X, Zhu Q, Chen Y et al (2016) CRISPR/Cas9 platforms for genome editing in plants: developments and applications. Mol Plant 9:961–974
26. Zhu L, Fernandez-Jimenez N, Szymanska-Lejman M et al (2021) Natural variation identifies SNI1, the SMC5/6 component, as a modifier of meiotic crossover in Arabidopsis. Proc Natl Acad Sci U S A 118:e2021970118
27. Pauwels L, De Clercq R, Goossens J et al (2018) A dual sgRNA approach for functional genomics in Arabidopsis thaliana. G3 (Bethesda) 8:2603–2615
28. Hyun Y, Kim J, Cho SW et al (2015) Site-directed mutagenesis in Arabidopsis thaliana using dividing tissue-targeted RGEN of the CRISPR/Cas system to generate heritable null alleles. Planta 241:271–284
29. Stuitje AR, Verbree EC, Van Der Linden KH et al (2003) Seed-expressed fluorescent proteins as versatile tools for easy (co)-transformation and high-throughput functional genomics in Arabidopsis. Plant Biotechnol J 1:301–309
30. Wise AA, Liu Z, Binns AN (2006) Three methods for the introduction of foreign DNA into Agrobacterium. Methods Mol Biol 343:43–54
31. Clough SJ, Bent AF (1998) Floral dip: a simplified method for Agrobacterium-mediated transformation of Arabidopsis thaliana. Plant J 16:735–743
32. Feng Z, Zhang B, Ding W et al (2013) Efficient genome editing in plants using a CRISPR/Cas system. Cell Res 23:1229–1232
33. Qi LS, Larson MH, Gilbert LA et al (2013) Repurposing CRISPR as an RNA-γuided platform for sequence-specific control of gene expression. Cell 152:1173–1183
34. Liu G, Zhang Y, Zhang T (2020) Computational approaches for effective CRISPR guide RNA design and evaluation. Comput Struct Biotechnol J 18:35–44
35. Alonso-Blanco C, Andrade J, Becker C et al (2016) 1,135 Genomes reveal the global pattern of polymorphism in Arabidopsis thaliana. Cell 166:481–491

Chapter 14

Generation of Deletion Lines in Allohexaploid Bread Wheat

Heïdi Serra, Radim Svačina, Jan Bartoš, and Pierre Sourdille

Abstract

Positional cloning in bread wheat (*Triticum aestivum* L.) remains a daunting task because of its large genome, high density of repeats, low recombination rate especially in pericentromeric regions and its allopolyploidy. One way to face this challenge is to decrease the size of the interval bearing the gene of interest both genetically and physically, in order to reduce significantly the number of potential candidate genes. In this chapter, we describe a technical approach to produce chromosome-specific deletion lines to locate precisely genes of interest onto wheat chromosomes, a step forward to their cloning.

Key words Wheat, *Triticum aestivum*, Deletion line, Gametocidal, Homeologs

1 Introduction

Bread wheat (*Triticum aestivum* L.) is an allopolyploid species belonging to the *Poaceae* family, derived from two successive interspecific crosses involving three different diploid species: *Triticum monococcum ssp. urartu* (AA genome), a species from the *Sitopsis* section (SS genome related to the BB genome of bread wheat, the closest known one being *Aegilops speltoides*) and *Aegilops tauschii* (DD genome) [1]. Each diploid species carries two sets of seven chromosomes (numbered from 1 to 7; $2n = 14$) and their hybridization resulted in allohexaploid bread wheat ($2n = 6x = 42$; AABBDD genome) [1].

Chromosomes with the same number from the A, B and D genomes are highly similar in terms of gene content and order but are highly different with regards to transposable elements (TE) [2], which constitute more than 85% of the wheat genome leading to genome obesity (1C ~16,000 Mb). Because of these two contrasting features, chromosomes from different genomes are called

Supplementary Information The online version of this chapter (https://doi.org/10.1007/978-1-0716-2253-7_14) contains supplementary material, which is available to authorized users.

Christophe Lambing (ed.), *Plant Gametogenesis: Methods and Protocols*, Methods in Molecular Biology, vol. 2484, https://doi.org/10.1007/978-1-0716-2253-7_14,

"homeologs" as opposed to true "homologues" originating from the two parents that are almost identical in stabilized cultivated varieties. The size of the wheat genome, its polyploidy and TE-content have long hampered the development of classical genomic tools useful to decipher genetics of complex traits.

Despite this, a gold-standard anchored and fully annotated wheat genome sequence [3] as well as high-density SNP arrays [4] or RNA-Seq data [5] have been recently produced bringing bread wheat on the same level as other crops and paving the way of cloning genes of agronomical interest. However, positional cloning of genes in wheat remains complicated especially because of low and uneven recombination rates along the chromosomes [6]. Indeed, chromosomes are almost deprived of recombination on roughly 70% of their length, especially in the pericentromeric regions while these latter bear almost 50% of the genes [7]. Cloning a gene that locates in these areas is thus almost impossible because of the difficulty to find appropriate recombinant individuals that break the gene-SNP linkage sufficiently close to the gene to allow its identification.

Interestingly, gene-content similarity between homeologs can be beneficial as pair of chromosomes, chromosome arms or fragments of chromosomes can be deleted without affecting wheat-plant development and fertility with only a few exceptions. Indeed, if the loss of a chromosome in a diploid species is usually too deleterious to be tolerated, this is not the case in polyploid species where homoeologous chromosome redundancy counteracts (at least partly) the effects of chromosome losses. This flexibility was a clue for the development of aneuploid stocks of wheat that were a main issue to understand the genetics of phenotypic traits.

Historically, Sears (1939) backcrossed a haploid wheat line—derived from a hybridization between wheat cultivar Chinese Spring and rye (*Secale cereale*)—with Chinese Spring and generated 11 lines with one or two univalents and for four of them, one or two trivalents [8]. After self-pollination of these lines, one progeny nullisomic for chromosome 3B exhibiting partially asynaptic behavior but a normal fertility was selected to produce additional monosomic and trisomic lines [9]. From these lines, the full set of nullisomic (missing a pair of chromosomes), nullisomic-tetrasomic (where one of the pairs of homeologous chromosomes replaces the missing pair), and ditelosomic (missing a pair of chromosome arms) lines was produced for the 21 chromosome pairs of Chinese Spring [10, 11]. This material was further used to locate genes of interest [12–14], as well as to assign molecular markers to their chromosomes [15, 16].

If locating a gene on a chromosome arm is already a considerable breakthrough in the large wheat genome, gene cloning is far to be achieved as each chromosome arm bears several thousands of functional genes [3]. As a comparison, a chromosome arm in wheat has a size equivalent to the entire rice genome. Development of additional stocks with shorter deletions was thus necessary [17]. Production of such material was based on the use of alien addition line in cultivar Chinese Spring with a chromosome 2C from *Ae. cylindrica* ($2n = 4x$; CCDD) or *Ae. triuncialis* ($2n = 4x$; CCUU) [18] or with a translocated-fragment from chromosome 2S of *Ae. speltoides* on the distal end of the long arm of chromosome 2B (T2BS-2BL-2SL) [17]. When introduced into CS cultivar of wheat, 2C gametocidal chromosome induces mostly terminal deletions in chromosomes of gametes to which it was not transferred. 436 (mostly homozygous) deletion lines were produced through the 2C gametocidal system and were useful to locate accurately molecular markers on physical regions and to establish relationships between physical and genetic distances, which is at the heart of positional cloning of genes [6, 19–21]. Such material remains a useful tool for positional cloning of genes in wheat [22, 23].

Here, we detail an approach based on the 2C gametocidal system to generate deletion lines in allopolyploid bread wheat, including the selection of plants carrying deletion on the chromosome of interest and their molecular characterization.

2 Materials

2.1 Plant Material

Two cultivars developed in hexaploid wheat landrace Chinese Spring (CS; $6x = 2n = 42$; AABBDD; 21″) background are needed to produce deletion stock.

1. A monosomic addition line of chromosome 2C from *Ae. cylindrica* (CS-2C; $2n = 43$; 21″ + 2C′) (*see* **Notes 1** and **2**).
2. A nullisomic-tetrasomic (NT) line lacking the chromosome of interest (where deletions aimed to be induced) with one of its homeologs in tetrasomic constitution. As an example, production of 3D deletion lines required NT line lacking 3D but with either 3A (N3DT3A; $2n = 42$; 19″ + 3A″″) or 3B (N3DT3B; $2n = 42$; 19″ + 3B″″) in tetrasomic constitution.

2.2 PCR

1. Master Mix AmpliTaq Gold 360.
2. Taq DNA polymerase.
3. 5× PCR buffer.
4. $MgCl_2$ solution.
5. dNTP.
6. M13 primers (*see* Table 1 for sequences).
7. Nuclease-free water.
8. 96-well PCR plates.
9. 96-well PCR plate seals.

Table 1
Primers used for identification of terminal deletions on chromosome 3D

Chromosome part	Oligo ID	Primer sequence 5′-3′	Position (bp)	Amplicon size (bp)
Short arm of 3D	3D_3.2Mb_F	**CACGACGTTGTAAAACGAC** CACAAATCAGTTTACTATCATACCG	3,222,209–3,222,463	274
	3D_3.2Mb_R	TAGAATTTATGTCGTAAATTTAATCTACC		
	M13_3DS	FAM-**CACGACGTTGTAAAACGAC**		
Centromere of 3D	3D_253.4Mb_F	**CACGACGTTGTAAAACGAC** TATGCGTTTGGAGTAGTTCTTGT	253,391,812–253,392,248	410
	3D_253.4Mb_R	CTCATCTCAGGCTGTCTAATTAA		
	M13_3DC	PET-**CACGACGTTGTAAAACGAC**		
Long arm of 3D	3D_598.5Mb_F	**CACGACGTTGTAAAACGAC** GTAATATGTAAAAACTGAACTTGTTCG	598,540,714–598,540,926	232
	3D_598.5Mb_R	AATCTACCGTGATAGCTTCCGT		
	M13_3DL	VIC-**CACGACGTTGTAAAACGAC**		

M13 extensions are in bold

10. 96-well PCR thermal cycler.
11. Agarose.
12. Ethidium bromide.
13. 100 bp DNA Ladder
14. 10× DNA loading buffer
15. Formamide Hi-Di.
16. GeneScan size standard.
17. 3730XL DNA Sequencer.
18. GeneMapper Software.

2.3 Droplet Digital PCR

1. ddPCR Supermix for Probes (No dUTP).
2. Droplet Generation Oil for Probes.
3. ddPCR Droplet Reader Oil.
4. PX1 PCR Plate Sealer.
5. Pierceable Foil Heat Seal.
6. DG8 Gaskets for QX200/QX100 Droplet Generator.
7. DG8 Cartridges for QX200/QX100 Droplet Generator.
8. QX200 Droplet Generator.
9. QX200 Droplet Reader.
10. ddPCR 96-well plates.
11. Nuclease-free water.
12. QuantaSoft Software.

3 Methods

3.1 Production of Deletion Lines Via the 2C Gametocidal System

1. Generate sufficient amount of seeds of wheat monosomic addition line of chromosome 2C from *Ae. cylindrica* and NT lines lacking the chromosome of interest. At least 50 seeds for each line are needed to produce ~3000 deletion lines.
2. Divide the seed stocks of both lines in batches of 10. Sow one batch per week of both lines to prolong the flowering stage of the population (*see* **Note 3**) and grow plants under standard conditions (16:8 h light–dark photoperiod, temperatures of 20 °C during the day and 16 °C at night, and 60% humidity). CS-2C addition line will be used as male while NT line will be used as female.
3. Select immature spikes of NT plants when plants reach the flowering stage.
4. Cut the top part of spike and remove spikelets of the bottom part (Fig. 1a).

Fig. 1 Emasculation procedure of spikes of NT lines. (**a**) Indicated area of removed spikelets. (**b**, **c**) Removal of middle florets of spikelets, leaving only one on each side. (**d**) Incision in two thirds of remaining florets. (**e**, **f**) Removal of the three anthers from cut florets. (**g**) Emasculated spike of wheat. Pictures edited from http:/www.wheat-training.com section "How to cross wheat" with permission from original authors

5. Remove the middle florets of each remaining spikelets using forceps (Fig. 1b, c).
6. Cut at one third of the length of the two remaining florets of each spikelet using scissors (Fig. 1d).
7. Carefully remove the three immature anthers from each floret (Fig. 1e, f).
8. Cover the spike with a crossing bag and mark date of emasculation (Fig. 1g).
9. Check the maturity of stigma 2 days after emasculation. The stigma is usually mature enough to receive pollen about 2–4 days after emasculation. The stigma should be feathered compared to immature one (Fig. 2a, b).

Fig. 2 Pollination procedure of emasculated spikes using CS-2C addition line pollen. Comparison of immature (**a**) and mature (**b**) stigma of emasculated NT lines. (**c**, **d**) Selection of mature anthers ready to release pollen from CS-2C lines. (**e**) Pollination of mature stigma. (**f**) Bagged pollinated spike of NT lines. Pictures edited from http:/www.wheat-training.com section "How to cross wheat"with permission from original authors

10. Select anthers from CS-2C plants ready to release pollen (Fig. 2c). The anthers should be yellow and slightly rounded (Fig. 2d).
11. Carefully remove the anthers from CS-2C line and place them to the base of your thumb. The warmth of your hand will trigger the anthers to reach maturity and open.
12. Hold open anthers using forceps and tap the mature stigmas of emasculated spikes of NT lines (Fig. 2e). Usually a single anther can be used to pollinate two to three florets. If enough maturing anthers are available, the chance of successful pollination can be increased by repeating the pollination step the next day.
13. Replace the crossing bag on the spike after each pollination (Fig. 2f) and keep it until harvest. Success of crosses can be checked by enlargement of ovary about 1 week after pollination.
14. Collect grains at complete maturity (1.5–2 months after pollination) and thresh.
15. Store the harvested seeds at 4 °C for at least 2 weeks to avoid dormancy.

3.2 Selection of Plants Carrying Deletion on the Chromosome of Interest

Majority of seeds generated by crossing CS-2C and NT lines will not carry deletion on the chromosome of interest. A screen of the full population is thus necessary to identify the lines carrying a deletion on the chromosome of interest. The screening procedure can be carried out by sequence-tagged sites (STS) molecular markers and gel separation (Subheading 3.2.1) or by PCR using M13 fluorescent marker and subsequent capillary electrophoresis for higher throughput workflow (Subheading 3.2.2). The first part of the protocol is identical for both approaches.

1. Sow the seeds resulting from the cross between CS-2C and NT lines in seeding trays and grow plants until three-leaf stage under standard conditions (*see* Subheading 3.1, **step 2**).
2. Collect ~50 mg of leaf-tips and isolate DNA from each plant (ideally in 96-well format to reduce the amount of work needed). Any standard protocol providing high-quality DNA (such as CTAB extraction) can be used.
3. Keep the plants in cool conditions (4–8 °C) until the end of screening for further multiplication of the selected individuals.
4. This procedure can be either followed by PCR screening and agarose gel electrophoresis (Subheading 3.2.1) or by PCR using M13 fluorescent markers and subsequent capillary electrophoresis (Subheading 3.2.2).

3.2.1 Screening Using PCR and Agarose Gel Electrophoresis

1. Choose one marker on each distal end of the chromosome of interest and one additional marker in centromeric area that will be used as a positive control. Alternatively, design new primers in these regions (*see* Subheading 3.3.1).
2. Perform PCRs using markers in terminal parts of desired chromosome arms and in centromeric area (positive control) with the following parameters: 20 μL of volume reaction containing 1× PCR buffer, 1.5 mM $MgCl_2$, 200 μM dNTPs, 1 μM primers, 20 ng of DNA and 0.4 U/20 μL Taq DNA polymerase. Conditions for amplification are as follows: initial denaturation at 95 °C for 10 min; 35 cycles of denaturation at 95 °C for 30 s, annealing at primer specific temperature for 30 s, and elongation at 72 °C for 50 s; followed by a final extension at 72 °C for 10 min.
3. Separate the PCR products using 1.2% agarose gel electrophoresis and score in presence/absence manner. Absence of PCR product on either terminal part of short/long arm of chromosome means a deletion present on respective arm.

3.2.2 Screening Using PCR with M13 Fluorescent Markers and Capillary Electrophoresis

1. Choose one marker on each distal end of the chromosome of interest and one additional marker in centromeric area and add an M13 extension on the 5′ end of forward oligos. Purchase these primers as well as M13 primers labelled with three different fluorescent dyes on their 5′ end, for example, first with

FAM, second with PET and third with VIC. Each of these fluorescent primers will be assigned to different PCR reactions, providing information which marker is being analyzed. Examples of primers used to detect terminal deletions on chromosome 3D are shown in Table 1.

2. For each marker, prepare a mix of M13-tailed forward primer and reverse primer at 1 and 10 μM, respectively.
3. Prepare PCR cocktail in 10 μL of volume reaction containing 5 μL of Master Mix AmpliTaq Gold 360, 0.5 μL of primer mix (previously prepared), 0.5 μM of labelled-primer and 25 ng of DNA. As an example, M13 primer labeled with FAM can be used with the marker specific to the short arm of chromosome, PET to the centromeric area and VIC to the long arm of chromosome.
4. Run PCR using the following parameters:

   1 cycle of,

   95 °C for 10 min.

   7 cycles of,

   95 °C for 30 s, 62 °C (decrease by 1 °C per cycle) for 30 s, 72 °C for 30 s.

   30 cycles of,

   95 °C for 30 s, 55 °C for 30 s, 72 °C for 30 s.

   8 cycles of,

   95 °C for 30 s, 56 °C for 30 s, 72 °C for 30 s.

   1 cycle of,

   72 °C for 5 min.

   Hold at 4 °C.
5. Dilute and multiplex PCR products from the same individual by transferring 3 μL of each PCR products (labelled with FAM, PET and VIC) in a new PCR well and by adding 190 μL of nuclease-free water.
6. Transfer 2 μL of the diluted mix in a PCR plate containing 5 μL of formamide and 0.02 μL of size standard.
7. Incubate the PCR plate at 95 °C for 5 min and then 4 °C for 5 min (minimum) to denaturate PCR products.
8. Separate PCR product on a 3730XL DNA Sequencer Instrument according to manufacturer's instructions.
9. Analyze capillary electrophoresis data with GeneMapper Software. Presence or absence of each marker in a given individual can be visually determined by a peak at the expected amplification size. Absence of a peak at the corresponding size indicates a deletion present on the respective arm. An example is provided in Fig. 3.

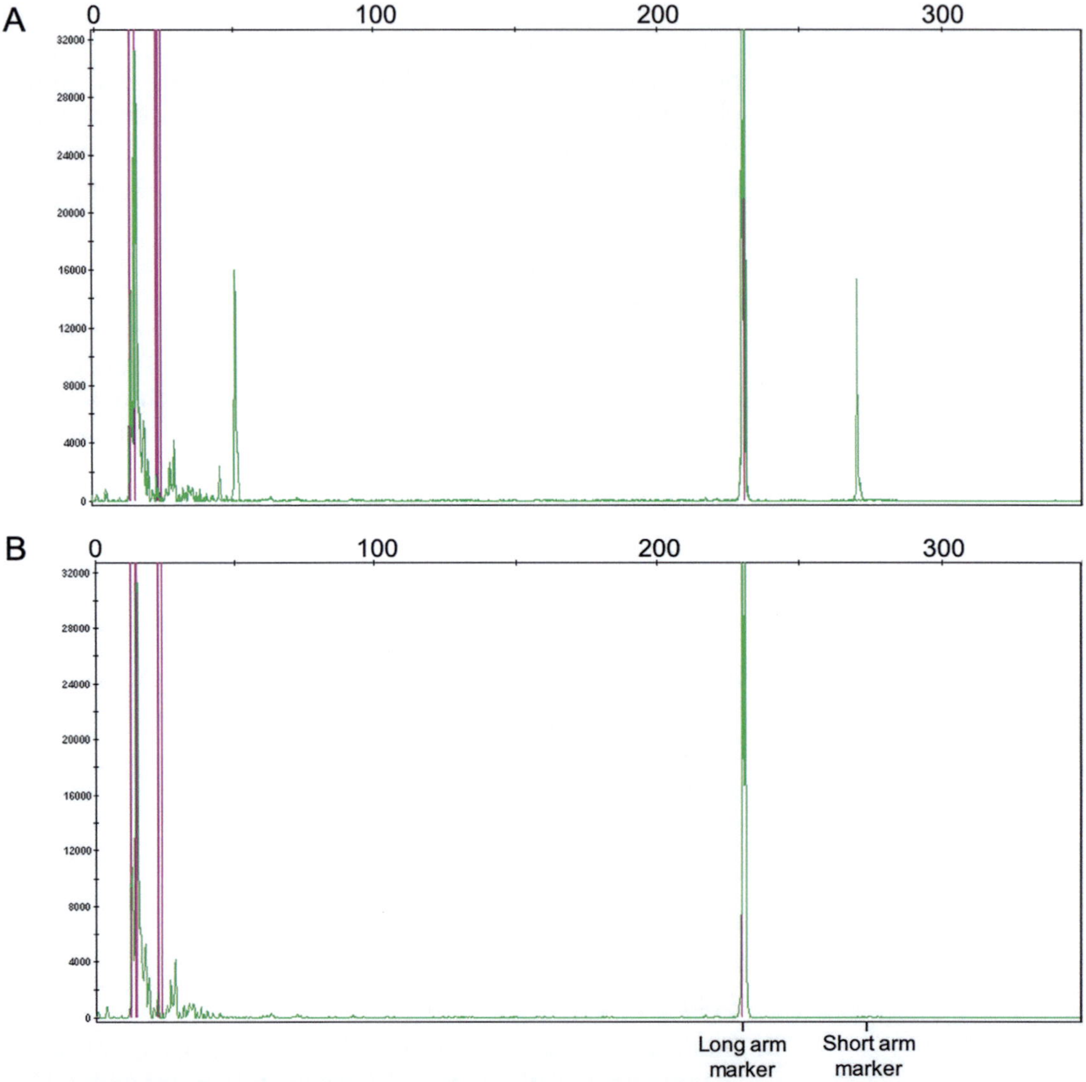

Fig. 3 GeneMapper electropherograms to detect deletions on chromosome 3D. Two individuals (A and B) were genotyped using 3D_3.2 Mb marker (specific to the short arm of chromosome 3D) and 3D_598.5 Mb marker (specific to long arm of chromosome 3D). Expected amplification sizes for both markers are 274 and 232 bp, respectively. The absence of a peak at 274 bp indicates a deletion on the short arm of chromosome 3D in individual B

3.3 Characterization of Deletion Sizes

Deletion breakpoint location for each selected deletion line can be precisely determined using sequence-tagged sites (STS) molecular markers (Subheading 3.3.1). In addition, other potential deletions located elsewhere in the genome can be easily detected via high-throughput SNP genotyping (Subheading 3.3.2).

3.3.1 Characterization Using Molecular Markers

1. Identify the region of interest for primer design using the reference sequence of the wheat genome [3] (*see* **Note 4**).
2. Align this region against the reference sequences of the corresponding homoeologous chromosomes using BLASTn and compare the three homoeologous sequences to depict SNPs specific to the chromosome of interest.
3. Design primers using the identified SNPs. Primers must strongly discriminate between the polymorphic homoeologous sequences during PCR amplification. They should be designed such that their 3′-ends anneal to polymorphic sites, where mismatches will strongly affect amplification. Examples of 3D-specific primers are shown in Supplementary Table 1.
4. Assess efficiency and specificity of the designed primers using template DNA from wild-type and nullisomic-tetrasomic lines lacking the chromosome of interest. PCRs are performed as in Subheading 3.2.1, **step 2**. Optimization of annealing temperatures and cycle numbers can be required for some primers. PCRs are scored for the presence/absence of specific products on 1.2% agarose gel. Successful genome-specific primers should amplify a single band from wild-type DNA (where the sequence of interest is present) but not from nullisomic-tetrasomic lines lacking the chromosome/sequence of interest.
5. Evaluate the presence or the absence of markers in deletion lines to determine the location of the deletion breakpoint. As deletions induced by 2C gametocidal system mostly affect the distal part of chromosome-arms, distal markers are expected to be negative (no PCR amplification) while more proximal ones, to be positives. An example is provided in Fig. 4. Additional primers can be designed to further refine breakpoint location.

3.3.2 Characterization Using High Throughput SNP Genotyping Array

1. Genotype DNA samples using the AXIOM BreedWheat 35 K Genotyping Array on a GeneTitan instrument. TaBW35K is derived from the AXIOM TaBW280K and allows the detection of deletions larger than a few megabases (variable depending on marker distribution). Alternatively, the AXIOM TaBW280K can be used to detect smaller deletions and achieve higher resolution of the deletion breakpoints. Affymetrix genotyping arrays generate files at .CEL format.
2. Analyze the output files using specific scripts available on request. An example of processed data is provided in Table 2. Dot plots of normalized signal intensities of individual markers along each chromosome can additionally be generated to visualize deletion breakpoints (Fig. 5). In the deletion lines produced via the above protocol, deletions were identified at the distal ends of either short or long chromosome arms, with a frequency of 1–2 per line (Table 2). In rare lines, both arms of the same chromosome carry a deletion.

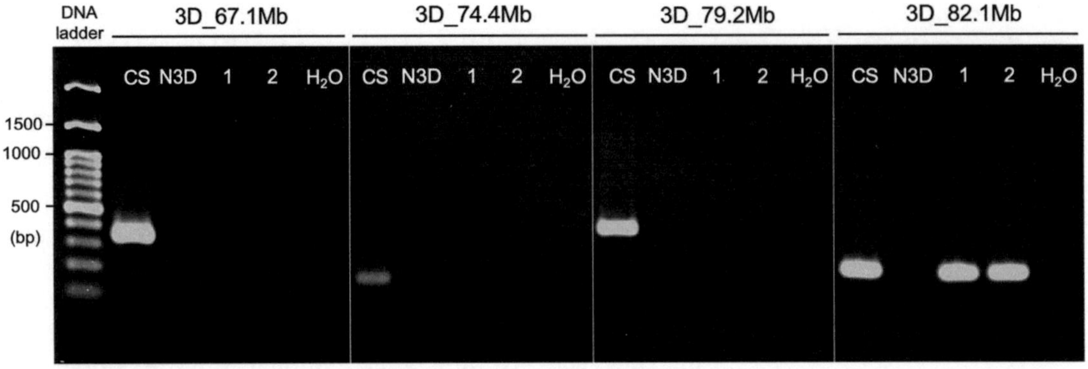

Fig. 4 Refining 3D deletion breakpoint location in the deletion line 2457 using molecular markers. Ethidium bromide stained gel showing PCR products of wild-type DNA (CS: Chinese Spring; positive control), DNA from nullisomic-tetrasomic line lacking chromosome 3D (N3D; negative control) and DNA from two individuals of the deletion line "2457" (1 and 2). Four 3D specific-markers located at positions 67.1, 74.4, 79.2, and 82.1 Mb on chromosome 3D were used for amplification. Expected size of the PCR products are 291, 138, 397, and 201 bp, respectively. Only the 3D_82.1 Mb primers allows amplification from 2457 DNA indicating that the deletion breakpoint in the line 2457 is located between positions 79.2 and 82.1 Mb on chromosome 3D

Table 2
Deletions detected in 3D deletion lines and other mutants using 35 K-SNPs genotyping array

Line	Chromosome	Segment	Coordinate of the first marker (bp)	Coordinate of the last marker (bp)	Min. deletion size (Mb)	Normalized signal intensity mean	Normalized signal intensity variance
A6S	3D	1	**318,330**	**65,230,227**	**64.91**	−1.376	1.068
		2	66,152,106	615,402,955	–	−0.498	0.097
B8S	3A	1	98,965	724,746,065	–	−0.064	0.112
		2	**724,870,059**	**750,322,878**	**25.45**	−1.321	1.025
	3D	1	**318,330**	**79,044,572**	**78.73**	−1.255	1.215
		2	79,246,431	615,402,955	–	−0.498	0.207
2457	3D	1	**318,330**	**81,040,484**	**80.72**	−1.428	1.146
		2	81,337,175	615,402,955	–	−0.012	0.060
ph1b	2D	1	278,775	621,558,871	–	−0.034	0.031
		2	**621,951,550**	**651,733,043**	**29.78**	−1.342	1.079
	5B	1	19,616	394,626,617	–	−0.039	0.029
		2	**395,692,680**	**454,386,489**	**58.69**	−1.384	0.935
		3	454,853,540	712,970,367	–	−0.044	0.028
	6D	1	**52,758**	**23,528,008**	**23.48**	−1.486	1.329
		2	24,003,386	473,347,817	–	−0.046	0.029
ph2a	3D	1	**318,330**	**120,722,379**	**120.40**	−1.352	1.125
		2	121,539,725	615,402,955	–	0.009	0.067

Coordinates (bp) of the first and last markers spanning the deleted fragment as well as the minimal size (Mb) of the deletion are in bold

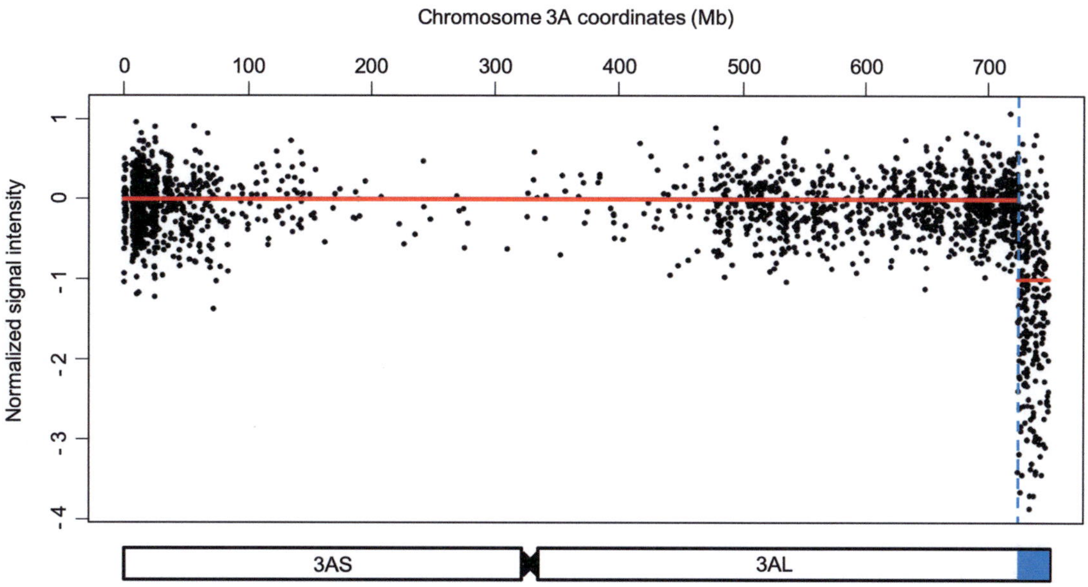

Fig. 5 Identification of a distal deletion on 3AL in B8S deletion line using 35 K-SNPs genotyping array. Dot plot showing the normalized signal intensities of 2040 markers from Affymetrix 35 K SNP array located along chromosome 3A. B8S signal intensities were normalized using wild-type Chinese Spring values for each marker. Each dot represents an individual marker. The deletion breakpoint (dotted blue line) is detected at ~724.8 Mb (between markers AX-89437774 and AX-89567713 at the coordinates 724,746,065 and 724,870,059, respectively) corresponding to a distal deletion of ~25.5 Mb. Red lines represent normalized signal intensity means of the markers located upstream (mean = −0.064) and downstream (mean = −1.321) of the deletion breakpoint

3.4 Selection of Deletion Lines with the Chromosome of Interest in Disomic Constitution

The selected plants in Subheadings 3.2 and 3.3 carry chromosome with deletion in a monosomic constitution, yielding unstable progeny regarding the number of chromosomes with deletion. To establish a population with stable inheritance of chromosome of interest carrying a deletion, a self-pollination of the F1 individuals followed by identification of lines with the chromosome in disomic constitution is necessary. Screening can be performed using cytogenetic approaches, however this approach is laborious and requires experienced staff to recognize chromosome of interest. Droplet digital PCR (ddPCR) is a reliable tool to solve this issue in high-throughput manner.

1. Transfer the selected plants (resulting from the cross between CS-2C and NT lines) into larger pots and let them grow until maturity.
2. Isolate young spikes using bags to ensure self-pollination.
3. Collect grains at full maturity and thresh them.
4. Sow F2 seeds and isolate DNA from each plant at three leaves stage. Any standard protocol providing high-quality DNA (such as CTAB extraction) can be used.

5. Perform PCR using markers outside of deletion area and separate the products using agarose gel electrophoresis. Absence of PCR product indicates lack of chromosome of interest in the analyzed individual, which is a frequent phenomenon in inheritance of monosomic chromosome. Discard the nullisomic plants for chromosome of interest.
6. Select or design primers and TaqMan probe in centromeric area on the chromosome of interest acting as a target and on different chromosome with known number of copies, which acts as a reference in ddPCR technique (*see* **Note 5**). See the supplier ddPCR Application guide for detailed instructions for probe design.
7. Prepare the PCR reaction in 20 μL (1× ddPCR Supermix for Probes (no dUTP), 900 nM primers for target, 900 nM primers for reference, 250 nM target probe, 250 nM reference probe, 100 ng DNA). Sequence of oligonucleotides used for ddPCR assay for determination of copy number of chromosome 3D is given in Table 3.
8. Create droplets using QX200 Droplet Generator with DG8 Cartridges and Gaskets and Droplet Generation Oil for Probes according to manufacturer's instructions.
9. Transfer the droplets to ddPCR 96-well plate and seal the plate by Pierceable Foil Heat Seal using PX1 PCR Plate Sealer. Perform PCR at ramp rate 2 °C/s under following conditions:

Table 3
Primers and probes used for determination of 3D chromosome number in ddPCR assay

Oligo ID	Sequence and modifications 5′-3′	Position (bp)	Amplicon length (bp)
Ta-4A_F	ATTTTGGGTCCTTG TTGTTATC	559,047,718–559,047,739	181
Ta-4A_R	ACACGCATGAAGTG TATAATGC	559,047,877–559,047,898	
Ta-4A_taq	FAM-AAGAACTTCA CACAC GAACTCGGA-QSY	559,047,745–559,047,768	
Ta-3D_F	CTCATCTCAGGCTG TCTAATTAA	253,384,668–253,384,690	167
Ta-3D_R	CATAGATCCCTCC TTGAAGGA	253,384,815–253,384,835	
Ta-3D_taq	VIC-CCTCACTCAAG CACCACATCG-QSY	253,384,692–253,384,712	

The TaqMan (taq) probes were either labelled by FAM (4A chromosome; used as a reference) or VIC (3D chromosome; target)

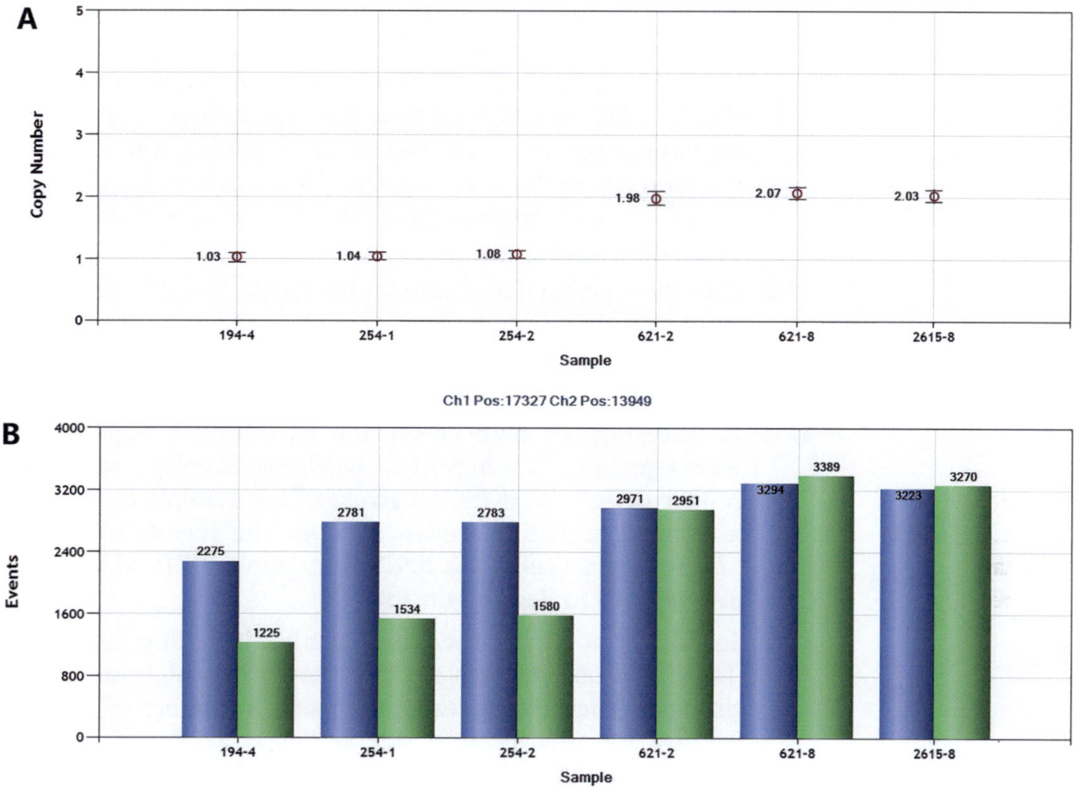

Fig. 6 Copy number variation analysis of chromosome of interest using ddPCR approach. Graph shows a number of copies of chromosomes 3D in four different lines in F2 generation. (**a**) Calculated number of copies based on droplets positive for reference (chromosome 4A) and target marker (chromosome 3D). (**b**) Numbers of droplets positive for target (green) and reference (blue) marker

enzyme activation at 95 °C for 10 min; 40 cycles of denaturation at 94 °C for 30 s; annealing/extension at 60 °C for 1 min; followed by enzyme deactivation at 98 °C for 10 min.

10. Transfer the plate to QX200 Droplet Reader and analyze the droplets under copy-number-variation (CNV) protocol.
11. Analyze the CNV data in QuantaSoft Software (Fig. 6).
12. Store the seeds of lines carrying chromosome of interest in disomic constitution and repeat self-pollination and ddPCR analysis in lines with no disomic plant detected.

4 Notes

1. Wheat monosomic addition line of chromosome 2C from *Aegilops cylindrica* can be obtained by crossing a disomic wheat line with CS-2C, backcrossing (usually five time) the hybrid with the recurrent wheat line and selecting at each round of backcross, plants with 43 chromosomes.

2. As offspring of wheat monosomic addition line of chromosome 2C from *Aegilops cylindrica* is not stable regarding the number of 2C chromosomes, the seed stock cannot be enlarged by simple self-pollination of this line. Extra step is crucial for increasing the seed stock, consisting of crossing wheat disomic addition line of 2C from *Aegilops cylindrica* with wheat, creating offspring of monosomic addition lines used in development of deletion lines.
3. Plan the crossing depending on the technical staff available for labor. If all the seeds were sown at once, the flowering stage of all plants would be similar, overwhelming the staff, resulting in self-pollination of most of flowers.
4. Considering the high proportion of repetitive sequences in wheat genome, it is important to choose a unique region for primer design. That can be achieved for example by masking annotated repetitive sequences from the reference genome [3]. Intergenic and genic SNPs from public AXIOM array can also be used to design primers.
5. In ddPCR, a polymerase-chain reaction including fluorescent TaqMan probes takes place individually in a high number of droplets. The droplet reader compares a number of droplets positive for two markers, one acting as a target (unknown number of copies) and the other one as a reference (known number of copies). Therefore, a set of primers and probe acting as a reference should be on a chromosome not affected by used NT line. Primers and probe targeting the chromosome of interest should be designed in the centromeric area to ensure amplification in all lines, regardless of deletion size.

Acknowledgments

We thank Clemence Marchal, Peter Scott, Tobin Florio, and James Simmonds for a permission to use figures from online manual "How to cross wheat" on website (http://www.wheat-training.com/wp-content/uploads/Wheat_growth/pdfs/How-to-cross-wheat-pdf.pdf).

References

1. Marcussen T, Sandve SR, Heier L et al (2014) Ancient hybridizations among the ancestral genomes of bread wheat. Science 345: 1250092
2. Wicker T, Gundlach H, Spannagl M et al (2018) Impact of transposable elements on genome structure and evolution in bread wheat. Genome Biol 19:103
3. Appels R, Eversole K, Stein N et al (2018) Shifting the limits in wheat research and breeding using a fully annotated reference genome. Science 361:6403
4. Rimbert H, Darrier B, Navarro J et al (2018) High throughput SNP discovery and genotyping in hexaploid wheat. PLoS One 13: e0186329

5. Ramírez-González RH, Borrill P, Lang D et al (2018) The transcriptional landscape of polyploid wheat. Science 361:6403
6. Saintenac C, Falque M, Martin O et al (2009) Detailed recombination studies along chromosome 3B provide new insights on crossover distribution in wheat (*Triticum aestivum* L.). Genetics 181:393–403
7. Choulet F, Alberti A, Theil S et al (2014) Structural and functional partitioning of bread wheat chromosome 3B. Science 345:1249721
8. Sears ER (1939) Cytogenetic studies with polyploid species of wheat. I. Chromosomal aberrations in the progeny of a haploid of *Triticum Vulgare*. Genetics 24(4):509–523
9. Sears ER (1944) Cytogenetic studies with polyploid species of wheat. II. Additional chromosomal aberrations in *Triticum Vulgare*. Nat Rev Genet 24(4):509–523
10. Sears ER (1966) Nullisomic-tetrasomic combinations in hexaploid wheat. In: Chromosome manipulations and plant genetics, pp 29–45
11. Sears ER (1978) The telocentric chromosomes of common wheat. In: Proceedings of 5th International. Wheat Genetics Symposium, New Delhi, vol 1, pp 389–407
12. Sears ER (1953) Nullisomic analysis in common wheat. Am Nat 87(835):245–252
13. Sears ER (1954) The aneuploids of common wheat. Univ Missouri Res Bull 572:1–58
14. Unrau J (1950) The use of monosomes and nullisomes in cytogenetic studies of common wheat. Sci Agric 30:66–89
15. Sharp PJ, Chao S, Desai S et al (1989) The isolation, characterization and application in the *Triticeae* of a set of wheat RFLP probes identifying each homoeologous chromosome arm. Theor Appl Genet 78(3):342–348
16. Anderson J, Churchill G, Autrique J et al (1993) Optimizing parental selection for genetic linkage maps. Genome 36(1):181–186
17. Endo T, Gill B (1996) The deletion stocks of common wheat. J Hered 87:295–307
18. Endo T (1988) Induction of chromosomal structural changes by a chromosome of *Aegilops cylindrica* L. in common wheat. J Hered 79:366–370
19. Gill K, Gill B, Endo T et al (1996) Identification and high-density mapping of gene-rich regions in chromosome group 1 of wheat. Genetics 144:1883–1891
20. Gill K, Gill B, Endo T et al (1996) Identification and high-density mapping of gene-rich regions in chromosome group 5 of wheat. Genetics 143:1001–1012
21. Sourdille P, Singh S, Cadalen T et al (2004) Microsatellite-based deletion bin system for the establishment of genetic-physical map relationships in wheat (*Triticum aestivum* L.). Funct Integr Genomics 4:12–25
22. Svačina R, Karafiátová M, Malurová M et al (2020) Development of deletion lines for chromosome 3D of bread wheat. Front Plant Sci 10:1–6
23. Serra H, Svačina R, Baumann U et al (2021) Ph2 encodes the mismatch repair protein MSH7-3D that inhibits wheat homoeologous recombination. Nat Commun 12:803

Chapter 15

Post-transcriptional Gene Silencing Using Virus-Induced Gene Silencing to Study Plant Gametogenesis in Tomato

Zhengming Wang, Xiao Yu Wang, Claudia Martinho, and David C. Baulcombe

Abstract

Loss-of-function analyses are essential to dissect the complex nature of biological processes, including gametogenesis. Virus-induced gene silencing (VIGS) has been widely used in crop species as an amenable and rapid way to generate gene knockdowns. As a transient assay, VIGS circumvents the generation of stable transgenic lines through laborious and time-consuming tissue culture techniques. VIGS involves inoculating plants during early development with genetically manipulated viral constructs carrying an endogenous gene target sequence. The viral infection triggers the host plant gene silencing machinery to process the viral genomic RNA into small RNAs (sRNAs) including the gene complementary region. The sRNAs with complementary sequences to the endogenous gene mediate posttranscriptional gene silencing of the targeted gene. Here, we provide a simple and reproducible VIGS protocol employing the tobacco rattle virus (TRV) in tomato (*Solanum lycopersicum cv. M82*). As it is stable at later developmental stages this approach is suitable for many traits in tomato including gametogenesis and it can be adapted to other crop species.

Key words Gene expression, Knockdown, Tobacco rattle virus, RNA silencing, VIGS, Tomato

1 Introduction

Manipulation of gene expression through gene knockout or knockdown is key to functional studies and enables a deeper understanding of complex biological processes. In the model plant *Arabidopsis thaliana*, there are T-DNA mutant collections which cover most of genes in Arabidopsis [1]. Other resources to address gene function in this and other species include knockout mutants through CRISPR-Cas9 [2] or RNAi and artificial miRNA strategies [3, 4]. Although these technologies are informative in gene functional studies, they all require a time-consuming and laborious tissue culture-mediated transformation.

Christophe Lambing (ed.), *Plant Gametogenesis: Methods and Protocols*, Methods in Molecular Biology, vol. 2484, https://doi.org/10.1007/978-1-0716-2253-7_15,

An approach that avoids the tissue culture step involves virus-induced gene silencing (VIGS)—a powerful tool to generate gene knockdowns in many species [5, 6]. Briefly, part of the gene of interest is cloned into a viral genome construct. Afterward, the cloned viral construct is transformed into *Agrobacterium tumefaciens* which is used to infect host plants. During agroinfection, transferred DNA (T-DNA) carrying viral sequences is delivered into host plant cells and viral proteins are expressed. These viral proteins, together with viral genomic RNA, form virus particles and they spread through the plant to cause systematic infection.

The RNA silencing machinery of host plants responds to the viral RNA [7] using the double-stranded forms of exogenous viral RNA that are produced by host or virus-encoded RNA-dependent RNA polymerases. The virus-derived sRNAs produced from the double stranded RNA by Dicer are bound by Argonautes (the effector proteins) that are then guided to target mRNAs through base pairing. Silencing is due to mRNA cleavage and/or translation inhibition and the sRNAs from inserted sequence target the corresponding endogenous mRNA. The systematic infection maintains gene silencing across different tissues and throughout development enabling studies of complex processes including gametogenesis.

Compared with gene editing and other methods of gene knockdown, VIGS has the advantage that it does not require tissue culture transformation. This makes it an easy and quick way to decipher gene function. Additionally, the sRNAs produced in VIGS can be designed to target paralogs in multigene families and thereby overcome the complication of functional redundancy in mutation analyses.

In tomato, tobacco rattle virus (TRV) induced VIGS is widely used [6]. There is evidence that TRV can enter meristems [8], which suggests the potential of TRV-VIGS in the study of gametogenesis. Recent evidence that TRV-VIGS mediated *MSH5* gene knockdown leads to meiotic crossover reduction, which further confirmed the availability of TRV-VIGS in study of gametogenesis [9]. Here, we describe the preparation of recombinant construct of TRV, Agrobacterium transformation and infiltration into tomato, and validation of systematic infection and knockdown of target gene, thus provide a detailed experimental protocol for TRV-VIGS in tomato, which has great potential to facilitate studies of gene function in tomato gametogenesis.

2 Materials

2.1 RNA Isolation and cDNA Synthesis

1. Liquid nitrogen.
2. Eppendorf Safe-Lock tubes, 1.5 mL.
3. Micropestles.

4. TRIzol reagent.
5. Direct-zol RNA MiniPrep.
6. Nuclease-free water.
7. RevertAid First Strand cDNA Synthesis Kit.
8. Thermocycler.
9. Low-temperature centrifuge.
10. PCR tubes.
11. Filter tips.

2.2 Construct Cloning

1. pTRV1.
2. pTRV2.
3. *Bam*HI and *Kpn*I restriction enzymes.
4. T4 DNA ligase.
5. *E. coli* DH5α competent cells.
6. Monarch PCR & DNA Cleanup Kit.
7. Monarch Plasmid DNA Miniprep Kit.
8. Heat block.
9. Incubator.
10. Centrifuge.
11. 1.5 mL Eppendorf tubes.
12. LB medium (Tryptone 1% (w/v), NaCl 1% (w/v), yeast extract 0.5% (w/v), adjust pH to 7.0 by 1 N NaOH).
13. Incubator shaker.
14. Nanodrop.

2.3 Agrobacterium Preparation

1. *Agrobacterium tumefaciens* strain GV3101.
2. IM buffer (10 mM MES pH 5.6, 150 μM acetosyringone, 10 mM $MgCl_2$).
3. Rifampicin.
4. Gentamicin.
5. Kanamycin.
6. Tetracycline.
7. Electroporation apparatus.
8. Incubator.
9. Incubator shaker.
10. LB medium (Tryptone 1% (w/v), NaCl 1% (w/v), yeast extract 0.5% (w/v), adjust pH to 7.0 by 1 N NaOH).
11. SOC medium (Tryptone 2% (w/v), yeast extract 0.5% (w/v), NaCl 10 mM, KCl 2.5 mM, $MgCl_2$ 10 mM, glucose 20 mM).
12. 1 mL syringe.

2.4 Plant Preparation

1. M3 soil.
2. Plant growth facilities.
3. M82 tomato seeds (accession: LA3475).

2.5 RNA Analyses for PCR/qPCR

1. Luna Universal qPCR Master Mix.
2. DreamTaq DNA Polymerases.
3. dNTPs 10 mM stock solution.
4. Equipment to perform DNA electrophoretic assays.
5. Agarose.
6. 1 × TBE buffer (0.13 M Tris pH 7.6, 45 mM boric acid, 2.5 mM EDTA).
7. Centrifuge.
8. RNase-free 1.5 mL Eppendorf tubes.
9. PCR tubes or 96-well PCR plates.
10. qPCR plates/sealers.
11. Filter tips.

3 Methods

3.1 Preparation of Recombinant Construct

3.1.1 Preparation of tomato RNA

1. Sow M82 tomato seeds into M3 soil, 20 °C 60% humidity, and 16 h day (300 μmol light), 8 h night and grow for 2 weeks.
2. Collect three new topical leaves per plant in 1.5 mL Safe-Lock Eppendorf tubes and flash freeze in liquid nitrogen. If necessary, samples can be stored at −80 °C until processing.
3. Cool down micropestle in liquid nitrogen and grind sample directly in the 1.5 mL Eppendorf until a homogenous powder is achieved. Thawing must be avoided until **step 4** to prevent RNA degradation.
4. Remove micropestle and add 500 μL of TRIzol reagent. Close the Safe-Lock 1.5 mL tube ensuring it is properly sealed. Vortex vigorously until a homogenous solution is formed.
5. Centrifuge at room temperature 5 min at 16,000 × *g* to remove debris.
6. Centrifuge a second time at room temperature 5 min at 16,000 × *g* to remove debris. Transfer the supernatant into a Zymo-Spin™ IIC Column in a Collection Tube without disturbing the debris pellet and centrifuge at 16,000 × *g* for 30 s. Transfer the column into a new collection tube and discard the flow-through.
7. Proceed to in-column DNase I treatment and add 400 μL RNA Wash Buffer to the column and centrifuge.

8. In an RNase-free tube, add 5 μL DNase I and 75 μL DNA Digestion Buffer per sample (add 10% extra volume to account for pipetting errors). Add 80 μL mix directly to the column matrix.
9. Incubate at room temperature (20–30 °C) for 15 min. Proceed to **step 10**.
10. Add 400 μL Direct-zol™ RNA PreWash buffer to the column and centrifuge 16,000 × *g* for 30 s. Discard the flow-through and repeat this step.
11. Add 700 μL RNA Wash Buffer to the column and centrifuge 16,000 × *g* for 30 s. Discard the flow-through, repeat this wash and centrifuge 16,000 × *g* for 1 min. Discard the flow-through and centrifuge 16,000 × *g* for 2 min to ensure complete removal of the wash buffer.
12. Transfer the column carefully to a RNase-free tube, add 40 μL of nuclease free water directly to the column matrix and centrifuge 16,000 × *g* for 1 min (*see* **Note 1**).

3.1.2 cDNA Synthesis

1. Use the RevertAid First Strand cDNA Synthesis Kit for cDNA synthesis,. The reverse transcription (RT) is carried out as described in the manual provided by the manufacture. Reaction should be set up using filter tips, nuclease-free PCR-tubes, gloves and on ice to avoid RNA degradation.
2. Add 1 μg total RNA and nuclease-free water up to 11 μL in a PCR tube or a 96-well PCR plate. This step has to be carried out in duplicates to include a RT-control for each sample.
3. Add 1 μL hexamer oligonucleotides (provided with the kit, *see* **Note 2**), mix well, spin down and incubate for 5 min at 65 °C. Immediately transfer samples to ice.
4. Prepare a cDNA reaction mixture and an RT-control mixture as follows: cDNA mixture (volume per sample—always add 10% extra to compensate for pipetting errors), 4 μL 5 × reaction buffer, 2 μL 10 mM dNTPs, 1 μL Ribo-lock, and μl Revert-Aid.

3.1.3 Amplification of Target Sequence of Selected Gene

1. Choose target sequence of selected gene using SGN VIGS Tool (https://vigs.solgenomics.net/). Design forward and reverse primers with appropriate restriction enzyme sites and protection bases. Here, we use *slPDS* gene (Solyc03g123760.3.1) as an example (Fig. 1).
2. Amplify the target sequence of *slPDS* using cDNA (prepared above) and primers with the following conditions. Per reaction mix: 2 μL 10 × DreamTaq green buffer, 0.5 μL dNTPs (10 mM each), 1 μL slPDS_F1 and slPDS_R1 primer mix (10 μM each), 0.2 μL DreamTaq, 1 μL cDNA, 15.3 μL

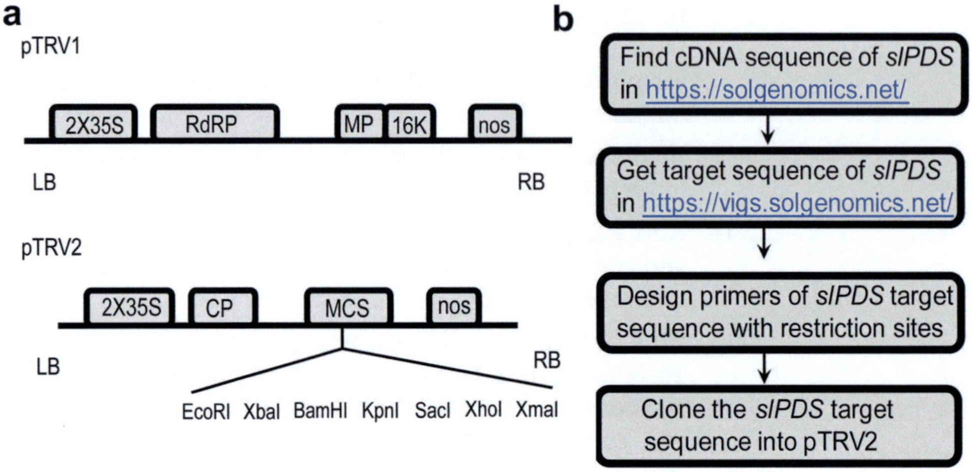

Fig. 1 VIGS plasmid diagrams and construct design. (**a**) Diagrams of TRV VIGS plasmids. Two plasmids (pTRV1 and pTRV2) are used for VIGS. PTRV1 contains dual promoter of 35S driven TRV genes (RNA dependent RNA Polymerase, RdDP; membrane protein, MP; 16 K protein, 16 K) and nos terminator. PTRV2 contains dual promoter of 35S driven coat protein gene (CP) and multiple cloning site (MCS) and nos terminator as well. The restriction sites of MCS are indicated. (**b**) Flowchart of VIGS construct design

H_2O. PCR conditions: 95 °C 3 min; 95 °C 20 s, 58 °C 20 s, 72 °C 30 s (30 cycles); 72 °C 5 min. (*see* **Note 3**).

3. Purify the PCR products by Monarch PCR purification kit following the manual protocol.

3.1.4 Clone Target Sequence into pTRV2

1. Take 100 ng pTRV2 and 100 ng target sequence PCR product, digest with selected restriction enzymes (e.g., *Bam*HI and *Kpn*I here) at 37 °C for 2 h, clean the digested vector and target sequence by Monarch PCR purification kit.
2. Setup a 10 μL ligation reaction with molar ratio of 1:3 (vector: insertion), for example, 20 fmol vector and 60 fmol insertion, using T4 DNA ligase at 16 °C overnight.
3. Add 2 μL of ligation product into a tube of DH5α chemically competent cells, 42 °C 45 s, chill on ice for 2 min. Add 0.9 mL SOC medium, shake at 225 rpm 37 °C for 1 h, plating on LB plate with 50 μg/mL of kanamycin.
4. Select single colony and culture it in 3–5 mL liquid LB medium with 50 μg/mL of kanamycin overnight at 37 °C and extract recombinant pTRV2-slPDS plasmids with Monarch Miniprep kit following the manufacturer's protocol. Measure the plasmid concentration using Nanodrop.

3.2 Agrobacterium Transformation, Culturing, and Induction

1. Take 100 ng of pTRV1, pTRV2, and recombinant pTRV2-slPDS plasmids, mix each with 50 μL Agrobacterium (GV3101) competent cells, mix and transfer into precooled 1 mm cuvette.

Table 1
Antibiotics for plasmids

Plasmid	Rifampicin	Gentamicin	Kanamycin	Tetracycline
pTRV1	50 μg/mL	25 μg/mL	50 μg/mL	NA
pTRV2	50 μg/mL	25 μg/mL	50 μg/mL	5 μg/mL
Recombinant pTRV2	50 μg/mL	25 μg/mL	50 μg/mL	5 μg/mL

2. Electroporation with 2400 V of voltage, 25 μF of capacitance, 200 Ω of resistance, add 800 μL SOC liquid medium immediately, transfer into an Eppendorf tube, and shake at 225 rpm 28 °C for recovery 3–4 h.
3. Plate 1/10 culture onto LB solid medium (*see* **Note 4**) with appropriate antibiotics (Table 1), and incubate 28 °C for 48 h.
4. Pick up single colony and culture it in 5 mL liquid LB medium with appropriate antibiotics, shake at 225 rpm 28 °C for 24 h (*see* **Note 5**).
5. Take 50 μL preculture into 50 mL liquid LB medium with appropriate antibiotics, 28 °C culture for 24 h.
6. Prepare IM buffer (*see* **Note 6**).
7. Spin down the Agrobacterium by centrifuging 2800 × *g* 10 min at room temperature, discard supernatant and resuspend the pellet in IM buffer by inverting of pipetting very gently up and down, adjust OD600 to 1 using IM buffer. Caution to use IM buffer as your blank for OD quantifications. Leave culture at room temperature for 2–3 h (*see* **Note 7**). Can proceed to infiltration.

3.3 Tomato Infiltration

1. Sow 3 trays (53 cm × 30 cm × 5 cm) of M82 tomato seeds with 15 pots, one seed in each pot, into M3 soil, 20 °C 60% humidity, and 16 h day (300 μmol light), 8 h night, grow for 1 week (*see* **Note 8**).
2. Mix up Agrobacterium of pTRV1 and pTRV2 or recombinant pTRV2-slPDS (OD600 = 1) by 1:1.
3. Use 1 mL needleless syringe to infiltrate the mixture into both cotyledons of each tomato plants through the abaxial side (Fig. 2), keep the infiltrated plants grow for two more weeks (*see* **Note 9**), keep checking phenotypes (Fig. 3).

Fig. 2 Tomato infiltration procedure. (**a**) M82 tomato seedlings (1-week-old) are ready for infiltration; (**b**) Agroinfiltration of cotyledons; (**c**) Tomato seedlings after infiltration. Dark areas of cotyledons are filled with Agrobacterium

Fig. 3 Phenotypes of *slPDS* knockdown by VIGS in tomato. *SlPDS* is a gene involved in carotenoid synthesis. Silencing of PDS causes photobleaching. (**a**) The *slPDS* knockdown effects have lasted for over 48 days since infiltration; (**b**) The VIGS effects happened at inflorescences (arrow) suggesting TRV can enter inflorescence, which assures the availability of TRV-VIGS in gametogenesis study

Table 2
Primers sequences

	Sequences (5′ to 3′)
TRV2_F1	ACTTAGATTCTGTGAGTAAG
TRV2_R1	AGACAATGAGTCGGCCAAAC
TRV2_F2	GAGTCCCACATATTCGCACG
TRV2_R2	CACAAGAACCGCCTGAACAT
slPDS_F1 slPDS_R1	CGCGGATCCTGGAAAGGTAGCTGCATGGA CGGGTACCCACCTCATCTGTCACCCTGT
slPDS_F2	AGTCGAGATGGTTGCTTGCA
slPDS_R2	ATCTTCTGGTCGTGGCATGG
slActin_F	GTCTGTTTCCCGGTTTTGCT
slActin_R	TCTGTTGTTTGCTACCGTGC

3.4 RNA Analysis to Confirm Viral Vector Expression and Target Gene Downregulation

3.4.1 Semiquantitative RT-PCR

1. Amplify the virus sequence using TRV2_F1 and TRV2_R1 (Table 2; Fig. 4a) with the following conditions. Amplify the *slActin* gene (Solyc03g078400.2) using slActin_F and slActin_R (Table 2) as internal control.
   (a) Per reaction mix:
   (b) 2 μL 10 × DreamTaq green buffer.
   (c) 0.5 μL dNTPs (10 mM each).
   (d) 1 μL primer mix (10 μM each).
   (e) 0.2 μL DreamTaq.
   (f) 4 μL cDNA (1:4 diluted) (*see* **Note 10**).
   (g) 12.3 μL H_2O.
   (h) Programme: 95 °C 3 min; 95 °C 20 s, 58 °C 20 s, 72 °C 30 s (29 cycles); 72 °C 5 min.
   (i) Run PCR products on 1.5% agarose gel (Fig. 4c).

3.4.2 Real Time Quantitative PCR

1. Amplify the virus sequence and *slPDS* gene sequence (Fig. 4d, e) using TRV2_F2/R2 (Table 2; Fig. 4a) and PDS_F2/R2 (Table 2; Fig. 4b) respectively, with the following conditions. Per reaction mix:
   (a) 4 μL cDNA (1:4 diluted).
   (b) 0.75 μL H_2O.
   (c) 5 μL Luna Universal qPCR Master Mix.

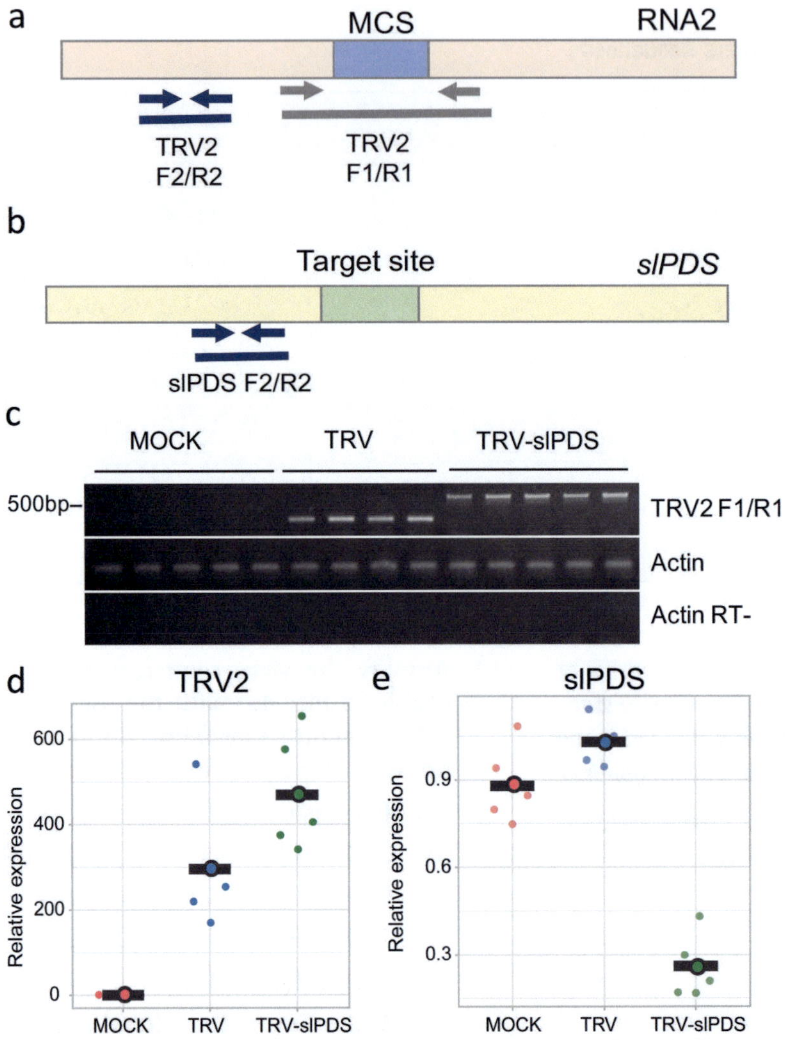

Fig. 4 RNA analyses of TRV2 and slPDS gene in VIGS samples. (**a**) Diagram of primer design of TRV RNA2. TRV2 F1 and R1 are designed in the flanking regions of multiple cloning site (MCS) to show the insertion of recombinant TRV2-PDS. TRV2 F2 and R2 are designed in the upstream region of MCS and are used for qPCR; (**b**) Diagram of primer design of *slPDS*. slPDS F2 and R2 are designed in the upstream region of VIGS targeting site; (**c**) RT-PCR of TRV2. TRV2 was amplified in the TRV and TRV-slPDS samples rather than the MOCK samples. Actin was used as loading control. RT minus was used as a negative control for RT-PCR; (**d**) qPCR results of TRV2 to further confirm the accumulation of TRV in TRV and TRV-slPDS samples rather than MOCK samples; (**e**) qPCR results of *slPDS* to show the downregulation of *slPDS* gene in TRV-slPDS samples than MOCK and TRV samples

(d) 0.25 μL Primer mix (10 μM each).

(e) Programme: 95 °C 3 min; 95 °C 10 s, 60 °C 30 s (40 cycles) + melt curve 65–95 °C, 0.5 °C increment per 5 s. (*see* **Note 11**).

4 Notes

1. Alternatively, for highly concentrated RNA use 25 μL of nuclease free water. RNA can be used immediately or stored frozen at −80 °C.
2. Hexamer oligonucleotides (rather than oligodT) must be used for cDNA amplification of viral RNA.
3. The temperature of annealing, extension time and cycle number need to be optimised for each primer and each gene depending on the primers, gene length and expression level.
4. This is to avoid overgrowth of transformed Agrobacterium.
5. At this point we can make a glycerol stock to store the Agrobacterium and restart from here when repeating the experiment.
6. Prepare IM buffer freshly every time.
7. Acetosyringone improves transformation efficiency by up-regulating the virulence of Agrobacterium.
8. The plant age is one of the most critical factors for the efficiency of the VIGS experiment. Infiltration should be carried out in plants before true leaves emerge. Later developmental stages should be avoided.
9. Keep the lid on to cover the infiltrated plants for 24 h, remove it afterward. This helps to maintain humidity at the beginning which is good for virus replication.
10. cDNA dilution allows pipetting of a larger cDNA volume (4 μL instead of 1 μL) decreases the impact of pipetting errors in quantitative analyses.
11. We recommend testing primer efficiency before qPCR and determining the best cDNA working concentration for each primer pair prior to this essay.

Acknowledgements

We thank the BBSRC grant BB/R022658/1. D. C. B. is the Royal Society Edward Penley Abraham Research Professor.

References

1. O'Malley RC, Ecker JR (2010) Linking genotype to phenotype using the Arabidopsis unimutant collection. Plant J 61:928–940
2. Zhu H, Li C, Gao C (2020) Applications of CRISPR–Cas in agriculture and plant biotechnology. Nat Rev Mol Cell Biol 21:661–677
3. Ossowski S, Schwab R, Weigel D (2008) Gene silencing in plants using artificial microRNAs and other small RNAs. Plant J 53:674–690
4. Small I (2007) RNAi for revealing and engineering plant gene functions. Curr Opin Biotechnol 18:148–153
5. Ratcliff F, Martin-Hernandez AM, Baulcombe DC (2001) Tobacco rattle virus as a vector for analysis of gene function by silencing. Plant J 25: 237–245
6. Dinesh-Kumar SP, Anandalakshmi R, Marathe R et al (2003) Virus-induced gene silencing. Methods Mol Biol 236:287–294
7. Baulcombe D (2004) RNA silencing in plants. Nature 431:356–363
8. Martín-Hernández AM, Baulcombe DC (2008) Tobacco rattle virus 16-Kilodalton protein encodes a suppressor of RNA silencing that allows transient viral entry in meristems. J Virol 82:4064–4071
9. Calvo-Baltanás V, Wijnen CL, Yang C et al (2020) Meiotic crossover reduction by virus-induced gene silencing enables the efficient generation of chromosome substitution lines and reverse breeding in Arabidopsis thaliana. Plant J 104:1437–1452

Chapter 16

Protocols for In Vivo Doubled Haploid (DH) Technology in Maize Breeding: From Haploid Inducer Development to Haploid Genome Doubling

Siddique I. Aboobucker, Talukder Z. Jubery, Ursula K. Frei, Yu-Ru Chen, Tyler Foster, Baskar Ganapathysubramanian, and Thomas Lübberstedt

Abstract

Doubled haploid (DH) technology reduces the time required to obtain homozygous genotypes and accelerates plant breeding among other advantages. It is established in major crop species such as wheat, barley, maize, and canola. DH lines can be produced by both in vitro and in vivo methods and the latter is focused here. The major steps involved in in vivo DH technology are haploid induction, haploid selection/identification, and haploid genome doubling. Herein, we elaborate on the various steps of DH technology in maize breeding from haploid induction to haploid genome doubling to produce DH lines. Detailed protocols on the following topics are discussed: in vivo haploid inducer line development, haploid selection using seed and root color markers and automated seed sorting based on embryo oil content using QSorter, artificial genome doubling, and the identification of genotypes with spontaneous haploid genome doubling (SHGD) ability.

Key words Doubled haploid, Genome doubling, Haploid induction, Haploid selection, Maize, Oil content, QSorter, Seed color

1 Introduction

Doubled Haploid (DH) technology is established in major crop species such as wheat, barley, maize, and canola [1]. The main purpose of DH technology is to accelerate inbred line development. DH lines are homogeneous lines (all genotypes in a line are identical) of completely homozygous genotypes. They have various advantages, such as compliance to variety registration regulations, more reliable phenotyping and genotyping compared to

Siddique I. Aboobucker and Talukder Z. Jubery have contributed equally to this work.

Christophe Lambing (ed.), *Plant Gametogenesis: Methods and Protocols*, Methods in Molecular Biology, vol. 2484, https://doi.org/10.1007/978-1-0716-2253-7_16,

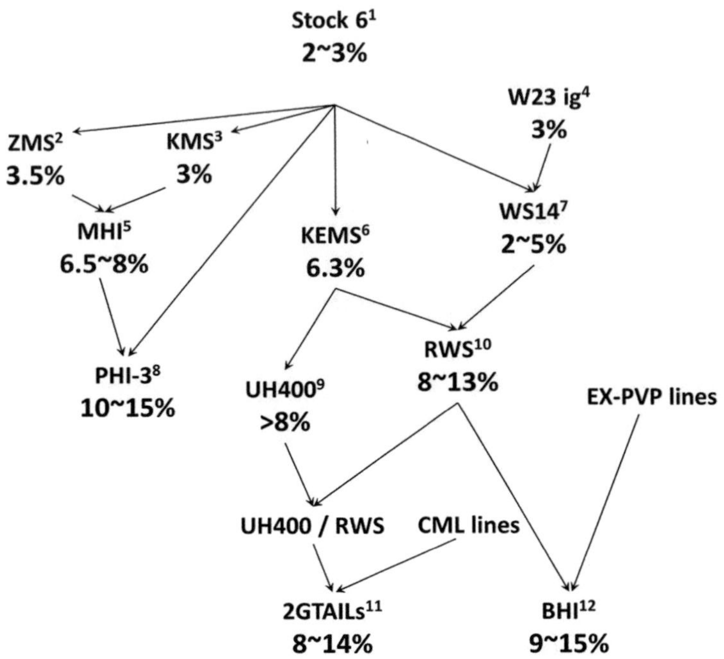

Fig. 1 The pedigree of published haploid inducers and their HIR (in percentage) are shown. The references for the respective haploid inducers are: [1] [5]; [2] [35]; [3] [36]; [4] [37]; [5] [38]; [6] [39]; [7] [40]; [8] [41]; [9] [19]; [10] [42]; [11] [15]; [12] [43]

segregating families, among others [2]. More recently, haploid induction has been combined with genome editing, to facilitate efficient production of homozygous edited genotypes without editing-related transgenes in their genomes [3].

DH lines can be produced by in vitro and in vivo methods, the latter involving haploid inducers within species, or from other species such as the wheat DH system using maize pollen for haploid induction [1]. Herein, we will focus on the predominant method of in vivo DH line production in maize, which has been conceived in the 1940s [4]. The first haploid inducer Stock 6 has been published in 1959 [5]. Since then, haploid inducers and the process of DH line development have been incrementally improved until it became a routine method in major maize breeding programs in the 1990s and 2000s (Fig. 1) [2].

Maternal haploid induction as a primary process to produce DH lines in maize requires the following steps: (1) a cross between a donor genotype (from which haploids / DH lines are derived) and a haploid inducer, (2) selection of kernels with haploid embryo, (3) genome doubling of haploid plants, at least of cell lines leading to fertile gametes, and (4) self-pollination of fertile haploid plants. Improvement of maize DH technology occurred for all steps in this process. The first haploid inducer Stock 6 had a haploid induction rate (HIR) of 2.3% [5]. Since then, haploid inducers have been developed with HIR exceeding 15%. Two major genes controlling maternal haploid induction, *ZmMTL* and *ZmDMP*, have been isolated [6–9], and this information can be used to develop novel inducers. This is important, as haploid induction must be combined with adaptation to the environment, where a haploid inducer will be used [10].

Since only a fraction of the kernels produced in a haploid induction cross carries a haploid embryo (ca. 15%), those "haploid kernels" need to be efficiently discriminated from the remaining 85% of regular diploid kernels. This is primarily done by visual and manual selection of kernels based on seed color marker, such as the dominant *R1-nj* gene, resulting in red coloration of the embryo and aleurone layer in diploid kernels. Haploid kernels from induction crosses are characterized by lack of red coloration of embryos but with red colored aleurone. Since expression of the *R1-nj* gene can be suppressed or masked in some genetic backgrounds, alternative haploid selection methods have been established, including one based on (purple plant) *Pl1*-based root coloration [10]. More recently, different methods have been evaluated to enable automated machine-based single kernel haploid selection, including Raman spectroscopy, NMR, kernel weight, image analysis, and NIRS. Respective high-throughput methods are expected to replace laborious manual selection, perhaps in conjunction with inducers facilitating machine sorting, such as high-oil inducers [10].

A major bottleneck for practical application of DH technology is the genome doubling step. This is routinely done by colchicine treatment of germinating plants or seedlings, which need to be transplanted into the field (rather than sown). Thus, this step is laborious and thus costly, and requires handling of toxic colchicine. Alternative treatments have been proposed, including different herbicides and N_2O. More recently, genetic mechanisms for spontaneous haploid genome doubling (SHGD) have been proposed, enabling direct sowing of haploid seed into the field without additional treatment [11]. In maize, haploid plants are displaying limited haploid female fertility (HFF), whereas male flowering organs of most genotypes are sterile. However, some genotypes have been identified with high level of haploid male fertility (HMF) as the main component of SHGD [12–14].

In this chapter, the objectives are to elaborate on the various steps involved to establish a successful in vivo maize DH program, including (1) haploid inducer development and improvement, (2) haploid induction (3) haploid selection by different methods (4) artificial genome doubling and (5) identifying genotypes with SHGD ability.

2 Materials

2.1 Haploid Inducer Development and Improvement

2.1.1 For Gel-Based Marker Analysis

1. Maternal haploid inducer maize lines can be obtained from Maize Genetics Cooperation Stock Center (MGSC, http://maizecoop.cropsci.uiuc.edu/) or through licensing from different organizations (Table 1, Fig. 1, *see* **Note 1**).
2. Field space or greenhouse space set to a temperature of 20 °C/25 °C and 8/16 h night/day.
3. Primers sequences for *ZmMTL (Matrilineal)* and *ZmDMP* markers are in Table 2 (*see* **Note 2**).
4. Genomic DNA extracted from individuals of a segregating breeding population by your method of choice.
5. PCR reaction components.
6. 4% Metaphor Agarose in 1× TBE buffer (100 mM Tris base; 100 mM boric acid; 2 mM EDTA disodium salt).

Table 1
Haploid inducer lines from different organizations

Haploid inducers	HIR	Adaptations	Haploid selectable marker	Additional-value traits	Licensing organizations
Stock 6	2–3%	Temperate			MGSC (Free access, Catalog No. M741F)
Improved Stock 6	N/A	Temperate			MGSC (Free access, Catalog No. M741H)
TAILs	6–8%	Tropical	*R1-nj*	–	CIMMYT
2GTAILs	8–14%	Tropical	*R1-nj*	–	CIMMYT
UH 400	8–15%	Temperate	*R1-nj*	–	University of Hohenheim
UH 600	8.5–12%	Temperate	*R1-nj*	High oil	University of Hohenheim
BHI201	9–15%	Temperate	*R1-nj*	–	Iowa State University
BHI306	9–15%	Temperate	*R1-nj, Pl1*	Ability to overcome partial dent sterility (*Ga1*)	Iowa State University
C1-I	8–10%	Temperate	*C1-I*	For haploid induction in *R1-nj* backgrounds; (*Ga1*)	Iowa State University

Table 2
PCR markers for two major QTLs for HIR introgression

Primer name	Primer sequence (5′ to 3′)	Expected size	Purpose
ZmMTL-HRM_F ZmMTL-HRM_R	AGCGGGTGTCCAGGG TCAAC TGCC TAGCGAGCCCACCGAG	98 bp	HRM marker to distinguish 4 bp difference (*see* **Note 2**) in *ZmMTL*
ZmMTL-SSR_F ZmMTL-SSR_R	TTGTTCTTGCATCCA TCCAG CATGCACTTGCCGTTG TACT	115 bp	SSR marker in the first intron of the *ZmMTL* gene (CGAT)n resolved in 4% metaphor
ZmDMP-HRM_4L ZmDMP-HRM_4R	CGTGCAGAAGACGCTC TCCA GTGGGGAGGAAG TTGGCCAG	50 bp	HRM marker to distinguish the SNP in *ZmDMP* for HIR
Umc1040-SSR_F Umc1040-SSR_R	CATTCACTCTC TTGCCAACTTGA AGTAAGAGTGGGATATTC TGGGAGTT	91 bp	SSR marker, located ca. 0.5 Mb downstream from *ZmDMP* gene, resolved in 4% metaphor

2.1.2 For High Resolution Melting (HRM) Analysis (Optional)

7. LC Green.
8. LightScanner Instrument.
9. Hard shell high profile 96-well semiskirted PCR plates for HRM.

2.2 Haploid Induction

1. Maternal haploid inducer (Table 1, *see* Subheading 2.1 and **Notes 3** and **4**).
2. Female donor population, to be induced for haploid production.
3. Shoot bags (Dry waxed white paper—2 1/2″ × 1″ × 8 1/2″) and tassel bags (Medium).

2.3 Haploid Selection

2.3.1 Haploid Selection Using Seed Color Marker

1. Potential haploid seed/kernel (PHKs) obtained from induction crosses between noncolored corn and haploid inducers with *R1-nj* as selectable marker or that obtained from crosses between colored corn (such as genetic backgrounds with *R1-nj*) and a *C1-I* (*colorless aleurone*) dominant haploid inducers (Table 1, Fig. 2, *see* **Notes 4** and **5**).
2. Well-lit workspace.
3. Razor blade or knife to cut a few kernels as a test to visualize the color difference between haploid and hybrids.

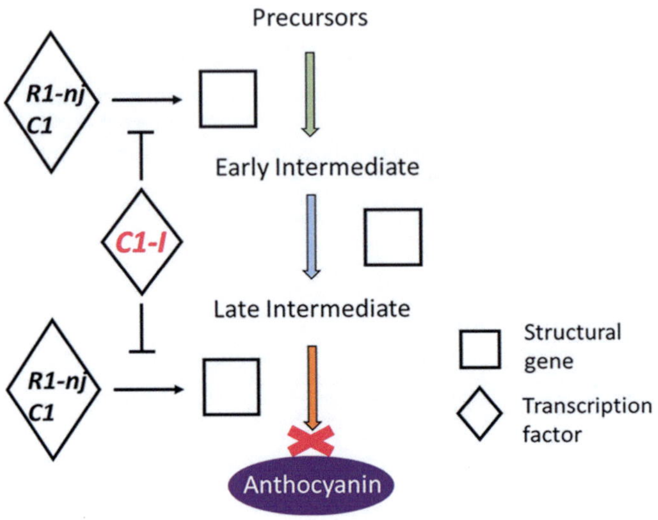

Fig. 2 The anthocyanin biosynthetic pathway in seed aleurone for haploid selection. R1-nj and C1 (colored aleurone1) are two transcription factors controlling the expression of structural genes in anthocyanin biosynthesis in seed aleurone and embryo. In the presence of *C1-I*, a negative dominant allele that represses the expression of structural genes, anthocyanin synthesis is blocked. These alleles are used as seed color markers to identify haploids in a mixture of seeds obtained from a haploid induction cross

2.3.2 Haploid Selection Using Root Color Marker

1. PHKs obtained from induction crosses between noncolored corn and *purple plant 1* (*Pl1*) dominant haploid inducers (Table 1, *see* **Note 6**).
2. Poly plug trays.
3. Soil mix suited for seeding.
4. Greenhouse space set to a temperature of 20 °C/25 °C and 8/16 h night/day.

2.3.3 Haploid Selection Using QSorter

1. Mixture of maize seeds with oil content ranging from 3 to 14% for calibration (e.g., exPVP lines and breeding lines extracted from the ALEXHO population—accessions are available through GRIN—https://npgsweb.ars-grin.gov/gringlobal/search).
2. 48-well tissue culture plates.
3. QSorter.
4. A computer with Microsoft windows operating system.
5. MATLAB 2018a, with image processing, machine learning, and deep learning toolboxes.
6. Bruker Minispec mq20 NMR analyzer.

2.4 Artificial Genome Doubling of Haploids Using Colchicine

1. Colchicine solution: 0.125% (w/v) colchicine dissolved in 0.5% DMSO; 0.05% Color Coat Blue (vegetable dye (blue), as used in seed treatment) (*see* **Notes 7** and **8**).
2. 1 mL surgical syringes.
3. Soil mix suited for seeding.
4. Peat pots (e.g., Jiffystrip).
5. Greenhouse space set to a temperature of 20 °C/25 °C and 8 h/16 h night/day.

2.5 Identifying Genotypes with SHGD Ability

1. PHKs of genotypes panel.
2. Field space.
3. Field hoe or tree lopper.
4. Stapler, staples, sharpie, labels, and knife.
5. Tassel bags.

3 Methods

3.1 Haploid Inducer Development

HIR and increased agronomic performance (resistance traits, seed production for maintenance of inducers, pollen production) of haploid inducers significantly enhance the efficiency of in vivo DH technology in maize. Poor agronomic performance and adaptation to the environment require the introgression of HIR into other genetic backgrounds. HIR can be conferred to noninducers by the introduction of mutations in two major genes, *ZmMTL* and *ZmDMP* by maker-assisted selection, and these alone may increase the HIR of noninducers to 6–12% [15] (*see* **Note 9**).

Subsequent inbreeding or DH production steps can help to obtain stable haploid inducer lines.

1. Plant haploid inducer line (Table 1) and a well-adapted genotype (for the region with good agronomic performance—"recipient line") side by side in the field or greenhouse. Several plantings of the inducer line are necessary to ensure simultaneous pollen and silk availability to cross the two lines.
2. Make crosses between the two lines to obtain F_1 seed. Use the haploid inducer as male and the recipient line as female in crosses. This allows to confirm, whether the selectable marker (such as *R1-nj* and/or *Pl1*) present in the haploid inducer is expressed in the recipient line. This is also recommended when the inducer source contains any of the pollen exclusion alleles (such as *Ga1-S* or *Ga2-S*) [16, 17]. No marker selection is necessary in this generation.

3. In the following growing season, either self-pollinate (to obtain F_2) or backcross the F_1 to the recipient line (to obtain BC_1F_1). No marker selection is necessary in this generation.
4. In the following generations, continue backcrossing (or self-pollinating) after marker selection for the two gene loci conferring haploid induction ability, and the presence of the selectable marker (*R1-nj* and/or *Pl1*) by visual observation (*see* **Note 10**).
5. Families that are fixed for *R1-nj* can be tested for haploid induction ability in test crosses onto a test donor (regional conventional hybrid—for example, Viking 60-01N or Viking 51-95UP). Grow inducer breeding families and several plantings of the test donor side by side in the field. Haploid inducer lines tend to flower early. Therefore, it is recommended to have more than one planting of inducers. During pollen shedding, collect a bulk pollen sample over all plants of the inducer family to be tested and pollinate 4–5 ears of the test donor, to obtain at least 1,000 induced seeds—this will allow a discrimination of HIR at the 1% level.
6. Determine HIR using the appropriate procedure as described (*see* Subheading 3.3.1). Select families with high HIR scores, the best expression of *R1-nj* in the offspring and good agronomic performance.
7. After 4–5 generations of self-pollination (or production of inducer DH lines), selected breeding lines will be tested on more than one test donor (hybrids in different heterotic backgrounds).

3.2 Haploid Induction

1. Estimate the number of haploid seeds needed, which also depends on the HIR of the haploid inducer used (*see* Table 1). In general, one inducer plant can be used to pollinate five donor plants on an average and one donor plant can produce at least 200 kernels under optimal conditions.
2. Plant donor parents and haploid inducers in the field. Silking time of donor parents need to match the tasseling time of haploid inducers. Thus, 2–3 staggered inducer plantings (1 week apart) are recommended.
3. Donor plants in the field should be well-isolated to avoid pollen contamination. In addition, detassel the donor plants before pollen shedding and cover the shoots (a.k.a shoot bagging) before silk emergence to avoid pollen contamination.
4. Cover tassels of haploid inducers with bags (a.k.a. tassel bagging) 1 day ahead of cross-pollination. The next day, pollen is collected by shaking the tassel while still covered in tassel bag. Remove the bag and dust collected pollen onto donor plants with at least 8 cm silk length, as these would be appropriate targets for haploid induction (*see* **Note 11**).

5. Alternatively, induction crosses can be done in an induction isolation block, in which haploid inducers and donor plants are grown side-by-side. Inducer plants are planted in several sets, to ensure pollen availability during the whole silking time of the donor plants. Great care must be taken to detassel the donor plants before pollen shedding to avoid contamination.
6. Harvest ears approximately 60 days after pollination. The seed should be fully mature (stage R6) for the seed color to be expressed by *R1-nj* and other color marker genes. Harvest time also depends on the donor and environment.
7. Dry ears in a dryer to 12% moisture or below. Several aborted kernels can be seen in a typical ear resulting from an induction cross (Fig. 3b).

3.3 Haploid Selection

Kernels obtained from a haploid induction cross typically consists of haploids, hybrids, deformed, and aborted kernels (Fig. 3). There are different methods available to identify haploids depending on the haploid inducer line used.

3.3.1 Haploid Selection Using R1-nj and C1 Seed Color Marker

After harvest and drying, haploid selection can be done either from the un-shelled cob or the shelled seeds.

1. All kernels on a maize cob are facing with their embryo side up to the tip of the cob. When selecting haploids hold the cob such that the tip faces you and remove seed one by one while making the decision. This is usually faster than selecting in shelled seeds, where each kernel must be turned embryo side up (Figs. 3 and 4).
2. In order to determine HIR, the number of hybrid (HYB) seeds and haploid seeds (HAP) is recorded and the HIR is calculated as follows.

$$\mathrm{HIR} = \mathrm{HAP}/(\mathrm{HYB} + \mathrm{HAP}) * 100$$

3. In test inductions made during haploid inducer development, the number of deformed and aborted kernels (Fig. 3) should also be recorded, to avoid selecting genotypes with a high percentage of this defect.

3.3.2 Haploid Selection Using Purple Plant 1 (Pl1) Root Color Marker

1. Stack two poly plug trays together such that one serves as bottom and the other as top (Fig. 5a, b). Thus, they form a stable tray, and it ensures that roots growing out of the small plugs are not right away exposed to light, which would lead to light-induced root coloring, a trait independent of the presence of *Pl1* in the seedlings (*see* **Note 7**).

Fig. 3 Haploid seed selection by *R1-nj* marker. (**a**) Schematic of haploid selection by *R1-nj* seed color marker. Donors carrying the *CI-I* allele, which represses *R1-nj*, will fail to express the *R1-nj* induced seed color. Haploid seeds are marked with a star. (**b**) The proportion of haploid seeds from a single cob is shown. The majority of seeds are diploid with purple crown endosperm and purple embryo (left). Haploid seeds with colorless embryo on average is 10% (bottom right), depending on the haploid inducer and donor genetic background. Some deformed kernels (middle right) and aborted kernels (top right) are also present

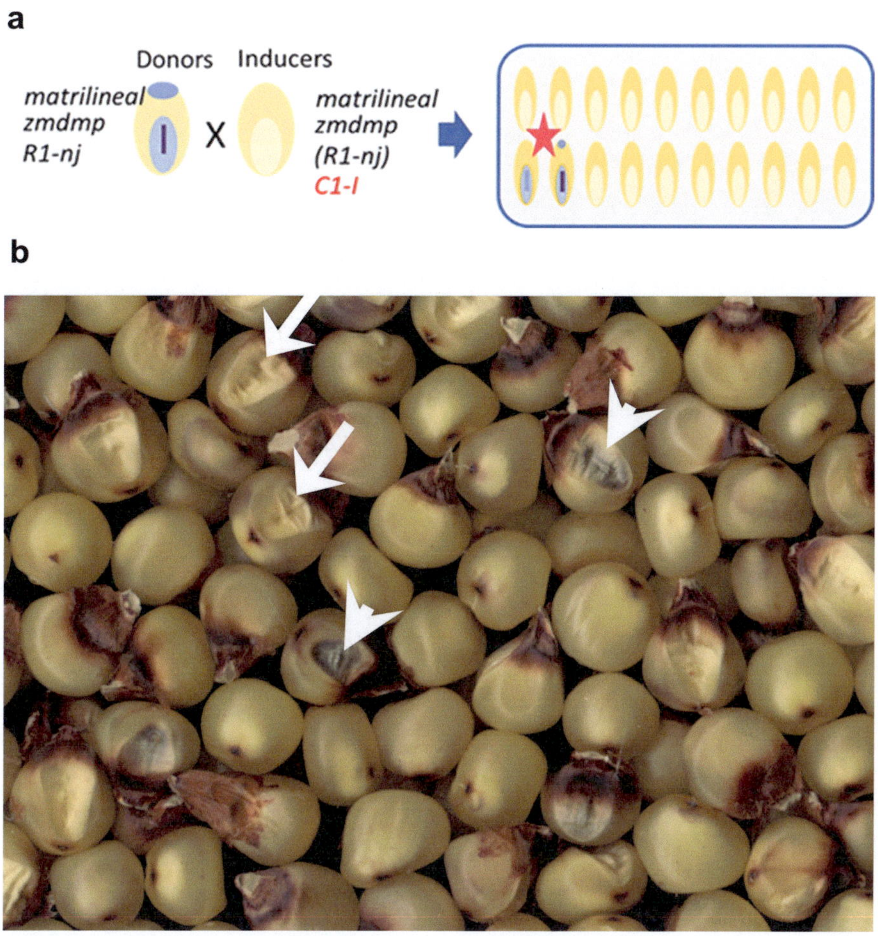

Fig. 4 Haploid seed selection by *C1-I* marker. (**a**) Schematic of haploid selection by *C1-I* seed color and it requires the presence of *R1-nj* marker in donor material. Haploid seeds are marked with a star. (**b**) Seeds resulting from a cross between *R1-nj* carrying donor and *C1-I* inducer are shown. Haploids are marked with an arrowhead and hybrids are marked with an arrow

2. Prefill the plugs by sprinkling soil over the tray—the bottom of the plugs should just be covered with soil, so that the seeds will not be located at the bottom of the plugs. Avoid using the rows/columns at the edges of the tray, where the bottom tray will not cover all seedling plugs.
3. Cover the seeds with soil and evenly press the soil down with a stack of empty trays—brush away excess soil and transfer to the greenhouse. Keep moist during germination. Warm, sunny days may require watering several times a day.
4. Haploid and hybrid seeds tend to germinate at different speed. Wait until most of the seeds germinated and developed 1–2 leaves before performing root color selection (Fig. 5a). This

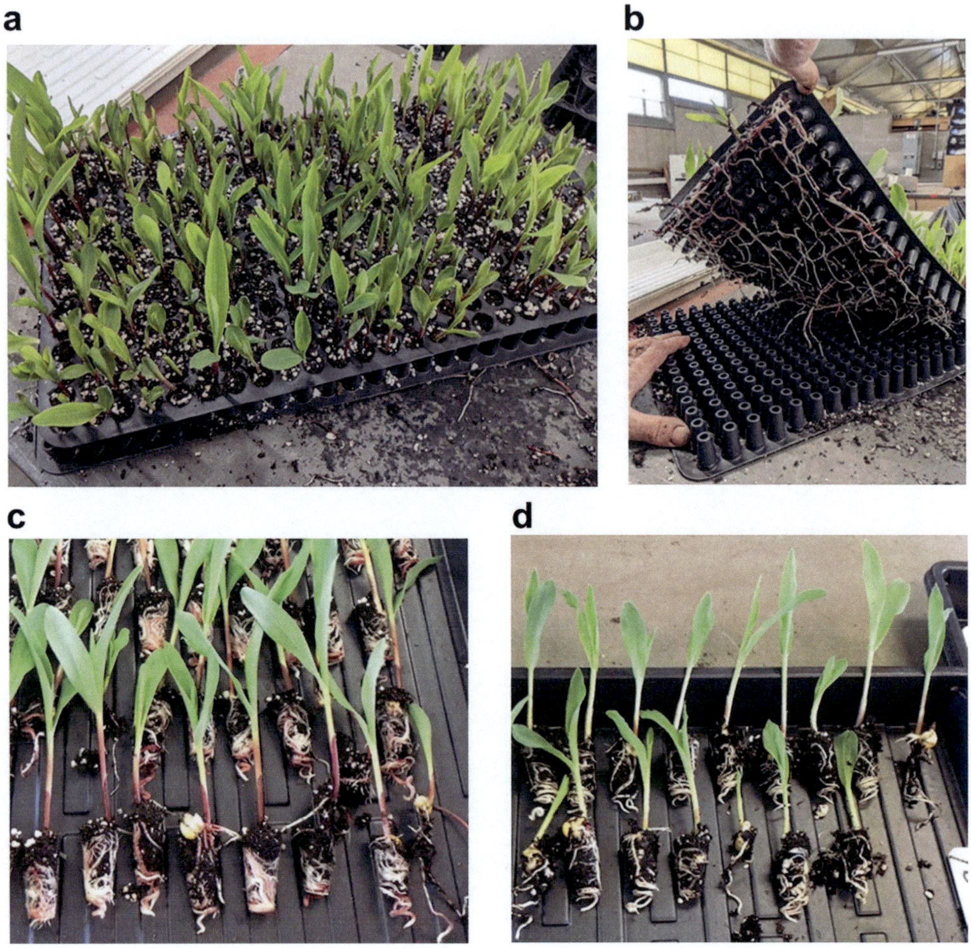

Fig. 5 Haploid selection using *Pl1* marker. (**a**) Germinated PHKs in #400 poly plug tray—note the uneven germination due to different germination speed of hybrid and haploid seeds. (**b**) Separating the two #400 poly plug trays before uprooting the seedlings. (**c**) Hybrid seedlings with clearly purple-colored roots. (**d**) Haploid seedlings with clearly white roots

usually takes 7–8 days from planting. When performing selection, the trays should be moist but not too wet, avoid watering immediately before the selection process.

5. Separate the two trays from each other and start to remove the seedlings with their roots from the tray by gently squeezing from below (Fig. 5b). The roots are usually curled up, which facilitates later repotting. Separate seedlings with red root color from white colored roots. To determine the percentage of seedlings from each class, lay them out in flat trays with 10 per row, to easily determine the total number of seedlings in each class (Fig. 5c, d).

6. After selection, haploid seedlings with white roots are repotted into peat pots (used for transplanting to the field) and grown for another 2 to 3 days in the greenhouse followed by colchicine application (*see* Subheading 3.4).

3.3.3 Haploid Selection Based on Embryo Oil Content Using QSorter

The embryo oil content of haploid seed (obtained from high oil inducer) is expected to be less than the diploid/hybrid counterpart's oil content [18]. For nondestructive prediction of the oil content, near infrared reflectance (NIR) spectrum has proven to be an economical alternative to nuclear magnetic resonance (NMR) imaging. However, prediction based on NIR is highly sensitive to seed position/orientation, which hinders its deployment to the widely used high-throughput seed sorters equipped with NIR spectrometer and RGB cameras. Here we provided a protocol to overcome this challenge using QSorter, a commercial high-throughput seed sorter and developing models for predicting oil content using both NIR spectrum and relative embryo size evaluated from RGB images.

QSorter produces one NIR spectrum (top view) and one composite RGB image consisting of top and two side views for every seed. Our protocol involves developing a deep learning-based classification model to identify seed orientation using the RGB images. Using the top view section of the RGB image and NIR spectrum, this protocol suggests developing models to sort seed based on oil content.

Select Seed Materials and Measure Oil Content

A prerequisite for using QSorter (or any other NIRS based device) for haploid seed sorting based on oil content is calibration of the device at hand. For calibration, a panel of seeds covering a wide range of oil content must be established. The total oil content of each seed of the panel must be determined in an established system, such as NMR spectroscopy, which has established nondestructive protocols. Seeds are arranged in 48-well plates, to be able to trace each individual kernel during the process.

1. Determine weight and oil content of the seeds (known or expected to cover a wide range of oil content, *see* Subheading 2.3.3) using Bruker Minispec mq20 NMR analyzer according to manufacturer's protocol (Fig. 6a).

NIR Spectrum and RGB Image Capturing Using QSorter

2. Select single kernel mode in QSorter.
3. Pass the seeds (from **step 1**) individually through QSorter. It will capture 512 bands NIR spectra within 900–1700 nm from top view (Fig. 6b, c). RGB images of the seeds from three views (i.e., one top and two side views) are also captured (Fig. 7).

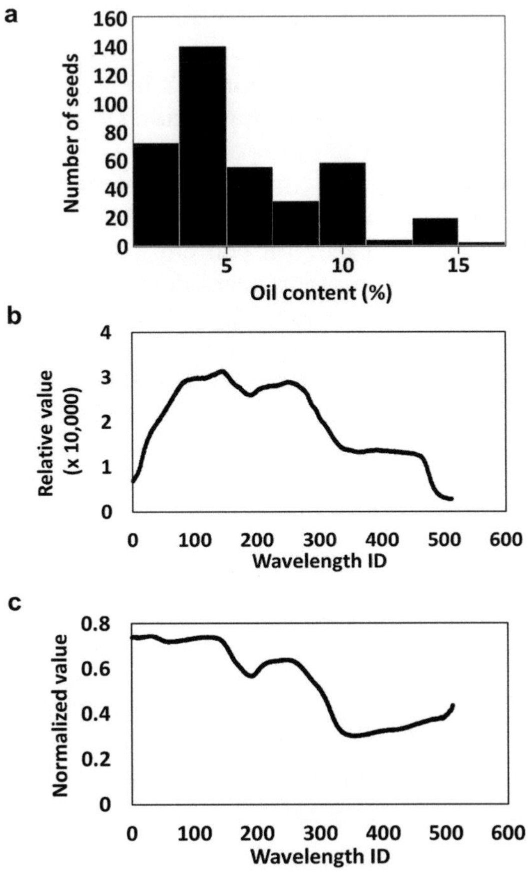

Fig. 6 Oil content of seeds obtained from NMR and NIRS. (**a**) Distribution of the seeds based on oil content measured by NMR (**b**) A typical NIR spectrum of a seed produced by the QSorter is shown (**c**) Normalized value of the NIR spectrum in (**b**)

4. Repeat the process for each seed 5 to 6 times to get a top view with correct orientation, i.e., embryo/scutellum fully visible on the top view (Fig. 7).
5. Save the collected data to a USB memory stick and transfer to a local computer.

Select Training Data and Train **Classification Model** in MATLAB to Identify Correct Seed Orientation

6. Inspect the RGB images and pick equal number of images (between 5–8% of the total number of acquired images) with correct and incorrect orientations (Fig. 7a, b) as training data.
7. Train a classification model to identify correct/incorrect orientations using MATLAB deep learning toolbox as in https://www.mathworks.com/help/deeplearning/ug/create-simple-deep-learning-network-for-classification.html

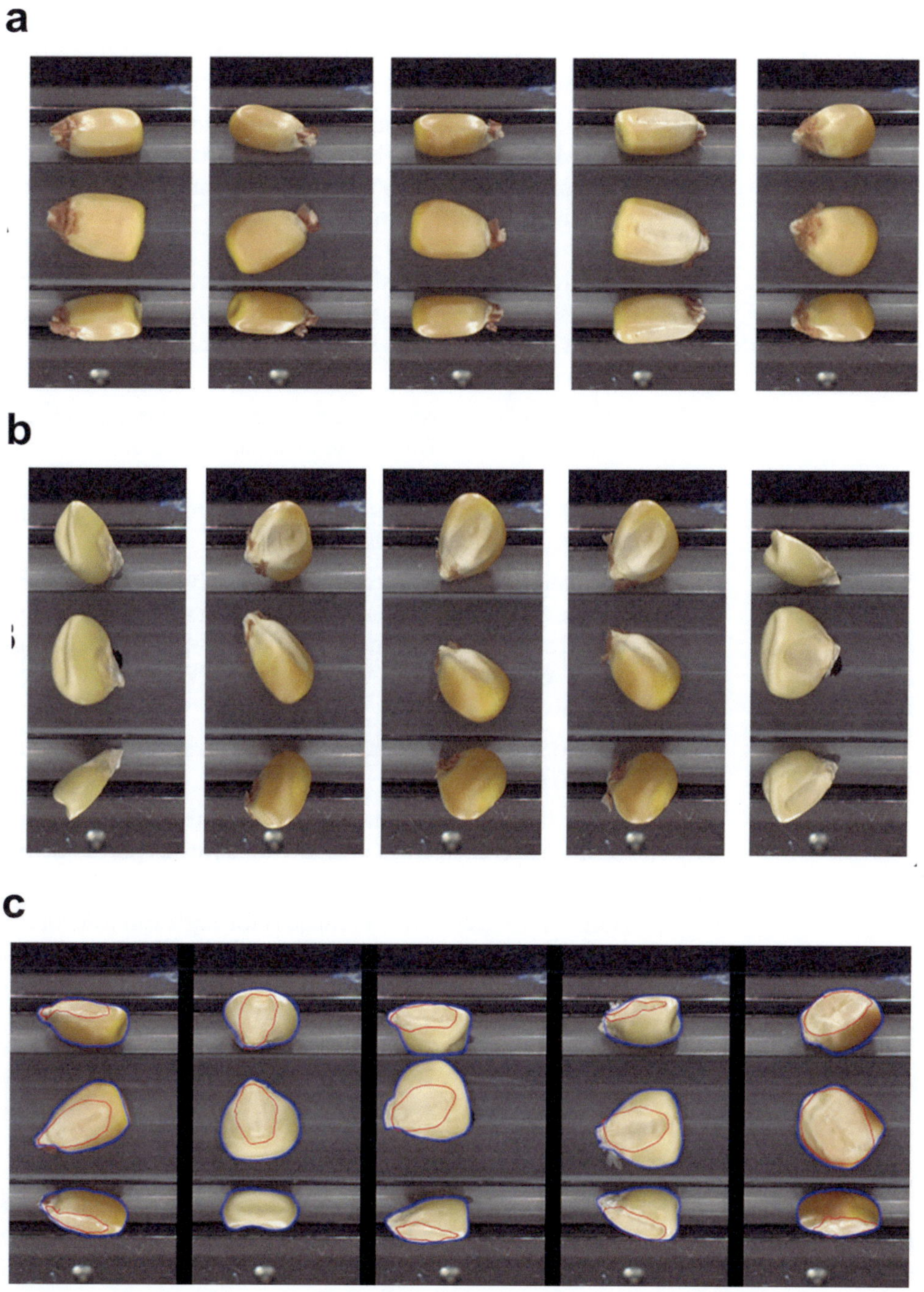

Fig. 7 Haploid selection using QSorter. Examples of (**a**) correctly and (**b**) incorrectly oriented seeds. (**c**) Examples of annotated images with traced outlines of embryo (red) and seed (blue). In all panels shown, the top and bottom rows show the two side views of the seeds and the middle row shows the top view of the seeds

Annotate Training Data and Train **Segmentation Model** in MATLAB to Identify Embryo

8. Trace outlines of embryo/scutellum and seed on the correctly oriented images for the training data (from **step 6**) as in Fig. 7c using the "Edit and Create" tool in Microsoft Photos.
9. Train a semantic segmentation model to identify the outlines using MATLAB deep learning toolbox as in https://www.mathworks.com/help/vision/ug/semantic-segmentation-using-deep-learning.html

Determination of Seed Orientation and Relative Embryo Size on the Rest of the Data (~92–95%)

10. Use the trained classification and semantic segmentation models (from **steps 7** and **9**) to identify "correctly" oriented images and the outlines of embryos and seeds on the rest of the images.
11. Evaluate relative area of the embryo with respect to the whole seed using MATLAB image processing toolbox.

Preprocessing NIR Data and Develop **Binary Classification Models** to Predict Oil Content

12. Collect QSorter generated NIR spectrum (from **step 3**) of the correctly oriented seeds.
13. Normalize the spectrum by scaling between 0 to 1 using the spectra values for dark and white objects captured by QSorter (QSorter suffixes file names containing spectra of the dark and white objects with "Dark" and "White," respectively) (Fig. 6b, c).
14. Discard the border (first and last five) bands from the normalized NIR data.
15. Assign the NIR and relative area of embryo data into groups according to their high and low oil content (as determined in **step 1**).
16. Train binary classification models for high and low oil content using classification learner app in MATLAB and pick the model with the best accuracy.

Model Deployment on QSorter

17. All the trained models (in **steps 7**, **9** and **16**) could be uploaded to QSorter and it can be run in batch mode for high-throughput sorting of haploid seeds based on embryo oil content.

3.4 Artificial Genome Doubling of Haploids Using Colchicine

1. Prefill peat pots with soil mix and bury haploid seeds (~ 2.5 cm) into the soil mix. Alternatively, transfer haploid seedlings identified from root color analysis (*see* Subheading 3.3.2) into peat pots and cover with soil.
2. Transfer to greenhouse and wait until the 2–3 leaf stage. This stage is the appropriate time for colchicine application. It is essential to make sure the seedlings are well-watered before treatment since watering should be avoided for 3–4 h after colchicine injection.

3. On the day of colchicine treatment, gently brush over the seedlings with arms to remove excess water from the leaf whorl.
4. Approximately 100 μL of 0.125% colchicine solution is injected per plant using surgical syringes (*see* **Notes 9, 12** and **13**).
5. After the treatment, seedlings are adapted in the greenhouse for 1 to 2 days and exposed to direct sunlight for 4 to 8 h before transplanting in the field.

3.5 Identifying Genotypes with SHGD Ability

1. Plant haploid seeds by either sown directly into the field or transplanted as seedlings (*see* Subheadings 3.3.1 and 3.3.2 and **Note 14**).
2. Regardless of the haploid selection method utilized, after a few weeks of growth, or approximately at V6 vegetative stage, false positives (diploids) can be discarded by using a field hoe or tree lopper (*see* **Note 15**).
3. Once anthesis begins, scoring for HMF can begin. Each haploid tassel must be checked independently and daily for male fertility traits, i.e., fertile anther emergence and pollen shedding (*see* **Note 16**).
4. Pollen quantity scores are recorded based on the presence of viable pollen and on a numeric scale of (1–5) for each trait (Fig. 8a). A score of "1" being the lowest and a score of "5" being the highest (*see* **Note 17**).
5. Anther quantity scores are recorded on the presence of viable anthers and on a numeric scale of (1–5) (Fig. 8b), with a score of "1" being the lowest and a score of "5" being the highest for fertile anthers. Sterile plants are scored "0." It is not abnormal for pollen/anther quantity scores for an individual tassel to change from day to day. It is thus necessary to score each tassel daily, with the highest score for each trait being the reported value for a given haploid plant.
6. Finally, it is possible to calculate the percentage of HMF for each genotype once field data have been collected. For a given genotype, count the number of haploids that were recorded as being male fertile and divide that number by the total number of haploids for that given genotype. When utilizing this equation, the HMF percentage reported for a given genotype should be expressed as the average of all replications across all environments. The abovementioned is expressed as,

$$\mathrm{HMF}_i = (\text{Number of Male Fertile Haploid Individuals}) / (\text{Number of Total Haploid Individuals}) * 100$$

where HMF_i is the calculated average for HMF percentage across replications and environments for the *i*th genotype.

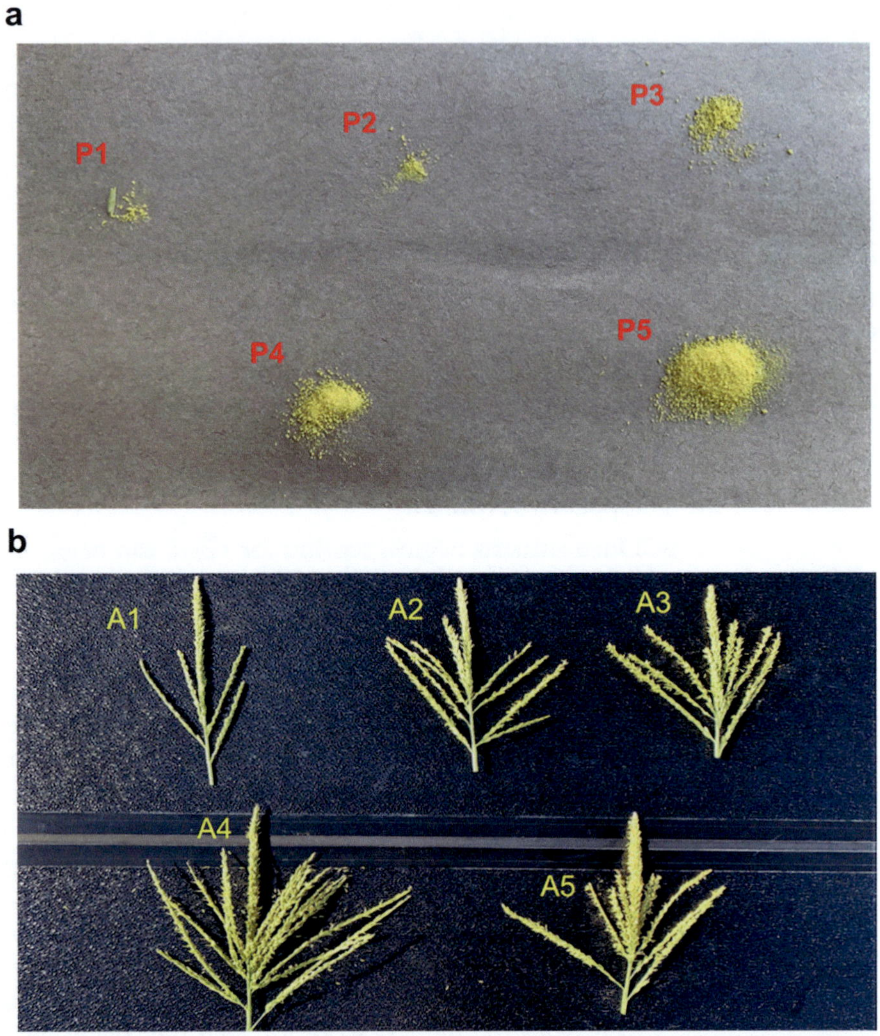

Fig. 8 Scoring for HMF traits. Pollen and anther quantity scores from various maize haploid genotypes. (**a**) Pollen scores recorded based on pollen presence on a numeric scale of 1 to 5. (**b**) Anther scores strictly recorded based on fertile anther presence on a numeric scale of 1 to 5. A score of "1" being the lowest and a score of "5" being the highest for fertile anthers. Sterile plants are scored "0"

4 Notes

1. Publicly available haploid inducer accessions are usually of poor agronomic performance and therefore, the HIR must be introgressed into more valuable and better-adapted genetic backgrounds. This can be accomplished through marker-assisted breeding schemes, such as marker-assisted backcrossing into the desired background followed by phenotypic selection for HIR to identify those few with an HIR of 10% or more.

2. The HIR trait is highly quantitative, but one major locus is already sufficient to explain more than 60% of the variability found in different studies [19], and a second locus enhances HIR by up to another 20%. The underlying genes for both loci have been cloned in recent years [6–9] and thus allow the development of molecular marker to trace them. *ZmMTL* located on chromosome 1 is needed for HIR. In this study, [7] showed that a 4 bp insertion toward the end of the translated sequence leads to an early stop codon and thus a truncated protein in haploid inducer lines. A single SNP in *ZmDMP* gene was found to enhance the HIR of *ZmMTL* by a third [9]. *ZmDMP* is located on the short arm of chromosome 9, thus segregating independently from *ZmMTL*.
3. Haploid inducer lines should have dominant color markers for haploid seed selection. The most typical phenotypic marker in haploid inducer lines is purple color observed in the embryo and endosperm, which is controlled by *R1-nj* transcription factor [20, 21]. It also requires that the female donors—for which haploids are to be obtained—normally have colorless embryo and endosperm in *c1* genetic background [22]. Thus, it becomes possible to select for colorless embryo and purple color endosperm of haploid seeds by in vivo conventional DH technology in maize.
4. The conventional haploid seed selection is based on the color on embryo. In diploid kernels, the red/purple color of anthocyanin, which is controlled by *R1-nj*, is exhibited in the aleurone (outermost layer of endosperm, 3n) and the scutellum (embryo, 2n) (Fig. 2). Because of single fertilization or chromosome elimination in the embryogenesis [23] the haploid kernels have a red/purple color of endosperm (3n), but the haploid embryo (1n), possessing only the donor genotype, is colorless. Note that if the donor parents contain dominant anthocyanin inhibitor genes, such as *C1-I*, *C2-Idf*, or *In1-D* [24], it would not be able to identify the haploid kernels by *R1-nj* characteristics [24].
5. We developed haploid inducing lines at Iowa State University that are fixed for the allele *C1-I* at the gene locus *C1*. C1 is a transcription factor, which interferes in the anthocyanin pathway upstream of *R1-nj* and leads to suppression of color in aleurone and embryo [24, 25] (Fig. 2). This method requires the presence of selectable marker *R1-nj* in the donor material.
6. *Pl1* is an alternative marker for haploid selection in genetic donor backgrounds, that have poor or completely suppressed expression of the kernel color marker *R1-nj*. *Pl1* encodes a transcription factor in the anthocyanin pathway and affects the coloring of various plant parts independent of light

exposure. It means that the typical purple color appears in organs such as coleoptile, roots, anthers, or pericarp even when the respective organ is not exposed to light.

7. A small batch of 1 g colchicine powder in general could be used to prepare an 800 mL standard solution for injection, which allows to treat 7 to 8 thousand haploid individual plants. Colchicine treatment by injection is relatively efficient and suitable for small, medium, and large scale (albeit laborious) in in vivo DH technology in maize. Colchicine solution can be stored at 4 °C refrigerator for 2 to 3 weeks, but it must be kept in a dark bottle as colchicine is susceptible to light.
8. Colchicine is toxic, and care should be taken when working with it. Addition of vegetable dye to colchicine solution does not only aid with the injection into seedlings but also aid the researcher in visualizing spills as a safety measure, to help avoid accidental contact with any spilled solution.
9. Genome-wide association study on HIR of Iowa State haploid inducers panel, in which some lines have HIR of >20%, reveals that there are other minor QTLs contributing to this trait [10]. In addition to breeding approaches based on MAS for the two major genes involved in HIR, cost and time intensive phenotypic evaluation of HIR in test inductions led over time to an improvement in the range of 12–15%. However, the breeding value predicted by genome-wide SNP markers could improve the genetic gain of the target population without phenotyping, once the training population is developed, which is called genomic selection [26, 27]. The inducer lines are all derived from Stock 6 [28, 29] (Fig. 1). The inducer lines in the training population and the superior individuals selected for next breeding cycle need to consider reduction of genetic variation for the long-term selection gains [30].
10. Four to five generations of inbreeding yield reasonable stable inducer lines. If using a backcrossing approach, it is recommended to go no further than BC2, under phenotypic selection for induction ability, as additional not yet identified factors contributing to HIR, might get lost.
11. In a large-scale induction nursery, tassel bagging can be omitted. Large amounts of pollen are collected openly from pollen shedding inducer plants, by tapping the tassels into a pollination bag. Donor plants showing good silking are prepared for pollination by cutting back the silk to about 1 cm, so they form a nice brush. The collected inducer pollen is dusted onto the silk and the ear covered with a dated tassel bag. This procedure can be performed very efficiently by two persons, one collecting and distributing the pollen, the other preparing the silks and covering pollinated ears.

12. The needle of the syringe is inserted a few millimeters above the basal meristem (but not in the meristem) of the seedlings, so the solution can contact the actively dividing meristematic cells, from which reproductive tissues will be differentiated [31]. It is vital that the progenitor cells of reproductive tissues have successfully doubled their genome. If the solution application was successful, the dye color can be seen in the whorl of leaf sheath.
13. Some research show that soaking haploid seedlings in colchicine solution result in a higher percentage of fertile haploids as compared to the described method of injection [32]. However, those methods have to deal with additional large volume of hazardous colchicine waste and need additional wash steps, which increase the chance to contact this toxic solution. There are some alternative chemicals which are less toxic such as Pronamid, APM, trifluralin, and oryzalin, but their abilities to inhibit the mitosis are less than colchicine [33].
14. In our experiments, we utilize a minimum of 40 seeds, 2 replications of 20 seeds, to determine the rate of HMF with the assumption that there will be false positives. In addition, there could be a few haploid seeds (directly sown) that do not emerge from the soil or a few haploid seeds that do not survive the process of transplanting.
15. There are many traits that can help to differentiate diploids and haploids in the field. For example, diploid plants tend to have wide leaf blades with a glossy sheen whereas haploid plants tend to have much more narrow and erect leaf blades, which can occasionally display white stripes [34]. Additionally, diploids often display purple pigmentation in the stalk, where this coloration is absent in haploids. Finally, diploid plants are significantly taller and are overall much more vigorous than their haploid counterparts. It is very important to rogue the diploid false positives as they can suppress the growth of haploid plants through resource competition.
16. Each day, before entering the field, images of scores for each of the traits should be reviewed to refresh one's mind of what constitutes a specific score for each trait. Once previous images have been analyzed, checking for the emergence of anthers and the ability of those anthers to shed pollen can be recorded. However, HMF will not be seen for each haploid plant. If, for instance, anthers emerge from the tassel, but no pollen is shed, then the haploid male is infertile. This implies that if there is no score recorded for pollen quantity, there is no need to record a score for the quantity of viable anthers—due to the anthers lacking pollen and the haploid being infertile.

17. A pollen score of "1" indicates that the tassel produces pollen, but the anthers must be cut to collect the pollen. In this case, the amount of pollen is not as important as how the pollen is collected. Although HMF is desired, having plants that shed pollen from anthers without having to cut anthers to obtain the pollen is preferred. The quantity of pollen is obtained by lightly tapping the tassel and collecting the shed pollen with a tassel bag.

Acknowledgments

This work was supported by the Foundation for Food & Agriculture Research under award number CA19-SS-0000000128. The content of this book chapter is solely the responsibility of the authors and does not necessarily represent the official views of the Foundation for Food & Agriculture Research.

References

1. Tuvesson S, Dayteg C, Hagberg P et al (2007) Molecular markers and doubled haploids in European plant breeding programmes. Euphytica 158(3):305–312
2. Geiger HH (2009) Doubled haploids. In: Bennetzen JL, Hake SC (eds) Maize handbook: genetics and genomics, vol 2. Springer-Verlag, New York, NY, pp 641–657
3. Kelliher T, Starr D, Su X et al (2019) One-stepgenome editing of elite crop germplasm during haploid induction. Nat Biotechnol 37(3):287–292
4. Chase SS (1947) Techniques for isolating monoploid maize plants. Am J Bot 34(10):582
5. Coe EH (1959) A line of maize with high haploid frequency. Am Nat 93(873):381–382
6. Gilles LM, Khaled A, Laffaire JB et al (2017) Loss of pollen-specific phospholipase *NOT LIKE DAD* triggers gynogenesis in maize. EMBO J 36(1460-2075 (Electronic)):707–717
7. Kelliher T, Starr D, Richbourg L et al (2017) *MATRILINEAL*, a spermspecific phospholipase, triggers maize haploid induction. Nature 542(7639):105–109
8. Liu C, Li X, Meng D et al (2017) A 4-bp insertion at *ZmPLA1* encoding a putative phospholipase A generates haploid induction in maize. Mol Plant 10(3):520–522
9. Zhong Y, Liu C, Qi X et al (2019) Mutation of*ZmDMP* enhances haploid induction in maize. Nat Plants 5(6):575–580
10. Trentin HU, Frei UK, Lübberstedt T (2020) Breeding maize maternal haploid inducers. Plants (Basel) 9(5):614
11. Boerman NA, Frei UK, Lübberstedt T (2020) Impact of spontaneous haploid genome doubling in maize breeding. Plants (Basel) 9(3): 369
12. Ren J, Boerman N, Liu R et al (2020) Mapping of QTL and identification of candidate genes conferring spontaneoushaploid genome doubling in maize (*Zea mays* L.). Plant Sci 293: 110337–110337
13. Trampe B, Dos Santos IG, Frei UK et al (2020) QTL mapping ofspontaneous haploid genome doubling using genotyping-by-sequencing in maize (*Zea mays* L.). Theor Appl Genet 133(7):2131–2140
14. Verzegnazzi AL, Dos Santos IG, Krause MD et al (2021) Major locus for spontaneous haploid genome doubling detected by a case-control GWAS in exotic maize germplasm. Theor Appl Genet 134:1423–1434
15. Chaikam V, Nair SK, Martinez L et al (2018) Marker-assisted breeding of improved maternal haploid inducers in maize for thetropical/subtropical regions. Front Plant Sci 9:1527
16. Bloom JC, Holland JB (2011) Genomic localization of the maize cross-incompatibility gene, gametophyte factor 1 (ga1). Maydica 56(4): 379–387
17. Wang M, Chen Z, Zhang H et al (2018) Transcriptome analysis providesinsight into the

molecular mechanisms underlying gametophyte factor 2-mediated crossincompatibility in maize. Int J Mol Sci 19(6):1757

18. Melchinger AE, Schipprack W, Würschum T et al (2013) Rapid and accurateidentification of *in vivo*-induced haploid seeds based on oil content in maize. Sci Rep 3(1):2129–2129
19. Prigge V, Schipprack W, Mahuku G et al (2012) Development of in vivohaploid inducers for tropical maize breeding programs. Euphytica 185(3):481–490
20. Chase SS, Nanda DK (1965) Screening for monoploids of maize by use of a purple embryo marker. Maize Genet Coop Newsl 39:59–60
21. Nanda DK, Chase SS (1966) An embryo marker for detecting monoploids of maize (*Zea mays* L.). Crop Sci 6(2):213–215
22. Greenblatt IM, Bock M (1967) A commercially desirable procedure for detection of monoploids in maize. J Hered 58(1):9–13
23. Zhao X, Xu X, Xie H et al (2013) Fertilization and uniparental chromosome elimination during crosses with maize haploid inducers. Plant Physiol 163(2):721731
24. Coe EH (1994) Anthocyanin genetics. In: Freeling M, Walbot V (eds) The maize handbook. Springer-Verlag, New York, pp 279–281
25. Ford RH (2000) Inheritance of kernel color in corn: explanations & investigations. Am Biol Teach 62(3):181–188
26. Meuwissen THE, Hayes BJ, Goddard ME (2001) Prediction of total genetic value using genome wide dense marker maps. Genetics 157(4):1819–1829
27. Bernardo R, Yu J (2007) Prospects for genomewide selection for quantitative traits in maize. Crop Sci 47(3):1082–1090
28. Liu Z, Wang Y, Ren J et al (2016) Maize doubled haploids. In: Janick J (ed) Plant breeding reviews, vol 40. Wiley-Blackwell, Hoboken, NJ, pp 123–160
29. Maqbool MA, Beshir A, Khokhar ES (2020) Doubled haploids in maize: development, deployment, and challenges. Crop Sci 60(6): 2815–2840
30. Almeida VC, Trentin HU, Frei UK et al (2020) Genomic prediction of maternalhaploid induction rate in maize. Plant Genome 13(1): e20014
31. Vanous K, Vanous A, Frei UK et al (2017) Generation of maize (*Zea mays*) doubled haploids via traditional methods. Curr Protoc Plant Biol 2:147–157
32. Eder J, Chalyk S (2002) *In vivo* haploid induction in maize. Theor Appl Genet 104(4): 703–708
33. Melchinger AE, Molenaar WS, Mirdita V et al (2016) Colchicine alternatives forchromosome doubling in maize haploids for doubled-haploid production. Crop Sci 56(2):559–569
34. Chase SS (1969) Monoploids and monoploid-derivatives of maize (*Zea mays* L.). Bot Rev 35(2):117–167
35. Zavalishina A, Tyrnov V (1984) Inducing high frequency of matroclinal haploids in maize. Dokl Akad Nauk SSSR 276:735–738
36. Hu H, Schrag TA, Peis R et al (2016) The genetic basis of haploid induction in maize identified with anovel genome-wide association method. Genetics 202(4):1267–1276
37. Kermicle JL (1969) Androgenesis conditioned by a mutation in maize. Science 166(3911): 1422–1424
38. Chalyk ST (1999) Creating new haploid-inducing lines of maize. Maize Genet Coop Newsl 73:53–54
39. Shatskaya OA (2010) Haploinductors isolation in maize: three cycles of selection on high frequency of induction of matroclinal haploids. Agric Biol 5:79–86
40. Lashermes P, Beckert M (1988) Genetic control of maternal haploidy in maize (*Zea mays* L.) and selection of haploid inducing lines. Theor Appl Genet 76(3):405–410
41. Rotarenco V, Dicu G, State D et al (2010) New inducers of maternal haploids in maize. Maize Genet Coop Newsl 84:21–22
42. Roeber FK, Gordillo GA, Geiger HH (2005) In vivo haploid induction in maize—performance ofnew inducers and significance of doubled haploid lines in hybrid breeding. Maydica 50(3):275–283
43. Iowa State University, Doubled Haploid Facility, 2015. https://www.biotech.iastate.edu/biotechnology-service-facilities/doubled-haploid-facility/

Chapter 17

Estimation of Heterografting Associated DNA Methylation Changes in Tree Crops by MSAP Analysis

Thomas Kadampanattu Uthup, Rekha Karumamkandathil, and Meenakumari Thuppale

Abstract

The genetic incompatibility of the seedlings which are used as rootstocks (stock–scion interactions) and the mechanical stress induced by grafting are two major factors responsible for the high intraclonal variations observed in tree crops which are propagated through bud grafting. Since stress-induced DNA methylation changes associated with heterografting is a major contributor of such variations in grafted tree crops, a proper assessment of this epigenetic phenomenon is inevitable to devise strategies for the development of more uniform planting materials with minimal intraclonal variations in the future. In order to evaluate and establish the effects of heterografting on the epigenome of plants, availability of ideal plant materials and a standard procedure for testing is very essential. Development of genetically uniform own-rooted seedlings through induction of cleavage polyembryony by a novel technique of *half ovulo* embryo culture is the first step. Grafting of buds from these genetically and epigenetically uniform plants to genetically divergent rootstock and identification of DNA methylation polymorphism among them forms the second part of the methodology for detecting epigenetic changes associated with grafting in tree crops. Methylation-sensitive amplification polymorphism technique (MSAP), a modified version of AFLP using a pair of methylation-sensitive and insensitive isoschizomers (such as HpaII and MspI), is an ideal methodology to assess DNA methylation polymorphisms on a genomic scale in such plants. Comparative analysis of two sets of restriction digestion products (EcoRI/HpaII and EcoRI/MspI) allows the identification of DNA methylation polymorphisms induced by grafting and will aid in the detection of differentially methylated regions (DMRs) among grafted plants. This chapter describes a detailed protocol for inducing multiple embryos of single zygotic origin and regeneration of seedlings in rubber tree (*Hevea brasiliensis*), grafting of buds from these genetically uniform own-rooted seedlings to divergent rootstocks, identification of epigenetic changes induced by grafting or stock–scion interactions through MSAP analysis, and locating the differentially methylated genomic region. The methodology described here could be applied to any tree species commercially propagated through grafting for detecting epigenetic changes putatively associated with intraclonal variability.

Key words DNA methylation polymorphisms, Epigenetic diversity, Bud grafting, *Half ovulo* embryo culture, *Hevea brasiliensis*, MSAP

Christophe Lambing (ed.), *Plant Gametogenesis: Methods and Protocols*, Methods in Molecular Biology, vol. 2484, https://doi.org/10.1007/978-1-0716-2253-7_17,

1 Introduction

Genetic integrity of elite varieties developed through cross pollination is preserved through bud grafting technique in several tree crops like rubber, walnut, sapota, mango, cocoa, cashew and apple. Seedlings are not preferred for the propagation of natural rubber producing tree crop *Hevea brasiliensis* given it is an outcrossing species with a very high degree of genetic heterogeneity. High-yielding and disease-tolerant rubber varieties are developed by hybridization programs and released for commercial planting after several years of extensive field evaluation trials. For preserving the genetic fidelity of these elite varieties, bud grafting is considered as the most preferred and commercially viable propagation technique. Even though the genetic composition of the scion remains the same in this vegetative mode of propagation, tree-to-tree variations in growth and yield are commonly observed in *Hevea* plantations. The responses induced by the mechanical stress of grafting procedure and the incompatibility of genetically divergent rootstocks and consequent changes in the physiology, gene expression, and protein function associated with it, widely known as stock–scion interactions, are primarily responsible for this phenomenon. The rootstock influences the growth as well as other properties of the scion using a series of intermediary means like signaling molecules, plant hormones and epigenetic factors like DNA methylation and small RNAs. Rootstock-induced epigenetic variation as indicated by variation in the methylation pattern was identified in *Hevea* recently [1]. In plants, DNA methylation is a major epigenetic modification that is important for genome stability and gene regulation. The aromatic ring in normal cytosine of the DNA molecule is modified by the covalent addition of a methyl group to the fifth position. The addition and removal of DNA methylation can be induced by both internal and external stimuli. DNA methylation and demethylation are used by plants to rapidly adapt to environmental changes or to counteract biotic and abiotic stresses. DNA methylation occurs commonly within three sequence contexts: CG, CHG, and CHH (where H is A, C, or T) in plants. Different molecular techniques have been developed to analyze DNA methylation either at the global level of methylated cytosines or at specific gene sequences. Methylation-sensitive amplification polymorphism (MSAP), a modification of the amplified fragment length polymorphism (AFLP), is one of the most widely used techniques to determine DNA methylation modulation at genome level. The AFLP technique was initially modified by Reyna-López et al. (1997) [2] to analyze fungal DNA methylation and later adapted to study genome methylation in different plant species [3, 4]. Currently, this modified AFLP protocol is widely used to analyze DNA methylation in several plant genomes of varying genome sizes. This technique is based on the use of isoschizomeric restriction endonucleases with different methylation sensitivities. Tetra cutter restriction enzymes such as HpaII and MspI are the frequently used isoschizomers detecting anonymous CCGG sites, in which cytosines are differentially methylated. Both restriction enzymes recognize the sequence 5′-CCGG. However, HpaII will not digest 5′-CCGG if one or both cytosines are methylated (both strands methylated), whereas MspI digests 5′-CCGG sequences even if the internal cytosine is methylated. Unlike normal AFLP technique, each sample is digested using both EcoRI/HpaII and EcoRI/MspI separately. A comparative analysis between EcoRI/HpaII and EcoRI/MspI fragments reveals genetic variability in terms of methylation insensitive polymorphisms and epigenetic variability in terms of methylation-sensitive polymorphisms (Fig. 1). This technique can be easily employed to assess DNA methylation polymorphisms associated with the grafting of genetically uniform scion of desired genotype onto genetically divergent rootstock in any vegetatively propagated crop plant. Genetically uniform own-rooted seedlings and their bud-grafted counterparts forms an ideal material for studying stock-induced epigenetic variation because they are identical except for the rootstock. For a highly heterozygous outcrossing tree species like *Hevea*, generating identical seedlings through conventional methods is nearly impossible. In this context, a novel strategy like the induction of genetically uniform seedlings through the cloning of zygotic embryos has much breeding significance. This technique assumes more relevance especially in perennial tree crops with long breeding cycles, as selection of plants generated after controlled breeding via hybridization of selected parents is based on single plant value. Cleavage polyembryony is a mechanism which can be induced in culture and uniform plantlets of single zygotic origin can be generated easily. This is achieved through a novel *half ovulo* embryo culture technique [5]. These uniform plants serve as the absolute control for all experiments using bud-grafted plants. By comparing the uniform seedlings with their bud-grafted counterparts, the changes induced by bud grafting can be specifically determined. For understanding the epigenetic variation among bud-grafted plants, these uniform plants of single zygotic origin grafted onto divergent rootstocks is the most ideal material. Since the polymorphic bands generated by MSAP arise randomly from any portion of the genome, direct estimation of the site of DNA methylation polymorphism is not possible through this analysis. However, it is possible to detect the genomic region harboring the DMRs and even locate the differentially methylated cytosine by follow-up techniques. The methylation-sensitive bands from the polyacrylamide gel can be eluted, reamplified, cloned, and sequenced to identify the genomic region where the change has happened. In order to determine the differentially methylated cytosine with single base pair resolution,

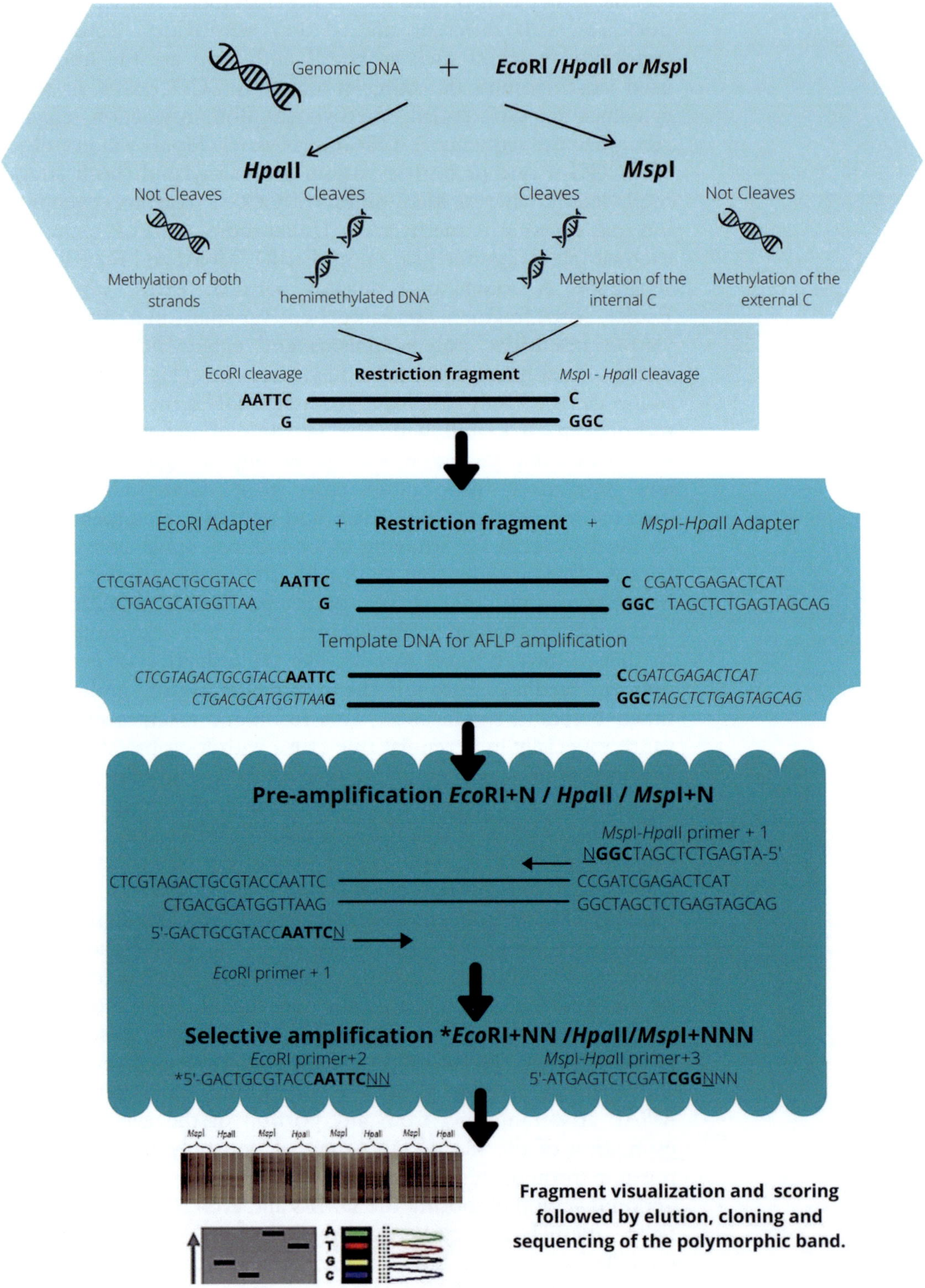

Fig. 1 Schematic representation of the MSAP technique

bisulfite sequencing of the genomic region can be performed using the original genomic DNA. Thus, the techniques described here lay the background for locating DNA methylation changes at single-base resolution and for follow-up studies on their functional relevance on gene expression and regulation.

The methodology described is broadly divided into four sections: (1) development of genetically uniform materials by polyembryony, (2) grafting of buds from these plants to genetically divergent rootstock, (3) MSAP analysis of the heterografted plants generated for locating DNA methylation polymorphisms, and finally (4) elution, cloning, and sequence analysis of the genomic region harboring the DMR.

2 Materials

2.1 Development of Genetically Uniform Seedlings Through Induction of Zygotic Polyembryony

1. *Hevea* fruits, 8–10 weeks old.
2. 80% alcohol.
3. Sterile blade and blade holder.
4. Sterile, glass petri plates.
5. Tissue culture medium: 720 mg/L NH_4NO_3, 950 mg/L KNO_3, 185 mg/L $MgSO_4$, 68 mg/L KH_2PO_4, 25 mg/L $MnSO_4$ $4H_2O$, 10 mg/L $ZnSO_4$ $7H_2O$, 0.25 mg/L Na_2MoO_4 $2H_2O$, 0.025 mg/L $CuSO_4$ $5H_2O$, 0.025 mg/L $CoCl_2$ $6H_2O$, 27.8 mg/L ferric sodium EDTA, 100 mg/L myoinositol, 30 g/L sucrose, 2 g/L Phytagel, 0.5 mg/L thiamine, 0.5 mg/L pyridoxine, 5 mg/L nicotinic acid, 2 mg/L glycine, 0.5 mg/L folic acid, 0.05 mg/L biotin, 3 mg/L benzyl adenine, 3 mg/L kinetin.
6. Germination medium: 500 mg/L NH_4NO3, 950 mg/L KNO_3, 220 mg/L $CaCl_2$, 185 mg/L $MgSO_4$, 85 mg/L KH_2PO_4, 3.2 mg/L H_3BO_3, 11.3 mg/L $MnSO_4$ $6H_2O$, 4.3 mg/L $ZnSO_4$ $7H_2O$, 0.083 mg/L KI, 0.25 mg/L $Na2MoO_4$ $2H_2O$, 0.025 mg/L $CuSO_4$ $5H_2O$, 0.025 mg/L $CoCl_2$ $6H_2O$, 37.3 mg/L ferric sodium EDTA, 100 mg/L myoinositol, 50 mg/L adenine sulfate, 30 g/L sucrose, 2 g/L Phytagel, 2 g/L activated charcoal, 0.5 mg/L thiamine, 0.5 mg/L pyridoxine, 5 mg/L nicotinic acid, 2 mg/L glycine, 0.5 mg/L folic acid, 0.05 mg/L biotin, 100 mg/L casein hydrolysate, 20% coconut water, 200 mg/L banana powder, 100 mg/L malt extract, 0.3 mg/L gibberellic acid, 0.3 mg/L benzyl adenine.
7. 0.2% activated charcoal.
8. Sand, soil, and Soilrite mixture.

2.2 Grafting

1. Clean sharp sickle/pruner for cutting budwood of stock plants.
2. Rootstock plant.
3. Scion bud patch from desired plant.
4. Budding knife.
5. Polythene tape for sealing the bud union.

2.3 MSAP Analysis for the Detection of DNA Methylation Polymorphisms in Heterografted Plants

1. 1.5 mL Eppendorf tubes.
2. PCR tubes or plates.
3. Thermocycler (PCR machine).
4. Agarose gel electrophoresis system.
5. Sequencing gel electrophoretic system.
6. Restrictions enzymes HpaII, MspI, EcoRI and restriction buffer.
7. BSA (10 mg/mL).
8. T4 DNA ligase and 10× ligation buffer.
9. Taq DNA polymerase and 10× reaction buffer.
10. dNTP set 10 mM, $MgCl_2$.
11. 10× PCR buffer: 100 mM Tris–HCl, pH 8.3, 25 mM $MgCl_2$.
12. Ethidium bromide.
13. Formamide buffer: 98% formamide (deionized and filtered), 10 mM EDTA, pH 8.0 (pH adjusted with 1 M HCl or 1 M NaOH), 0.025% of xylene cyanol.
14. STE buffer: 0.1 M NaCl, 10 mM Tris HCl (pH 8.0), 1.0 mM EDTA.
15. TE buffer: 10 mM Tris–HCl (pH 7.5), 0.1 mM EDTA.
16. A double-stranded EcoRI adapter 5 pmol/L. It is made of two primers: [(5′-CTC GTA GAC TGC GTACC-3′/3′-CATCTG ACG CAT GGT TAA-5′).
17. A double stranded Hpall/MspI-adapter (50 pmol/μL) It consists of the combination of primers [(5′-ACG ATG AGT CTA GAA-3′/3′-CTA CTC AGA TCTTGC-5′)].
18. EcoRI primer +1 (50 ng/μL; 5′-GAC TGC GTA CCA TTCN-3′).
19. EcoRI primer +3 (12 ng/μL; 5′-GAC TGC GTA CCA TTCNNN-3′).
20. HpaII/MspI primer +1 (50 ng/μL; 5′-ACG ATGAGT CTA GAA CGG N-3′).
21. HpaII/MspI primer +3 (50 ng/μL; 5′-ACG ATGAGT CTA GAA CGG N NN-3′).
22. Polyacrylamide gel solution (40%).

23. TEMED.
24. Ammonium persulfate (APS: 100 mg/mL).
25. Loading buffer: 0.1% of bromophenol blue, 0.1% of xylene cyanol FF, 10 mM EDTA pH 7.5 (pH adjusted with 1 M HCl or 1 M NaOH), and 98% of deionized formamide.
26. DNA extraction kit.
27. Fix/stop solution for silver staining: 10% glacial acetic acid.
28. Oxidizing solution for silver staining: 1.5% nitric acid.
29. Staining solution for silver staining: 0.2% silver nitrate.
30. Developing solution for silver staining: 3% sodium carbonate, 0.054% formaldehyde (37%).

2.4 Identification and Annotation of the Differentially Methylated Genomic Region

1. Genetic Analyzer 3500xL.
2. Thermal cycler.
3. Microcentrifuge.
4. Scalpel.
5. Sterile Milli Q water.
6. Parafilm.
7. 3 M sodium acetate.
8. Glycogen (10 mg/mL).
9. Absolute alcohol.
10. *Escherichia coli* DH5α competent cells.
11. TA cloning kit.
12. Plasmid miniprep kit.

3 Methods

3.1 Development of Genetically Uniform Seedlings Through Induction of Zygotic Polyembryony

1. Identify trees of desired clone from a well-managed plantation having an age of 10–20 years (*see* **Note 1**).
2. Collect immature fruits of 8–10 weeks old from the selected trees (*see* **Note 2**, Fig. 2.1).
3. Wash the fruits thoroughly with tap water, then keep the fruits in dry filter paper for 10–15 min in a laminar flow until all water droplets dry off (*see* **Note 3**).
4. Immerse the fruits in 80% alcohol for 15 min. Then keep in sterile petri plates until all the alcohol is evaporated (*see* **Note 4**).
5. Dissect out the developing ovules from the fruits using a sterile blade and forceps (*see* **Note 5**, Fig. 2.2 and 3).

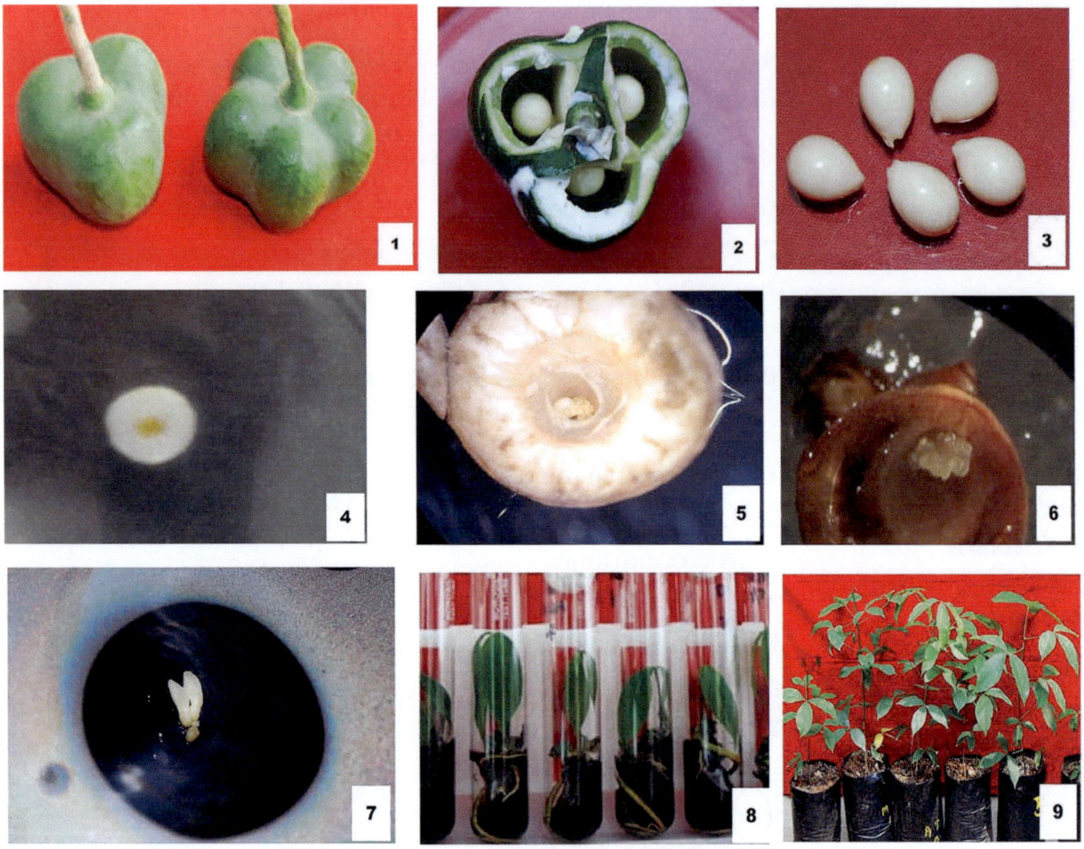

Fig. 2 Development of uniform seedlings of single zygotic origin in *Hevea brasiliensis*. Various developmental stages of genetically uniform own-rooted seedlings right from fruits to hardened pants are shown. (1) Immature fruit, (2) transverse section of the fruit, (3) separated ovules, (4) *half ovulo* culture, (5) initiation of multiple embryos, (6) proliferation of multiple embryos, (7) maturation of separated multiple embryos from the zygote, (8) regenerated plantlets, (9) hardened plants

6. Cut the ovules in 2 transverse halves and keep the micropylar halves in another petri plate (*see* **Note 6**).
7. Culture the micropylar halves in culture medium for embryo development (*see* **Note** 7, Fig. 2.4).
8. Keep the cultured plates in the culture room with standard tissue culture conditions under dark incubation for 1–3 months.
9. Observe the plates under the microscope at weekly intervals. After 1–2 months, small globular structures will be visible (*see* **Note 8** Fig. 2.5 and 6).
10. Sub-culture the half ovules in the same medium at regular intervals until the proliferation of globular structures (*see* **Note 9**).

11. After 2–3 months, the embryogenic mass formed from the globular structures will be proliferated. Separate them and culture directly in the same media composition with addition of 0.2% activated charcoal (*see* **Note 10** Fig. 2.7).
12. After 2 weeks, separate the visible embryos and culture individually for maturation. Maturation of the embryos needs 4 to 6 weeks under dark incubation (*see* **Note 11**).
13. Separate enlarged, ivory colored embryos with opaque cotyledons from the previous medium and culture in the germination medium. Keep under 16 h photoperiod (*see* **Note 12**).
14. After germination and plant regeneration (Fig. 2.8), carefully take out the plantlets with forceps. Remove adhering media carefully and plant in small polythene bags filled with sand, soil, and Soilrite mixture. Keep the polybags in environmentally controlled growth chamber for 2–3 weeks. Transfer the hardened plants to the nursery for further experiments (*see* **Note 13**, Fig. 2.9).
15. For maintenance, irrigate the plants at regular intervals and provide fertilizers as per recommendation.
16. Ensure the genetic and epigenetic uniformity of the plants using already established genetic markers (*see* **Note 14**, see also Subheading 3.3).
17. Use one selected plant from the above for raising bud-grafted counterparts as per the standard procedure described in Subheading 3.2.2.

3.2 Grafting

3.2.1 Establishment of Seedling Nursery

1. Collect fresh open pollinated seeds from monoclonal or polyclonal plantations (*see* **Note 15**).
2. Prepare germination bed of size 90 cm wide and convenient length covered with a layer of sand to a depth of 4–5 cm. The germination bed should be raised ideally 10–15 cm above ground.
3. Spread the seeds over the bed in single layer and gently press so that the surface of seed is level with the ground. (*see* **Note 16**, Fig. 3.1).
4. Collect sprouted seeds soon after radicle emergence on alternative days and transfer to seedling nurseries (*see* **Note 17**, Fig. 3.2).
5. Sow the seeds in the ground nurseries or directly in polybags (*see* **Note 18**, Fig. 3.3).

3.2.2 Establishment of BudWood Nursery

1. Plant the genetically and epigenetically tested uniform own-rooted seedlings in the nursery (*see* **Notes 19** and **20**).
2. One year after planting, cut the shoot at a height of 75 cm from the base. Subsequently, allow 2–3 shoots to grow by cutting

Fig. 3 Bud-grafting technique in *Hevea brasiliensis*. Different stages of bud grafting in *Hevea*. (1) Seed germination bed, (2) sprouted seeds, (3) seedling nursery, (4) budwood nursery, (5) green bud sticks, (6) insertion of bud patch, (7) bud-grafted seedling, (8) two whorl polybag plants

back every year leaving enough length for regeneration of budwood so that a bush of branches is established (*see* **Note 21**, Fig. 3.4).

3. Harvest 6- to 8-week-old bud shoots with one mature whorl on top for green budding (*see* **Note 22**, Fig. 3.5).

3.2.3 Grafting Procedure (Green Budding)

1. Preparation of stock: Clean the basal region of the stock plant. Make two vertical incisions 5 cm long and 1 cm apart, and 2.5 cm above ground level with the sharp budding knife. Horizontal cuts are made connecting the lower and upper end of these cuts (*see* **Note 23**).
2. Preparation of scion: Carefully remove bud patch from fresh green shoots of the same size from the budwood (*see* **Note 24**).

3. Budding procedure: Peel the cut panel on the stock plant and keep the bud patch on the budding panel. Wrap firmly with a polythene film starting from the bottom and move upward in a spiral above the bud patch (*see* **Note 25**, Fig. 3.6, 7).
4. Examination of budding success: Bud union takes place within 21 days from grafting when the polythene tape can be removed. If the bark is green, initial budding success can be ascertained (*see* **Note 26**).
5. Uproot the stock plants and cut back them about 7.5 cm from the bud patch. Prune the taproot and laterals to a convenient length before transplanting to polybags and allow the scion shoot to grow (*see* **Note 27**, Fig. 3.8).

3.3 MSAP Analysis for the Detection of DNA Methylation Polymorphisms in Heterografted Plants

1. Genomic DNA is extracted with a DNA extraction kit according to the manufacturer's recommendations. Digest genomic DNA with two different restriction enzymes followed by ligation of the double-stranded MSAP adapters to the sticky ends. The adapter and restriction site motifs serve as primer binding sites in the subsequent amplification steps. Complete digestion of genomic DNA is very important to prevent the amplification of undigested fragments at a later stage.
2. Perform double digestion of approximately 1.5 μg of genomic DNA in two sets. One set with EcoRI (15 U)/MspI (30 U) and the other set with EcoRI (15 U)/HpaII (15 U). The restriction digestion is performed in 1× NEB buffer 4 in a final volume of 50 μL with 0.5 μL BSA (10 mg/mL) and incubated at 37 °C overnight. Heat inactivation of the enzyme is done at 70 °C for 15 min.
3. Adapter preparation and ligation: EcoRI and HpaII-MspI oligos are used for adapter preparation. Mix the two oligonucleotides together at 1:2 ratios. For each reaction, 5 pmol of oligo-1 and 10 pmol of oligo-2 are mixed with STE buffer. Prepare each adapter in a total volume of 50 μL (reaction contents are 12.5 μL of oligo-1, 25 μL of oligo-2, 5.0 μL of 10× STE buffer, and 7.5 μL of water). Mix the reagents and place the tubes at 70 °C for 10 min and allow it to cool to room temperature slowly. Ligate both EcoRI and Hpa-MspI adapters onto the restricted DNA fragments by keeping the mix at 16 °C overnight using T4 DNA ligase in a final volume of 50 μL.

Adapter ligation	
EcoRI Adapter	2.0 μL
HpaII-MspI Adapter	2.0 μL
10× T_4 DNA Ligation Buffer	5.0 μL

(continued)

Adapter ligation	
T_4 DNA Ligase	5.0 μL
Nuclease free water	11.0 μL
DNA	25.0 μL
	50.0 μL

Dilute the adapter ligated product to 1:10 with 10:0.1 TE buffer and use it as template for the preamplification PCR reactions (*see* **Note 28**).

4. Perform two successive selective PCR reactions. Use primers with increased number of selective nucleotides at their 3′ end in the PCR reactions in order to selectively amplify a single set of digested DNA fragments at the end of each analysis. The selectivity can be enhanced and the complexity can be reduced by using one, two or three selective nucleotides at the 3′ end of each primer. It can be either EcoRI +1, EcoRI +2 or HpaII/MsPI +1, HpaII/MsPI +2. The first PCR reaction is named pre amplification. It is performed with a single selective nucleotide at the 3′ end of both EcoRI and HpaII/MsPI primers. The second PCR or selective amplification, is carried out using more than one selective nucleotide at the 3′ end of both EcoRI and HpaII/MsPI primers.
5. Set up the preamplification reactions in a total volume of 25 μL. Major components are 0.2 μM of each primer that is complementary to the EcoRI and Hpa-MspI adapter sequences with an additional selective 3′ nucleotide (e.g., EcoRI + A and Hpa-MspI +C) and 1 U of Taq DNA polymerase. Set the thermal cycler for 20 cycles at the temperature cycling profile of an initial denaturation of 94 °C for 30 s, annealing at 56 °C for 1 min and a final extension of 72 °C for 2 min (*see* **Note 29**).
6. In order to check the efficiency of preamplification reaction, electrophorese 5 μL of the PCR product on a 1% agarose gel and visualized it using gel documentation system. Appearance of a thick smear after electrophoresing for 15 to 20 min indicates successful preamplification step (*see* **Note 30**). Dilute the preamplified DNA 50 fold with sterile water and use it as template for selective amplification (*see* **Note 31**).
7. Set up the selective PCR reactions using EcoRI and Hpa-MspI primers having three additional nucleotides at their 3′ ends (EcoRI-ANN and Hpa-Msp I-CNN) where N represents A, C, G, or T. Perform the reactions in a total volume of 25 μL containing 5 μL of 50 fold diluted preamplified template DNA, 2 mM of $MgCl_2$, 200 μM of each dNTPs, 1.2 μM of MseI, 0.2 μM of EcoRI primer and 1 U of Taq DNA polymerase in

the same thermal cycler. The PCR cycling conditions are as follows: an initial denaturation at 95 °C for 5 min followed by a touch down PCR program. The temperature profile includes: one cycle of initial denaturation at 94 °C for 30 s, annealing at 65 °C for 30 s and extension at 72 °C for 1 min. The annealing temperature is lowered down by 0.7 °C at each cycle for 12 cycles with the touch down PCR program. This is followed by a normal PCR of 35 cycles of 94 °C for 30 s, annealing at 56 °C for 30 s, extension at 72 °C for 1 min and a final elongation step at 72 °C for 10 min (*see* **Note 32**).

8. Add equal quantity of the loading buffer to the PCR product and store at −20 °C until the electrophoresis step.
9. Prepare 6% polyacrylamide gel using the composition given below for 60 mL (*see* **Note 33**).

Preparation of the gel	
Reagents	**Required quantity**
Acrylamide 40%	9 mL
TBE + urea	51 mL
10% APS	270 μL
TEMED	27 μL

Percentage of gel (PAGE) concentration can be changed by changing the quantity of 40% acrylamide used.

10. Deaerate the acrylamide solution prior to adding ammonium persulfate (APS) and TEMED. Gently swirl the flask during deaeration until no air bubble is released.
11. Allow the acrylamide to polymerize for 60 min at room temperature. Attach the gel to the electrophoresis tank and fill the reservoirs of the electrophoresis tank with 0.6× TBE. Use a syringe to flush out the wells with 1× TBE.
12. Denature the PCR products at 95 °C for 10 min and quickly chill on ice. Load 3–5 μL of the sample into the well (*see* **Note 34**). Run the gel at a constant power of 60 watts until the marker dye has migrated to the desired distance. The temperature of the gel should be below 55 °C for best results.
13. After electrophoresis, carefully separate the glass plates and place the gel (attached to the glass plate) in a shallow plastic tray.

14. For silver staining, follow the steps listed below.

Step	Solution	Time
(a) Fixing	Fix/stop solution	20 min
(b) Wash	dH_2O	2 × 2 min
(c) Oxidizing	1.5% nitric acid	3 min
(d) Rinse	dH_2O	10 s
(e) Staining	Staining solution	60 min
(f) Rinse	dH_2O	10 s
(g) Developing	Developer solution	2–5 min
(h) Stopping	Fix/stop solution	5–6 min
(i) Wash	dH_2O	2–3 min

All steps should be done with constant shaking. The volume of the solutions used in each step is typically 1 L per gel. Position the gel upright and allow it to dry overnight and the gel should be scanned or photographed for identifying and scoring the alleles (Fig. 4).

15. Analyze the results. In the gel, each DNA sample is represented by two lanes of fragments resulting from EcoRI/MspI and EcoRI/HpaII-digested DNA (M and H lanes, respectively) (*see* **Note 35**). Since the isoschizomers MspI and HpaII recognize the same tetra-nucleotide sequence, 5′-CCGG, but have different sensitivities to methylation of the cytosines. HpaII will not cut if the cytosines in the double strand is methylated, whereas MspI will not cut if the external cytosine is fully or hemi methylated (single-strand). Thus, for a given DNA sample in the MSAP analysis, CG methylation or CHG hemimethylation at the assayed 5′-CCGG sites are discovered with the presence of a band in only one of the enzyme digests and absence from the other. The (+, +) pattern is attributed to digestion by both enzymes at a nonmethylated CCGG site. The inhibition of digestion with both enzymes at a fully methylated mCmCGG site when another treated sample shows the presence of a fragment at that position is indicated as (−, −) pattern. This pattern may also represent a mutated site when genetically distinct samples are compared. The (+, −) pattern corresponds to digestion with MspI but not HpaII and refers to the presence of a CmCGG site. The (−, +) pattern is designated as CHG hemimethylation which is common in plant genomes. Score the bands observed in the MSAP gels as described in Fulneček and Kovařík (2014) [6]. Determine the relative CG, CHG, and total methylation levels by tabulating the number of bands representing the various types of MSAP patterns (Figs. 4 and 5).

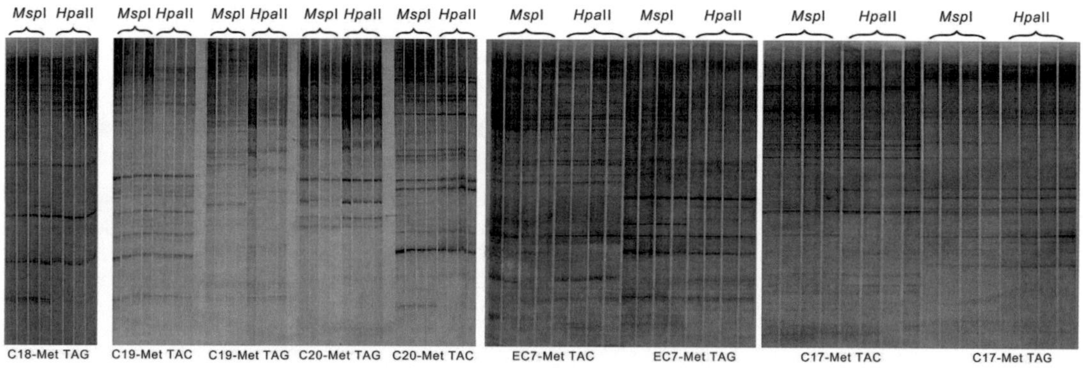

Fig. 4 Epigenetic profiling by MSAP analysis. Gel picture depicting MSAP patterns generated using different selective primer combinations. Variations in banding pattern between the MspI set and HpaII set indicate the presence of methylated regions in *Hevea* genome whereas variations among the HpaII set indicate differentially methylated regions. (Reproduced from Rekha et al., 2015 [9] with permission from Springer Nature)

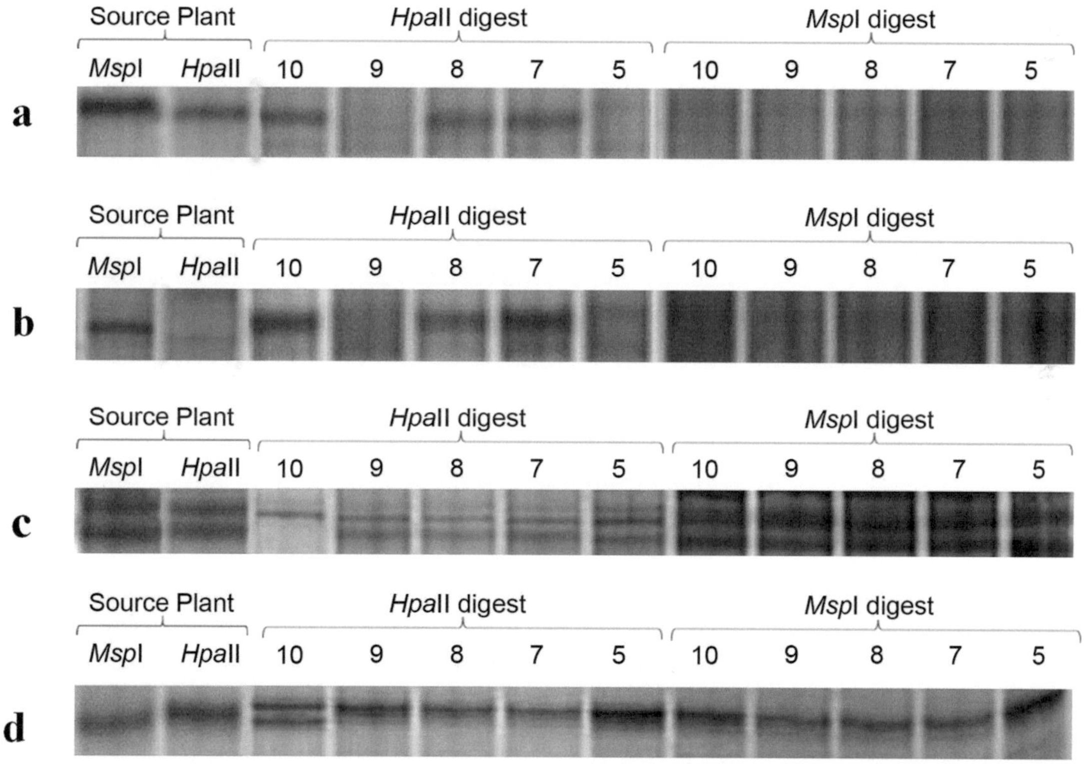

Fig. 5 Polymorphic methylation-sensitive bands in MSAP gel. MSAP gel picture highlighting prominently polymorphic methylation-sensitive bands in the source own-rooted plant and the five bud-grafted plants. (Reproduced from Uthup et al. 2018 [1] with permission from Springer Nature)

3.4 Identification and Annotation of the Differentially Methylated Genomic Region

1. Cut the selected bands showing clear DNA methylation polymorphisms from as many samples as possible and transfer them to an eppendorf tube (*see* **Note 36**).
2. Soak the gel slice in 100 μL of autoclaved Milli Q water for 10 min in a microcentrifuge tube.
3. Boil the tube with the cap tightly closed with parafilm for 15 min.
4. Spin for 2 min to collect the gel debris at the bottom of the tube.
5. Transfer the supernatant to a new microcentrifuge tube. Add 10 μL of 3 M Sodium acetate, 5 μL of glycogen (10 mg/mL) and 450 μL of 100% ethanol (*see* **Note 37**). Keep the samples for 30 min at 80 °C in a dry bath or water bath.
6. Spin for 10 min at 4 °C to pellet the DNA. Remove the supernatant and rinse the pellet with 200 μL ice cold 85% ethanol. Spin briefly and remove the residual ethanol.
7. After the pellet is properly dried, dissolve it in 10 μL of autoclaved Milli Q water and use it for reamplification with the same primer combinations used earlier (*see* **Note 38**).
8. Run the reamplified PCR products on a 3% agarose gel and elute and purify the PCR product using any gel band purification kit.
9. Quantify the purified PCR product and clone it to any TA cloning vector following the manufacturer's instructions. Follow standard TA cloning protocol for ligation, and transformation of *E.coli* DH5α strain. Extract plasmids from the transformed E.coli cells using any standard plasmid isolation kit.
10. Sequence the cloned fragments in the TA cloning vector from both directions using universal primers by Genetic Analyzer 3500xL.
11. Using BLASTN and BLASTX programs from NCBI (http://www.ncbi.nlm.nih.gov/) perform functional homology analysis of the DNA sequence obtained from the methylation-sensitive fragment for locating and annotating the genomic regions harboring each DMR (*see* **Note 39**).
12. Bisulfite sequencing of the methylation-sensitive polymorphic fragments can be performed by standard protocols to determine the exact location, class, and extent of methylation in the samples analyzed (*see* **Note 40**).

4 Notes

1. In tree tissue culture, the age of the plant is an important factor. Usually too young and too old trees are not used. Selection of trees is also an important factor. Selected trees need to be maintained without disease and with proper nutrient management.
2. For embryo cloning, immature embryos are preferred in order to exploit the advantage of juvenility. Rubber is a perennial tree with seasonal flowering character. Immature fruits are available from the month of March until the end of May. In rubber tree, growth of the fruit completes first, then the endosperm growth and only after that the embryo starts growing [7]. The embryos are not visible until the fruits reach 14 weeks of age. The ideal stage of the fruit for induction of multiple embryos is 8–10 weeks after pollination.
3. Since the fruits are collected from field grown trees, the dust and other particles are to be removed by washing in clean water. Since the sterilization process is done in alcohol, the adhering water particles if remained will reduce the efficiency of sterilization and hence it is recommended to dry the fruits before sterilization.
4. After sterilization with alcohol, there is no need to wash with sterile water. The alcohol is allowed to evaporate in the laminar flow hood. Instead of alcohol, mercuric chloride can also be used. However, handling of the sterilized fruits will be difficult while using mercuric chloride.
5. *Hevea* fruits are trilocular and each fruit contain three ovules. Since the inner cavity of the fruits are always sterile, ovules need not be sterilized before culturing. Ovules are dissected aseptically from the sterilized fruits using a sterile blade and forceps.
6. The ovules are cut in two equal halves. The embryo will be embedded in the micropylar half and that half needs to be cultured with the micropyle touching the medium. Care should be taken to identify the micropylar region and for proper culturing.
7. Nitsch basal medium [8] fortified with 3 mg/L GA and 3 mg/L kinetin along with organic supplements as reported by Rekha et al. (2015) [9] is used for culturing. Media is prepared in 60 mm sterile petri plates.
8. As in all tissue culture experiments, observation at regular intervals is the most important point to be noted. Plates with contamination should be removed periodically. The emergence of globular embryos is expected usually around 2 months of

culture. However, early emergence can also happen due to the genotype effect. Regular observation will ensure timely identification of positive response and proper subculture.

9. Initial medium will be depleted by 1–2 months and for continued growth, the cultures should be transferred to new media with the same composition. At early stages of growth, the division of zygotic cells needs the support of ovular tissues and hence the half ovules are cultured as such without disturbing the embryonic cells.
10. Isolation and culture of embryonic mass or cluster of embryos from the ovular tissues without damage is an important step. This is cultured in same media with activated charcoal. Activated charcoal will ensure the slow release of nutrients and the growth of individual embryos will be promoted.
11. For maturation, half-strength MS medium [10] fortified with organic supplements such as CW (20%), BP (200 mg/L), ME (100 mg/L), and casein hydrolysate (CH) (100 mg/L) devoid of growth regulators is ideal. 0.2% charcoal is also added in the maturation medium. The cultures are incubated in the dark.
12. For germination, half-strength Murashige and Skoog (MS) medium along with 0.3 mg/L benzyl adenine (BA), 0.3 mg/L GA_3 [10] and organic supplements is used [11].
13. Hardening is the most difficult step in rubber tissue culture. Careful handling is important for rescuing the in vitro developed seedlings. The humidity of the hardening chamber is reduced gradually so as to acclimatize the plants. However, embryos generated through cleavage of zygotic embryo will have a capacity to thrive better. Hence, more than 50% of the plants can be hardened.
14. Since the embryos developed did not involve a callus phase, the plants generated through this technology are expected to be uniform in all aspects. However, their genetic uniformity needs to be ensured before going for epigenetic studies so that the variations, if any observed among them, can be solely attributed to epigenetic changes. We have already established markers for assuring uniformity [12].
15. Collection of seeds soon after dehiscence is an important criterion as *Hevea* seeds' viability is too short.
16. The seed bed should be irrigated twice daily and covered with coir mat or gunny bags.
17. Germination starts within 6–7 days after sowing. Sprouted seeds are to be handled with utmost care and should be carried in a bucket of water. Late germination seeds (after 3 weeks) are to be discarded.

18. Weaker plantlets in a nursery are removed at interval of 2–3 months. Seedlings are grown at a spacing of 23 × 23 cm for green budding. 2–8-month-old seedlings with a collar girth of 2.5 cm are used for budding. Proper agromanagement practices including regular removal of weeds, spraying, fertilizer application, and irrigation are needed for uniform growth and buddability in the seedling nurseries.
19. It is very important to ensure the uniformity of the plants which is to be used as scion, in order to rule out rootstock influence. Seedlings are planted at a spacing of 90 × 90 cm for green budwood.
20. Two whorl polybag plants are the commonly used planting materials. Based on the color and age of buds, two common budding techniques, namely, brown and green are practiced in rubber. For the development of bud-grafted counterparts of uniform seedlings, green budding is adopted [13, 14].
21. The scale buds of fresh green shoots are ideally suited. One green shoot yields 4–6 buds.
22. Green shoots are collected early in the morning and used immediately.
23. Time is given for the latex to ooze out and is wiped off when the latex flow ceases.
24. As separation of buds is difficult from a green shoot, remove the bud patch with a section of the wood and carefully separate the bark from the wood ensuring that the core of the bud is intact.
25. The cambium tissue of stock plant and bud patch should be in tight contact for effecting budding success. 25 cm × 2 cm with 200 gauge polythene tape is found to be suitable for green budding.
26. Examine the bud patch 10 days after the first examination. If the bark is still green then the bud grafting is successful. Green budding gives good success from November to March.
27. The cut end of the seedling is sealed with molten paraffin wax. Alternately, seeds can be directly sown in polybags and budding done in the stock seedlings raised in polybags.
28. Reconstruction of restriction sites is avoided by the design of adapters. Due to the presence of restriction enzymes in the ligation step, primer concatemers that may be generated by ligation are restricted, thereby ensuring almost complete adapter/fragment ligation.
29. MSAP reaction mixes should be prepared for a minimum of ten different DNA samples to minimize pipetting error discrepancies.

30. Electrophoresis after preamplification is performed for short duration (10–15 min) to compare intensities among amplified samples. Smears will be too faint to detect if longer runs are performed. The dilution of preamplified DNA fragments depends on the intensity of smears visualized on the agarose gels.
31. At this stage, the PCR product can be stored for more than a year at -20 °C.
32. The start of the PCR at a very high annealing temperature allows optimal primer selectivity and by decreasing gradually the annealing temperature, we could increase the efficiency.
33. The AFLP reaction products are analyzed on 4.5% denaturing polyacrylamide gels (acrylamide-bisacrylamide mix containing urea) or 6–8% Long Ranger gels (Long Ranger™ 50% Gel Solutions, Lonza) polyacrylamide gels containing urea). Gels should be casted at least 2 h before use to ensure sufficient time for gel polymerization and can be stored for 24 h at 4 °C.
34. After the pre-run, remove the urea precipitate or the pieces of gel from the well forming surface of the gel with a syringe before loading.
35. Fragments are scored visually or using different AFLP scoring software. In the MSAP gel, methylation-sensitive fragments showing different intensities are usually observed which can be considered as an indirect quantification of methylated sites.
36. It is better to cut the polymorphic band from the gel before fully drying the gel. After drying and documenting the gel for the first time, the same samples can be electrophoresed again for eluting the selected fragments.
37. In ethanol and isopropanol, glycogen remains insoluble and forms a precipitate that traps nucleic acids. A glycogen concentration of 20 mg/mL may be used for recovery of low amounts of nucleic acids and short oligonucleotides.
38. For the reamplification of the eluted fragment, selective PCR amplification step can be repeated with the eluted DNA as template and the same primer combination used.
39. The sequences can be blasted against the whole genome shot gun sequences or transcriptome sequences available in public databases to identify the location of these methylation-sensitive regions. The contigs which harbored these portions may be extracted and further analyzed to annotate the region of interest.
40. Bisulfite sequencing of the same region is performed following standard procedure using the same genomic DNA that was used for MSAP. The bisulfite sequencing data obtained from

multiple genetically uniform grafted plants are aligned with genomic DNA sequence of the same region to identify the differentially methylated cytosine.

Acknowledgments

We thank Dr. James Jacob, Director, Rubber Research Institute of India for his constant encouragement and support at every stage of this work. Thanks are due to Dr. Mercykutty V C, Senior Scientist (Rtd.), Rubber Research Institute of India for providing the photographs on grafting method in *Hevea*. We also acknowledge the help rendered by Ms. Minimol Ravindran in preparing the manuscript and Ms. Krishna Priya S for the artwork.

References

1. Uthup TK, Rekha K, Ravindran M et al (2018) Heterografting induced DNA methylation polymorphisms in *Hevea brasiliensis*. Planta 248:579–589
2. Reyna-Lopez GE, Simpson J, Ruiz-Herrera J (1997) Differences in DNA methylation patterns are detectable during the dimorphic transition of fungi by amplification of restriction polymorphisms. Mol Gen Genet 253:703–710
3. Marfil CF, Camadro EL, Masuelli RW (2009) Phenotypic instability and epigenetic variability in a diploid potato of hybrid origin, *Solanum ruiz-lealii*. BMC Plant Biol 9:21
4. Verhoeven KJF, Jansen JJ, van Dijk PJ et al (2010) Stress-induced DNA methylation changes and their heritability in asexual dandelions. New Phytol 185:1108–1118
5. Rekha K, Lincy A, Jayashree R et al (2012) Half ovulo embryo culture –an ideal method for raising true-to-type seedlings in *Hevea brasiliensis*. In: International rubber conference, Oct. 29–31, Kovalam, Kerala, India, pp 89–90
6. Fulnecek J, Matyasek R, Kovarik A et al (1998) Mapping of 5-methylcytosine residues in *Nicotiana tabacum* 5S rRNA genes by genomic sequencing. Mol Gen Genet 259:133–141
7. Muzik TJ (1956) Studies on the development of the embryo and seed of *Hevea brasiliensis* in culture. Lloydia 19:86–91
8. Nitsch P, Nitsch C (1969) Haploid plants from pollen grains. Science 163:85–87
9. Rekha K, Uthup TK, Sobha S et al (2015) Genetic and epigenetic uniformity of polyembryony derived multiple seedlings of *Hevea brasiliensis*. Protoplasma 25:783–796
10. Murashige T, Skoog F (1962) A revised medium for rapid growth and bioassays with tobacco tissue cultures. Physiol Plant 15: 473–497
11. Rekha K, Jayashree R, Gireesh T et al (2010) Embryo rescue and plant regeneration in *Hevea brasiliensis*. Nat Rubber Res 23: 147–154
12. Saha T, Roy B, Nazeer MA (2005) Microsatellite variability and its use in the characterization of cultivated clones of *Hevea brasiliensis*. Plant Breed 124:86–92
13. Mercykkutty VC (2008) Rubber nursery: a practical guide. Malayalam book. Rubber Research Institute of India, Rubber Board, Kottayam, p 95
14. Marattukalam JG, Mercikutty VC (2000) Propagation techniques. In: George PJ, Jacob CK (eds) Natural rubber agromanagement and crop processing. Rubber Research Institute of India, Rubber Board, Kottayam, pp 75–96

Chapter 18

In Vitro Fertilization System Using Wheat Gametes by Electric Fusion

Tety Maryenti, Norio Kato, Masako Ichikawa, and Takashi Okamoto

Abstract

In vitro fertilization (IVF) systems using isolated gametes have been used to dissect post-fertilization events in angiosperms, as female plant gametophytes are deeply embedded within the ovaries. In addition, hybrid and polyploid zygotes can be produced by using IVF systems. Complete IVF systems of maize and rice, two out of three major energy-providing crops, have been established in order to acquire detailed knowledge of mechanisms of fertilization and early embryogenesis. Following in the footsteps of previous success, a wheat IVF system was developed to introduce the advantages of this technology to wheat research. Fusion of gametes was performed via a modified electrofusion method, and the zygote formed a cell wall and two nucleoli. The zygotes divided into symmetric two-celled embryos, globular-like embryos and multicellular club-shaped embryos which are mostly consistent with those in the embryos in planta. IVF-produced club-shaped embryos developed into compact embryonic calli and subsequently regenerated into fertile plants. In this chapter, we provide a detailed description of wheat IVF system that might become an important technique for generating new genotypes of wheat and/or new hybrids as well as for investigating fertilization-induced events in wheat.

Key words Egg cell, Embryogenesis, In vitro fertilization, Sperm cell, Wheat, Zygote

1 Introduction

Fertilization consists of plasmogamy between parental gametes and subsequent karyogamy in fused gamete, and is the central event in the life cycle of most organisms to generate progeny with parental genetic materials. In most animals and cryptogamic plants, flagellated sperms swim toward female gametes in water or body fluids. While in the angiosperms, male gametes lacking motility are delivered into the female gametophyte via pollen tube. The female gametophyte of angiosperms, which is also referred to as the embryo sac or megagametophyte, develops in the ovule, and fertilization and subsequent events such as embryogenesis and endosperm development also occur in the embryo sac, which is profoundly implanted in ovular tissue [1, 2]. Inaccessibility to the

Christophe Lambing (ed.), *Plant Gametogenesis: Methods and Protocols*, Methods in Molecular Biology, vol. 2484, https://doi.org/10.1007/978-1-0716-2253-7_18,

embedded female gamete, zygote, and early embryo has been a hurdle for the investigations into the molecular mechanisms of fertilization and embryogenesis. Therefore, such studies have been carried out predominantly through analyses of *Arabidopsis* mutants or transformants coupled with live-imaging [3–6]. Alternatively, direct analyses using isolated gametes or zygotes are possible because protocols for isolating viable gametes have been established. The initial IVF system using plant gametes was developed by Kranz and Lörz (1993) in maize [7] and a second precedent of in vitro fertilization in rice was eventually developed by Uchiumi et al. (2007) [8], demonstrating the chance that led to the sophisticated progression of IVF method as a mediator to observe and analyze fertilization and post-fertilization processes directly [9].

The IVF system used for angiosperms involves a combination of three basic microtechniques: (1) the isolation and selection of male and female gametes, (2) the fusion of pairs of gametes, and (3) single-cell culture [10]. For the isolation of viable gametes, procedures have been established in a wide range of plant species, including both monocotyledonous and dicotyledonous plants [10, 11]. Fusion of isolated gametes can be performed electrically [12, 13] or chemically via calcium [14–16], polyethylene glycol [17, 18], or bovine serum albumin [19], since plasma membranes of isolated gametes are largely exposed, and plasma membranes between the isolated gametes can be attached each other. Among these four kinds of fusion procedures, only zygotes produced by electrofusion are known to divide and develop into embryo-like structures and plantlets.

The electrofusion-based IVF systems have been successfully utilized to observe and investigate post-fertilization events such as karyogamy [20–23], egg activation and zygotic development [24–26], paternal chromatin decondensation in zygote nuclei [27], microtubular architecture in egg cells and zygotes [28], fertilization-induced/suppressed gene expression [29], contribution of parental genomes to zygotic development [30–32], initiation of zygotic development by male allele-derived factor [33], and epigenetic resetting in early embryos [34].

In addition to the aforementioned valuable works, a genome-editing system via direct delivery of Cas9-gRNA ribonucleoprotein into plant zygotes was recently achieved [35]. This newfangled technology was adapted by the combination of IVF method and polyethylene glycol calcium (PEG-Ca^{2+})-mediated transfection system with zygotes and egg cells [36]. With this PEG-Ca^{2+} transfection protocol, macromolecules such as DNA, RNA, and proteins can be delivered into zygotes, and transient expression of genes and proteins of interest in the target cells is possibly conducted [33]. Taken together, the IVF system can be considered as an important technique for creating new genotypes and/or hybrids that have desirable characteristics as well as for investigating post-fertilization events.

Complete IVF systems have been established in maize and rice [7, 8]. Notably, these plant species are two of three major crop species, maize, rice, and wheat, that provide the majority of the human energy intake worldwide. Among those crops, functional analyses of genes of interest have not been intensively conducted for wheat, since bread wheat is a hexaploid with a large genome size (17 Gbp). However, a draft sequence of the wheat genome has been sequenced [37, 38], and assembly of the sequences is mostly complete [39–41]. Moreover, state-of-the-art genome/molecular biology technologies enable high-throughput and high-resolution analyses. These provide opportunities for the use of wheat plants as experimental materials similar to those for maize and rice, which are monocot model plants. Therefore, the establishment of a wheat IVF system has been awaited. In wheat, there are reports of the isolation and subsequent culture of zygotes and of the production of zygotes by gamete fusion. Kumlehn et al. (1998) and Bakos et al. (2003) mechanically isolated zygotes from pollinated wheat ovaries, and the zygotes were successfully cultured into embryo-like structures that could be regenerated into wheat plants [42, 43]. For wheat zygotes produced by gamete fusion in vitro, Kovács et al. (1995) reported that egg cells isolated from wheat ovules that were treated with a cell wall-degrading enzyme mixture were successfully fused with isolated wheat sperm cells by electric pulses, however, development of the resulting zygotes was arrested at a multicellular stage [44]. Therefore, a complete IVF system using wheat gametes was still lacking, despite two decades having passed after those valuable reports. We recently established a complete wheat IVF system to utilize the valuable resources rooting in wheat [45]. In the protocol, we describe the procedures for mechanical isolation of gametes from unpollinated wheat flowers, electrofusion of isolated gametes, culture of zygotes into embryonic calli, and regeneration of wheat plants from the calli.

2 Materials

2.1 Isolation and Transfer of Gametes

1. Unpollinated wheat spikes (*see* **Note 1**).
2. Straight dissection forceps with fine tips (No. 5 and 5A) (*see* **Note 2**).
3. Laminar air flow chamber.
4. Stereomicroscope.
5. Inverted microscope.
6. Nontreated plastic dishes with a diameter of 3.5 cm.

7. Coverslips (24 × 40 mm) siliconized at the edges with 5% dichlorodimethylsilane in 1,1,1-trichloroethane (*see* **Note 3**).
8. Mineral oil (embryo culture tested grade).
9. Mannitol solution (330 mM) adjusted to 370 mOsmol/kg H_2O and autoclaved.
10. Sliding stage for the insertion of a coverslip and a plastic dish.
11. Glass capillaries made from 50 μL aspirator tubes, tip openings 150–250 μL (drawn by hand).
12. Glass needles with fine and blunt tips (*see* **Note 7**).
13. Manual handling injector.

2.2 Fusion of Gametes

1. Mannitol solution (330 mM) adjusted to 370 mOsmol/kg H_2O and autoclaved.
2. Mannitol solution (430 mM) adjusted to 520 mOsmol/kg H_2O and autoclaved.
3. Inverted microscope.
4. Electrofusion generator.
5. Manipulator with a double pipette holder.
6. Electrodes fixed to the pipette holder.

2.3 Culture of Zygotes into Compact Embryonic Calli and Regenerated Plantlets

1. Nontreated 3.5-cm plastic dishes.
2. Millicell-CM cell culture inserts, diameter 12 mm.
3. Feeder cells: wheat suspension cells culture (line HT3–1) (*see* **Note 4**).
4. Medium for zygote culture (Fig. 1b) (N6Z-medium [42] with modifications): 2 g/L CHU (N6) basal salt mixture, 0.025 mg/L $Na_2MoO_4 \cdot 2H_2O$, 0.025 mg/L $CoCl_2 \cdot 6H_2O$, 0.025 mg/L $CuSO_4 \cdot 5H_2O$, 0.01 mg/L retinol, 0.01 mg/L calciferol, 0.01 mg/L biotin, 1 mg/L thiamine hydrochloride, 1 mg/L nicotinic acid, 1 mg/L pyridoxine·hydrochloride, 1 mg/L choline chloride, 1 mg/L Ca-pantothenate, 0.2 mg/L riboflavin, 0.2 mg/L 2,4-D, 0.02 mg/L cobalamine, 0.02 mg/L *p*-aminobenzoic acid, 0.4 mg/L folic acid, 2 mg/L ascorbic acid, 40 mg/L malic acid, 40 mg/L citric acid, 40 mg/L fumaric acid, 20 mg/L Na-pyruvate, 1000 mg/L glutamine, 250 mg/L casein hydrolysate, 100 mg/L myoinositol, 150 mg/L xylose. Osmolality, 450 mOsmol/kg H_2O adjusted with glucose. Adjust pH to 5.7 with NaOH and filter sterilized.
5. Medium for callus induction (Fig. 1c) (N6Z-medium [42] with modifications): 4 g/L CHU (N6) basal salt mixture, 0.05 mg/L $Na_2MoO_4 \cdot 2H_2O$, 0.05 mg/L $CoCl_2 \cdot 6H_2O$, 0.05 mg/L $CuSO_4 \cdot 5H_2O$, 0.02 mg/L retinol, 0.02 mg/L

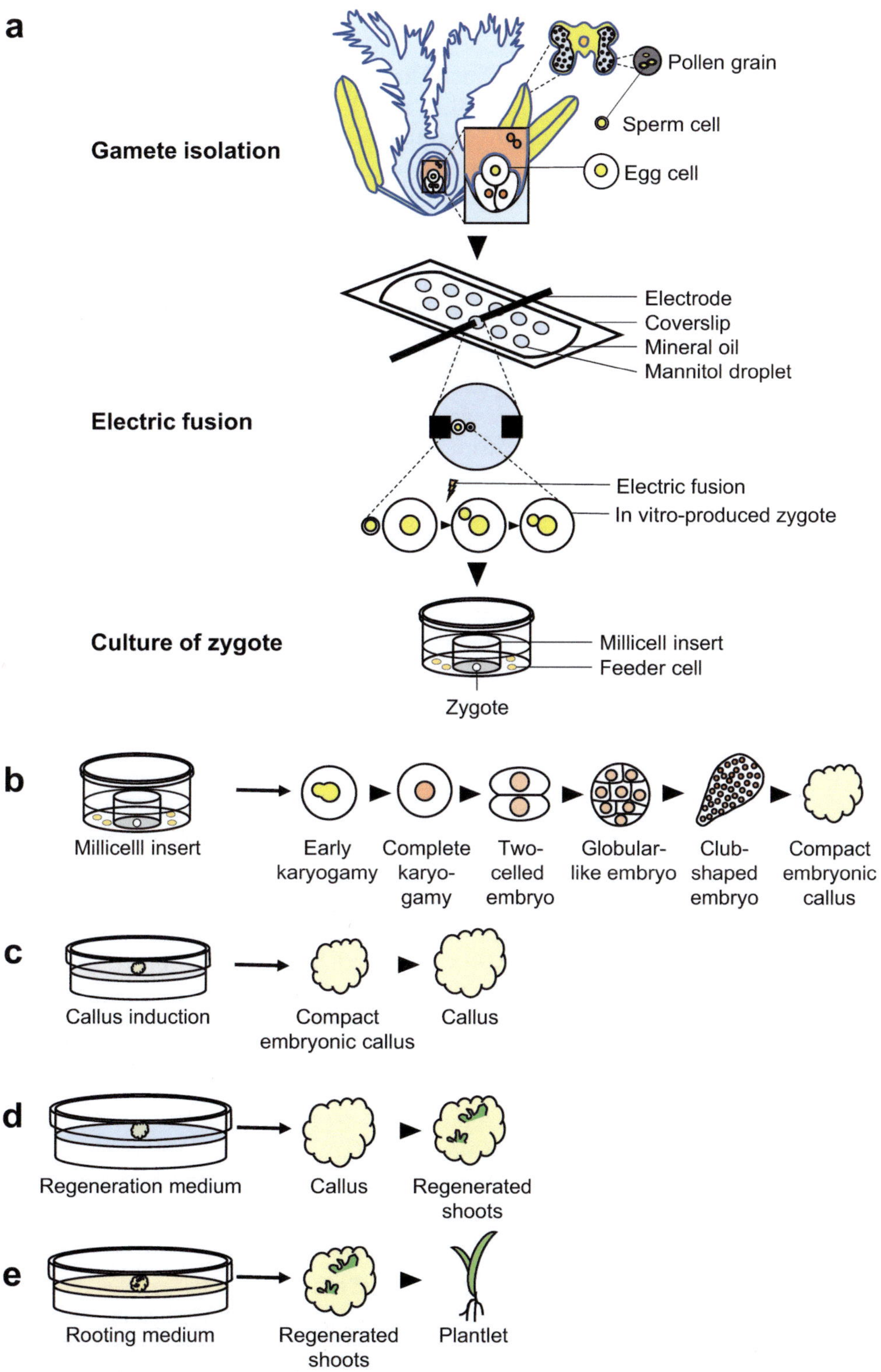

Fig. 1 Schematic illustration of procedures for in vitro production of wheat zygotes via electric fusion and for culture of the zygotes. (**a**) Gametes, an egg cell and a sperm cell, are isolated from unpollinated flower and transferred into droplet of mannitol solution overlaid with mineral oil on a coverslip. The gametes are

calciferol, 0.02 mg/L biotin, 2 mg/L thiamin hydrochloride, 2 mg/L nicotinic acid, 2 mg/L pyridoxine·hydrochloride, 2 mg/L choline chloride, 2 mg/L Ca-pantothenate, 0.4 mg/L riboflavin, 0.4 mg/L 2,4-D, 0.04 mg/L cobalamine, 0.04 mg/L p-aminobenzoic acid, 0.8 mg/L folic acid, 4 mg/L ascorbic acid, 80 mg/L malic acid, 80 mg/L citric acid, 80 mg/L fumaric acid, 40 mg/L Na-pyruvate, 2000 mg/L glutamine, 500 mg/L casein hydrolysate, 200 mg/L myo-inositol, 300 mg/L xylose, 54 g/L maltose. After filter sterilization of the modified N6Z medium, mix with equal volume of 1.6% autoclaved gelrite solution.

6. 10× LS major salts: Dissolve 19.0 g KNO_3, 16.5 g NH_4NO_3, 4.4 g $CaCl_2{\cdot}2H_2O$, 3.7 g $MgSO_4{\cdot}7H_2O$, and 1.7 g KH_2PO_4 in 900 mL deionized distilled water (DDW) and fill to 1000 mL. Store at 4 °C.
7. 100× ferric ethylenediaminetetraacetic acid (FeEDTA): Dissolve 2.78 g $FeSO_4{\cdot}7H_2O$ in 900 mL of hot DDW and add 3.73 g ethylenediamine-*N, N, N′, N′*-tetraacetic acid, disodium salt. Fill to 1000 mL. Store at 4 °C.
8. 100× LS minor salts: Dissolve 2.23 g $MnSO_4{\cdot}5H_2O$, 1.06 g $ZnSO_4{\cdot}7H_2O$, 620 mg H_3BO_3, 83 mg KI, 25.0 mg $Na_2MoO_4{\cdot}2H_2O$, 2.5 mg $CuSO_4.5H_2O$, and 2.5 mg $CoCl_2{\cdot}6H_2O$ in 900 mL of DDW and fill to 1000 mL. Store at 4 °C.
9. 100× Modified LS vitamins: Dissolve 10 g myoinositol, 100 mg thiamine hydrochloride, 50 mg pyridoxine hydrochloride, and 50 mg nicotinic acid in 900 mL of DDW and fill to 1000 mL. Store at 4 °C.
10. Regeneration medium (Fig. 1d): 100 mL of the 10× LS major salts, 10 mL of 100× FeEDTA, 10 mL of 100× LS minor salts, 10 mL of 100× Modified LS vitamins, 50 mL of 100 mg/L zeatin, 20 g sucrose, 0.5 g MES, 2.5 mg $CuSO_4{\cdot}5H_2O$ were added to 700 mL of DDW and make up the volume to 1000 mL. After adjusting pH to 5.8 with NaOH, add 8 g agar and autoclave at 121 °C for 15 min.
11. Rooting medium (Fig. 1e): 100 mL of the 10× LS major salts, 10 mL of 100× Fe-EDTA, 10 mL of 100× LS minor salts, 10 mL of 100× Modified LS vitamins, 2 mL of 100 mg/l IBA,

Fig. 1 (continued) positioned in the center of a droplet, fixed at one of the electrodes under an AC field and fused by multiple DC pulses (5–7 times). The in vitro-produced zygote is cultured on the membrane of a Millicell insert containing N6Z medium and feeder cells. (**b**) Zygote divides into compact embryonic callus through early embryonic development in the Millicell insert. (**c**) Compact embryonic callus develops to callus on the callus induction medium. (**d**, **e**) Thereafter, the callus is transferred onto regeneration and rooting medium to obtain regenerated shoots and plantlet, respectively

15 g sucrose, and 0.5 g MES were added to 700 mL of DDW and make up the volume to 1000 mL. Adjust pH to 5.8 with NaOH and add 3 g Gelrite. Autoclave at 121 °C for 15 min.

3 Methods

A complete flow of wheat gametes isolation, electrofusion and its subsequent culture are demonstrated in Fig. 1. Egg cells and sperm cells are isolated from mature unpollinated wheat spikes that have yellowish anthers (Fig. 1a). Thereafter, these gametes are transferred into freshly made fusion droplets (0.5–1.0 μL) on a coverslip, fused electrically and cultured (Fig. 1a). The zygote is developed into compact embryonic callus on a Millicell membrane insert in N6Z medium with feeder cells (Fig. 1b), and the compact embryonic callus is transferred onto callus induction medium (Fig. 1c). The resulting callus is further transferred onto regeneration medium (Fig. 1d), and the regenerated shoots are transferred eventually to the rooting medium to induce root formation (Fig. 1e). Culture steps in Fig. 1c–e are conducted as described by Ishida et al. (2015) [46].

3.1 Isolation of Wheat Gametes

1. Collect the wheat spikes that have emerged from the culm. Pick up the unpollinated spikelets from the bottom to middle part of the wheat spikes (*see* **Note 5**) and dissect them under the stereo microscope using straight dissection forceps with fine tips. Isolate the pistils and stamens, and transfer them separately into 3.5 cm plastic dishes containing 3 mL of mannitol solution (370 mOsmol/kg H_2O) for isolating the egg and sperm cells, respectively.
2. For egg cell isolation, sink the pistils to the bottom of new 3.5 cm plastic dishes filled with 3 mL of the mannitol solution using straight fine forceps and cut them diagonally with a razor blade, leaving the cut ovaries alone without stigmas and styles (*see* Fig. 2a–d, **Note 6**). Gently push the chalazal part (basal part) of the cut ovaries using a blunt glass needle (*see* **Note** 7). Approximately 5–7 egg cells are transferred into fresh mannitol droplets on coverslip using glass capillary connected with manual handling injector under an inverted microscope (*see* Fig. 2e–f, **Note 8**).
3. For sperm cell isolation, immerse the anthers in mannitol solution and roughly break them with forceps to let the anthers release its pollen grains (*see* Fig. 2g). Use the sperm cells released from the burst pollen grains for electrofusion (*see* Fig. 2h, **Note 9**).

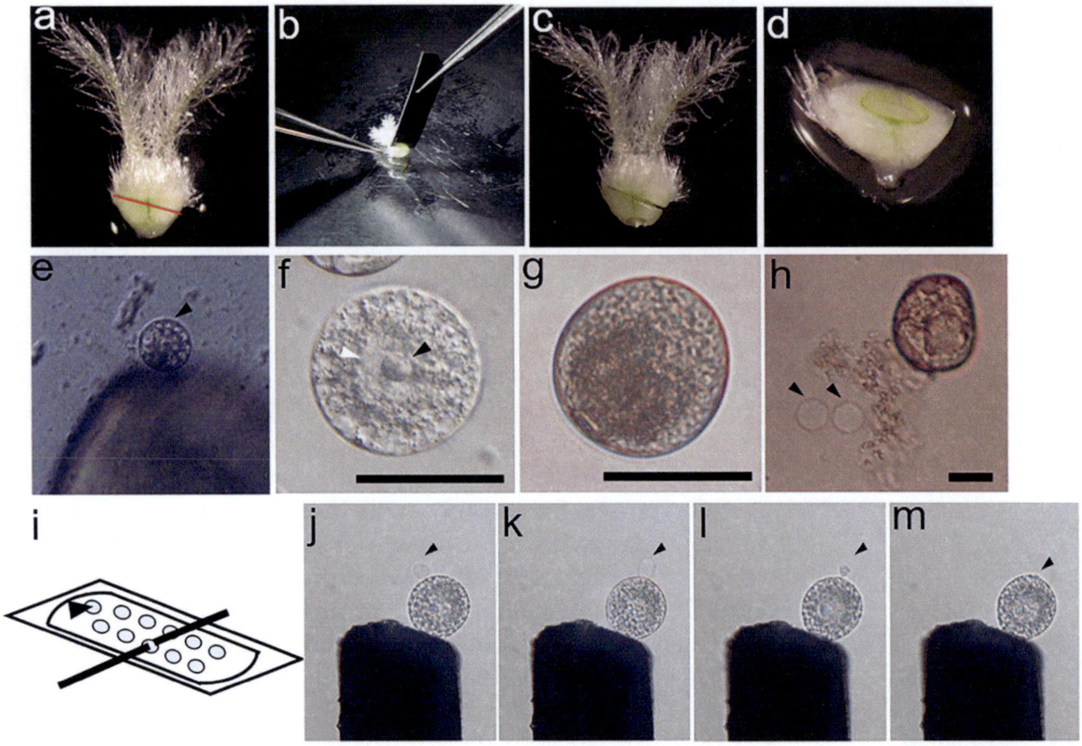

Fig. 2 Procedures for isolating the egg cells and sperm cells and for in vitro fusion of wheat gametes. (**a**) A mature pistil with developed stigma harvested from an unpollinated wheat flower. The red line indicates the incision line on the ovary for egg cell isolation. (**b**) The stigma of the pistil is fixed with forceps, and the fixed ovary is cut with a small blade using the other forceps. (**c**) A diagonally cut wheat ovary. (**d**) Side view of cut ovary. (**e**) An egg cell released from the basal portion of a dissected ovary. The arrowhead indicates the released egg cell. (**f**) An isolated wheat egg cell. The white- and black-colored arrowheads indicate the nucleus and nucleolus, respectively. (**g**) A pollen grain from wheat anther. The pollen grains are immersed in 370 mOsmol kg/H_2O mannitol solution to induce bursting. (**h**) A burst pollen grain releasing its contents. The arrowheads indicate two sperm cells released from a pollen grain. (**i**) An illustration of the coverslip containing fusion droplets overlaid with mineral oil. The black arrowhead indicates a mannitol droplet for tentative storage of isolated egg cells for subsequent fusion. (**j**) Alignment of wheat gametes at one of the electrodes under alternating current (AC) field. An arrowhead indicates the sperm cell which is attached to the egg cell. (**k**) Aligned egg cell and sperm cell after the addition of mannitol solution with a relatively high osmolality (520 mOsmol kg/H_2O) to the fusion droplet. The attachment of the gametes to the electrode becomes more stable and sperm cell becomes elongated (arrowhead). (**l**) Fusion of gametes following multiple negative direct current (DC) pulses. The arrowhead indicates a fusing sperm cell. (**m**) A zygote at 1 min after fusion. The arrowhead indicates the point of fusion. Bars = 50 μm in (**f**, **g**) and 20 μm in (**h**)

3.2 Fusion of Gametes

1. Overlay a siliconized coverslip with 0.28 mL mineral oil. For tentative storage of isolated egg cells, make 0.5–1.0 μL mannitol droplets (370 mOsmol/kg H_2O) using glass capillary connected with manual handling injector. In addition, make 2 μL mannitol droplets (370 mOsmol/kg H_2O) in two rows, each with five droplets (*see* Fig. 2i, **Note 10**).

2. Transfer one egg cell to each mannitol droplet and distribute one sperm cell to each of the droplets (*see* **Note 11**).
3. Set up the fusion generator and fine-tune the position of electrodes according to gametes positions in the droplet.
4. Align and fix the two gametes at one electrode under an alternating current (AC) field (1 MHz, 5 V rms). By adjusting the position of electrodes using manipulator, first attach an egg cell to the electrode. Thereafter, fix a sperm cell to the female gamete by a gentle movement of manipulator (*see* Fig. 2j). Adjust the final distance of the electrodes to approximately half to one time the sum of the diameter of the cell.
5. Add approximately 0.5–1.0 μL of mannitol solution (520 mOsmol/kg H_2O) gently to the fusion droplet using a thin glass capillary (*see* Fig. 2k, **Note 12**).
6. Induce cell fusion by applying 5–7 multiple negative direct current (DC) pulses (40 μs, 12–15 kV/cm) (*see* Fig. 2l, m, **Note 13**).
7. Detach the fusion products from the electrode by gently tapping the manipulator. Move the electrodes out of the droplet and conduct the next gamete fusion (*see* Subheading 3.2, **step 4**, **Note 14**).

3.3 Culture of Zygotes into Embryo-Like Structure and Plantlets

1. Place 0.2 mL zygote culture medium in a Millicell insert and put the insert into a 3.5 cm plastic dish containing 2 mL of the medium. Add 40–60 μL of wheat suspension cells culture into the dish as feeder cells.
2. Sterilize the microcapillary by washing it with absolute ethanol and sterilized water and transfer the fusion products (zygotes) by using sterilized microcapillary into fresh mannitol droplets (450 mOsmol/kg H_2O) three times. Then transfer those zygotes onto the membranes of a Millicell insert carefully (*see* Fig. 1a).
3. Culture the zygotes at 26 °C in the dark overnight without shaking, and continue the subsequent culture with gentle shaking (30 rpm) (*see* Fig. 3a–i, **Notes 15** and **16**).
4. Five to seven days after fusion, remove feeder cells by transferring the Millicell dishes containing the embryos into new 3.5 cm plastic dishes filled with 2 mL of fresh zygote culture medium (*see* Fig. 3h, **Note 17**). Continue culturing as above.
5. After 21 days of culture, subculture the compact embryonic calli from the IVF-produced zygotes onto callus induction medium by the use of sterilized forceps. Incubate under continuous light at 30 °C for 2 weeks (*see* Fig. 1c, Fig. 3j).

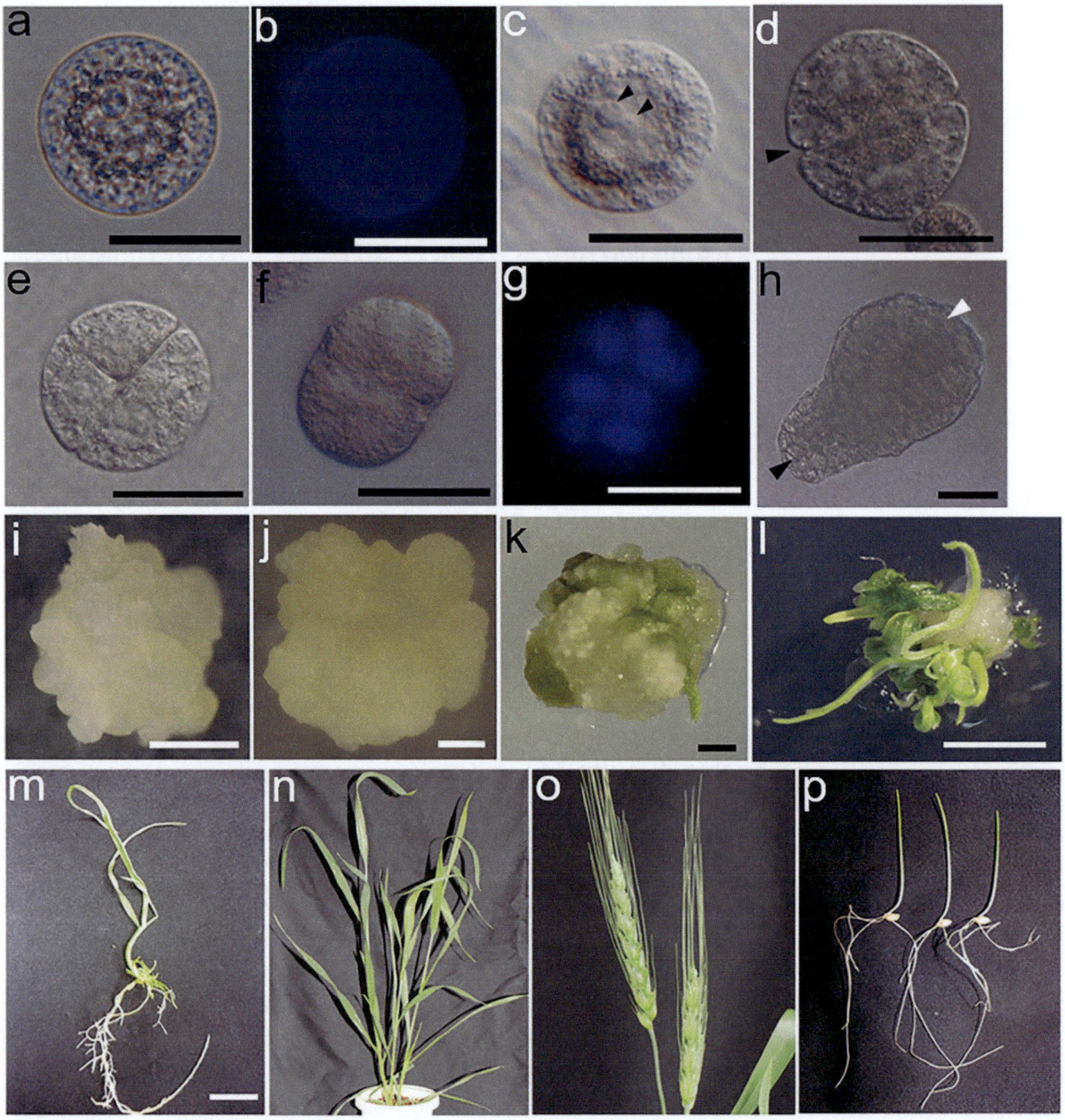

Fig. 3 Development of IVF-produced wheat zygotes into early embryos and regeneration of plantlets derived from the IVF-produced club-shaped embryos. (**a**, **b**) Cell wall staining of an IVF-produced zygote at 5 h after gamete fusion. Panels **a** and **b** are bright-field and fluorescence images, respectively. (**c**) A zygote at 5 h after fusion. Two nucleoli are indicated by arrowheads. (**d**) A two-celled embryo at 25 h after fusion. The arrowhead indicates the cleavage furrow. (**e**) An early embryo at 44 h after fusion. (**f**, **g**) Nuclear staining of a globular-like embryo at 78 h after fusion. Panels **f** and **g** are bright-field and fluorescence images, respectively. (**h**) A club-shaped embryo at 7 days after fusion. The white- and black-colored arrowheads indicate a possible embryo proper and suspensor, respectively. (**i**) A compact embryonic callus (21 days after fusion). (**j**) A callus that is developed from the compact embryonic callus on the callus induction medium (35 days after fusion, Fig. 1c). (**k**) A regenerating callus 4 days after transferring the callus in panel **j** to regeneration medium (39 days after fusion). Green-colored parts are visible in/on the callus. (**l**) Regenerated shoots. Generation of shoots is observed after 14 days of subculturing the callus on regeneration medium (49 days after fusion, Fig. 1d). (**m**) A regenerated plantlet (63 days after fusion, Fig. 1e). (**n**) A regenerated plant at 41 days after being transferred to a soil pod (104 days after fusion). (**o**) Flowers of a regenerated plant (111 days after fusion). (**p**) Five-day-old seedlings germinated from seeds that were harvested from regenerated plant. Bars = 50 μm in (**a–h**), 1 mm in (**i–k**) and, and 1 cm in (**l**, **m**)

6. After 14 days of incubation, subculture the calli from the IVF-produced zygotes onto regeneration medium by the use of sterilized forceps. Incubate under continuous light at 30 °C for the other 2 weeks (*see* Fig. 1d, Fig. 3k, l, **Note 18**).
7. Transfer the regenerated shoots onto a rooting medium and culture them under a 13 h/11 h light/dark cycle at 30 °C for 2 weeks (*see* Fig. 1e, Fig. 3m).
8. Transfer the resulting plantlets to soil pods and grown in environmental chambers as described in **Note 1** (*see* Fig. 3n, *see* **Note 19**). If needed, harvest the seeds from the regenerated plants and germinate them (*see* Fig. 3o, p).

4 Notes

1. Wheat plants (*Triticum aestivum* L. cv. Fielder) are grown in an environmental room under conditions of 23 °C under light (18 h)/20 °C dark (6 h) cycle with a photosynthetic photon flux density of 100–150 mmol photons/m^2/s. Spikes can be obtained throughout all seasons under these growth conditions. The maturity of the spikes can be judged by the well-developed stigma of the ovaries and the yellowish anthers. The harvested spikes can be immediately used for gamete isolation or stored at 4 °C for 7 days before use.
2. Straight dissection forceps with fine tips are needed to easily remove the palea, lemma, lodicule part and stamen of the flower, leaving the pistil alone.
3. Coverslips should be noncoated, as using coated coverslips will result in attachment of the cells to the surface of the coverslip.
4. Wheat suspension cells (line HT3–1) provided by KANEKA Corp. (Tokyo, Japan) are subcultured once weekly. No difference has been observed in feeder effects between freshly subcultured cells and 1-week-cultured cells.
5. To collect the unpollinated spikelets, the middle part of the spikes is preferable because most of the pistils have appropriate size. Since the pollination of wheat spikes is not synchronous, if the middle part is already pollinated, the egg cells can be isolated from the bottom part of wheat spikes. Refrain from taking the spikelets that have small size of wheat pistils and green anthers. Although the egg cells are possible to be isolated, the size of the egg cells would be small and immature.
6. The procedure of cutting the pistils (*see* Fig. 2b–d) must be conducted in mannitol solution. In addition, since the cut

ovaries often levitate on the mannitol solution, sinking the cut ovaries into the mannitol solution is essential in order to isolate the egg cells efficiently. Usually, 15–25 cut ovaries are put into a dish.

7. Two types of glass needles are necessary to efficiently isolate the egg cells. A fine tip of glass needle is used to hold the left side of cut ovaries and another blunt tip of glass needle is used to gently squash the cut ovaries to obtain the egg cells. Sometimes 1–2 egg cells are automatically released from approximately 20 cut ovaries. Gently push the bottom to the middle position of lower part of the cut ovaries with a glass needle to release additional egg cells. Breaking the cut ovaries results in the damage of egg cells and inefficient egg cells isolation.
8. Egg cells can be kept in the mannitol droplet at room temperature up to 6 h after isolation for conducting electrofusion without decrease in fusion efficiency.
9. Sperm cells should be used for IVF within 15 min after isolation to increase the possibility of a successful gamete fusion. Using isolated sperm cells 20 min after isolation is also possible, although the fusion efficiency will be slightly reduced. Sperm cells appear to degenerate around 25–30 min after isolation and such cells cannot be fused with the egg cells. Repeated isolation of the sperm cells from the fresh anthers is preferable to conduct a successful gamete fusion.
10. Volume of mannitol droplets can vary between 0.5 and 2 μL. The mannitol droplets should be located inside the oil without the access to the air, since access of the droplets to the air results in increase of osmolality via water evaporation and in degeneration of the gametes in the droplets.
11. At each round of fusion procedures, five sets of gamete fusions are recommended because conducting many sets of gamete fusion takes time and sperm cell will degenerate during the course of the experiments.
12. The addition of mannitol solution with a higher osmolality changes the shape of the sperm cell to oblong and makes the attachment of the egg cell to the electrode more stable (*see* Fig. 2j, k). Without this treatment, egg cells are often released from the electrode upon fusion induced by a DC pulse, and fusion efficiency is greatly reduced.
13. If no cell fusion occurs, reduce the distance between the two electrodes and pulse again. Alternatively, transfer the egg cell and a newly isolated sperm cell into a new droplet and fuse again.

14. The efficiency of successful electrofusion is approximately 65% under optimal conditions. Three egg cells can be isolated routinely from approximately 10 excised ovaries, and 10–15 egg cells can be fused with the sperm cells by 1 experimenter per day.
15. Gamete fusion usually takes approximately 1–2 min since plasma membrane of wheat egg cell is physically rigid (*see* Fig. 2l, m) and 30–60 min settling the fused product (zygote) on the fusion droplet is required to allow the zygote to form stable cell wall before transferring to the culture medium. The IVF-produced wheat zygote starts to form cell wall (*see* Fig. 3a, b) and two nucleoli can be observed in a zygote within 4–5 h after fusion (*see* Fig. 3c). The first symmetric cell division of the zygote is observed at 25 h after fusion (*see* Fig. 3d). After the first division, the two-celled embryo continues to develop into early and late globular-like embryo at 44 and 78 h after fusion, respectively (*see* Fig. 3e–g).
16. Approximately 90% IVF-produced zygotes divide into two-celled embryos, and approximately 90% of IVF-produced two-celled embryos develop into globular-like embryos.
17. After 5–7 days of culture of the IVF-produced zygotes, cocultivation with feeder cells is not needed.
18. Normally, after 4 days of subculture of the callus (*see* Fig. 3j) on a solidified regeneration medium, green spots become visible (39 days after fusion, *see* Fig. 3k) and the emergence of multiple shoots is observed after 14 days of subculturing on the rooting medium (49 days after fusion, *see* Fig. 3l).
19. A regenerated plant is obtained approximately at 41 days after being transferred to a soil pod (104 days after fusion, *see* Fig. 3n). Normal flowers of a regenerated plant are observed (*see* 111 days after fusion, Fig. 3o) and are consistent with the flowers of wild-type wheat plant.

References

1. Russell SD (1992) Double fertilization. Int Rev Cytol 40:357–390
2. Raghavan V (2003) Some reflections on double fertilization, from its discovery to the present. New Phytol 159:565–583
3. Berger F (2011) Imaging fertilization in flowering plants, not so abominable after all. J Exp Bot 62:1651–1658
4. Hamamura Y, Nagahara S, Higashiyama T (2012) Double fertilization on the move. Curr Opin Plant Biol 15:70–77
5. Hamamura Y, Nishimaki M, Takeuchi H et al (2014) Live imaging of calcium spikes during double fertilization in *Arabidopsis*. Nat Commun 5:4722
6. Maruyama D, Völz R, Mori TH et al (2015) Rapid elimination of the persistent synergid through a cell fusion mechanism. Cell 161: 907–918
7. Kranz E, Lörz H (1993) In vitro fertilization with isolated, single gametes results in zygotic embryogenesis and fertile maize plants. Plant Cell 5:739–746
8. Uchiumi T, Uemura I, Okamoto T (2007) Establishment of an in vitro fertilization system in rice (*Oryza sativa* L.). Planta 226:581–589

9. Wang YY, Kuang A, Russell SD et al (2006) In vitro fertilization as a tool for investigating sexual reproduction of angiosperm. Sex Plant Reprod 19:103–115
10. Kranz E (1999) In vitro fertilization with isolated single gametes. Methods Mol Biol 111:259–267
11. Okamoto T (2011) In vitro fertilization with isolated rice gametes: production of zygotes and zygote and embryo culture. Methods Mol Biol 710:17–27
12. Kranz E, Bautor J, Lörz H (1991) In vitro fertilization of single, isolated gametes of maize mediated electrofusion. Sex Plant Reprod 4:12–16
13. Uchiumi T, Komatsu S, Koshiba T et al (2006) Isolation of gametes and central cells from *Oryza sativa* L. Sex Plant Reprod 19:37–45
14. Faure JE, Digonnet C, Dumas C (1994) An in vitro system for adhesion and fusion of maize gametes. Science 263:1598–1600
15. Kranz E, Lörz H (1994) In vitro fertilization of maize by single egg and sperm cell protoplast fusion mediated by high calcium and high pH. Zygote 2:125–128
16. Khalequzzaman M, Haq N (2005) Isolation and in vitro fusion of egg and sperm cells in *Oryza sativa*. Plant Physiol Biochem 43:69–75
17. Sun M, Yang H, Zhou C et al (1995) Single-pair fusion of various combinations between female gametoplasts and other protoplasts in *Nicotiana tabacum*. Acta Bot Sin 37:1–6
18. Tian HQ, Russell SD (1997b) Micromanipulation of male and female gametes of *Nicotiana tabacum*: II. Preliminary attempts for in vitro fertilization and egg cell culture. Plant Cell Rep 16:657–661
19. Peng XB, Sun MX, Yang HY (2005) A novel in vitro system for gamete fusion in maize. Cell Res 15:734–738
20. Faure JE, Mogensen HL, Dumas C et al (1993) Karyogamy after electrofusion of single egg and sperm cell protoplasts from maize: cytological evidence and time course. Plant Cell 5:747–755
21. Ohnishi Y, Hoshino R, Okamoto T (2014) Dynamics of male and female chromatin during karyogamy in rice zygotes. Plant Physiol 165: 1533–1543
22. Ohnishi Y, Okamoto T (2017) Nuclear migration during karyogamy in rice zygotes is mediated by continuous convergence of actin meshwork toward the egg nucleus. J Plant Res 130:339–348
23. Ohnishi Y, Kokubu I, Kinoshita T et al (2019) Sperm entry into the egg cell induces the progression of karyogamy in rice zygotes. Plant Cell Physiol 60:1656–1665
24. Kranz E, von Wiegen P, Lörz H (1995) Early cytological events after induction of cell division in egg cells and zygote development following in vitro fertilization with angiosperm gametes. Plant J 8:9–23
25. Nakajima K, Uchiumi T, Okamoto T (2010) Positional relationship between the gamete fusion site and the first division plane in the rice zygote. J Exp Bot 6:3101–3105
26. Sato A, Toyooka K, Okamoto T (2010) Asymmetric cell division of rice zygotes located in embryo sac and produced by in vitro fertilization. Sex Plant Reprod 23:211–217
27. Scholten S, Lörz H, Kranz E (2002) Paternal mRNA and protein synthesis coincides with male chromatin decondensation in maize zygotes. Plant J 32:221–231
28. Hoshino Y, Scholten S, von Wiegen P et al (2004) Fertilization induced changes in the microtubular architecture of the maize egg cell and zygote—an immunocytochemical approach adapted to single cells. Sex Plant Reprod 17:89–95
29. Okamoto T, Scholten S, Lörz H et al (2005) Identification of genes that are up- or down-regulated in the apical or basal cell of maize two-celled embryos and monitoring their expression during zygote development by a cell manipulation- and PCR-based approach. Plant Cell Physiol 46:332–338
30. Toda E, Ohnishi Y, Okamoto T (2016) Development of polyspermic rice zygotes. Plant Physiol 171:206–214
31. Toda E, Ohnishi Y, Okamoto T (2018) Effects of an imbalanced parental genome ratio on development of rice zygotes. J Exp Bot 69: 2609–2619
32. Deushi R, Toda E, Koshimizu S (2021) Effect of paternal genome excess on the developmental and gene expression profiles of polyspermic zygotes in rice. Plan Theory 10:255
33. Rahman MH, Toda E, Kobayashi M (2019) Expression of genes from paternal alleles in rice zygotes and involvement of *OsASGR-BBML1* in initiation of zygotic development. Plant Cell Physiol 60:725–737
34. Jahnke S, Scholten S (2009) Epigenetic resetting of a gene imprinted in plant embryos. Curr Biol 19:1677–1681
35. Toda E, Koiso N, Takebayashi A et al (2019) An efficient DNA- and selectable-marker-free genome-editing system using zygotes in rice. Nat Plants 5:363–368

36. Koiso N, Toda E, Ichikawa M (2017) Development of gene expression system in egg cells and zygotes isolated from rice and maize. Plant Direct 1:e00010
37. Brenchley R, Spannagl M, Pfeifer M et al (2012) Analysis of the bread wheat genome using whole-genome shotgun sequencing. Nature 491:705–710
38. International Wheat Genome Sequencing Consortium (2014) A chromosome-based draft sequence of the hexaploid bread wheat (*Triticum aestivum*) genome. Science 345: 1251788
39. Clavijo BJ, Venturini L, Schudoma C et al (2017) An improved assembly and annotation of the allohexaploid wheat genome identifies complete families of agronomic genes and provides genomic evidence for chromosomal translocations. Genome Res 27:885–896
40. Zimin AV, Puiu D, Luo MC et al (2017) Hybrid assembly of the large and highly repetitive genome of *Aegilops tauschii*, a progenitor of bread wheat, with the mega-reads algorithm. Genome Res 27:213405–213116
41. International Wheat Genome Sequencing Consortium (2018) Shifting the limits in wheat research and breeding using a fully annotated reference genome. Science 361:eaar7191
42. Kumlehn J, Lörz H, Kranz E (1998) Differentiation of isolated wheat zygotes into embryos and normal plants. Planta 205:327–333
43. Bakos F, Darko E, Ponya Z et al (2003) Application of wheat (*Triticum aestivum* L.) microspore culture and ovaries to raise wheat zygotes in vitro. Acta Biol Cracov Bot 45:107–110
44. Kovács M, Barnabás B, Kranz E (1995) Electro-fused isolated wheat (*Triticum aestivum* L.) gametes develop into multicellular structures. Plant Cell Rep 15:178–180
45. Maryenti T, Kato N, Ichikawa M et al (2019) Establishment of an in vitro fertilization system in wheat (*Triticum aestivum* L.). Plant Cell Physiol 60:835–843
46. Ishida Y, Tsunashima M, Hiei Y et al (2015) Wheat (*Triticum aestivum* L.) transformation using immature embryos. Methods Mol Biol 1223:189–198

Part IV

Bioinformatics and Data Analysis

Chapter 19

High-Throughput Fluorescent Pollen Tetrad Analysis Using DeepTetrad

Dohwan Byun and Kyuha Choi

Abstract

Meiotic recombination initiates from ~100–200 s of programmed DNA double stranded breaks (DSBs) in plants. Meiotic DSBs can be repaired using homologous chromosomes to generate a crossover. Meiotic crossover is critical for chromosomal segregation and increasing genetic variation. The number of crossovers is limited to one and three per chromosome pair in most plant species. Genetic, epigenetic, and environmental factors control crossover frequency and distribution. Due to the limited number of crossovers it is challenging to measure crossover frequency along chromosomes. We adapted fluorescence-tagged lines (FTLs) that contain *quartet1* mutations and linked transgenes expressing dsRed, eYFP, and eCFP in pollen tetrads into the deep learning-based image analysis tool, DeepTetrad. DeepTetrad enables the measurement of crossover frequency and interference by classifying 12 types of tetrads from three-color FTLs in a high-throughput manner, using conventional microscope instruments and a Linux machine. Here, we provide detailed procedures for preparing tetrad samples, tetrad imaging, running DeepTetrad, and analysis of DeepTetrad outputs. DeepTetrad-based measurements of crossover frequency and interference ratio will accelerate the genetic dissection of meiotic crossover control.

Key words Meiosis, Crossover, Tetrad analysis, QRT1, FTLs, DeepTetrad

1 Introduction

Meiosis is a particular type of cell division that occurs in sexually reproducing organisms to produce haploid gametes [1, 2]. Meiosis consists of a single round of DNA replication and two consecutive cell divisions, producing four daughter cells from a single progenitor cell [1, 2]. During meiosis prophase I, homologous chromosomes undergo pairing and alignment by programmed DNA double-strand breaks (DSBs) and homology-directed repair (HDR), which can produce a reciprocal crossover between nonsister chromosomes [1, 2]. Meiotic recombination is essential for genetic diversity and proper segregation of homologous chromosomes via at least one crossover per chromosome pair [1, 2].

Christophe Lambing (ed.), *Plant Gametogenesis: Methods and Protocols*, Methods in Molecular Biology, vol. 2484, https://doi.org/10.1007/978-1-0716-2253-7_19,

Meiotic recombination is initiated by the formation of programmed DSBs by the SPO11 topoisomerase-like protein complex [3]. Meiotic DSBs are repaired by the HDR pathway and only a subset of DSBs are repaired using homologous chromosomes as templates and proceed into crossover products, indicating that the crossover number is severely restricted via diverse antirecombination pathways [1, 4]. In *Arabidopsis*, two pathways are known to cause crossovers. The major crossover pathway, known as the class I pathway, is governed by a group of ZMM proteins, MLH1 and MLH3. Class I crossovers are responsible for 80–85% of crossovers [1, 5, 6]. Class I crossovers are subjected to interference, which prevents crossovers from being spaced close to another crossover [7]. Class II crossovers represent approximately ~15–20% of crossovers. Class II crossovers are noninterfering and are partially dependent on the MUS81 endonuclease [1, 8].

Cytological analysis and meiotic product analysis are the two major experimental methods used to observe crossover patterns and frequency [9]. Cytological methods rely on microscopy to identify the chromosome features related to crossover formation in addition to axis formation, DSB formation, and synapsis [9]. The morphology of bivalent chromosomes and chiasmata when applied with DAPI staining provides evidence of the number and position of crossovers [10, 11]. Immunohistochemistry of MLH1 foci at late pachytene, diplotene, or diakinesis can identify the number and position of Class I crossovers [12, 13]. The main advantage of cytological analysis is that it allows for observation and quantification of crossover events in single meiocytes as well as the correlation between cytological traits and physiological characteristics, such as fertility and viability [12]. However, cytological methods are limited in their ability to identify the exact number and location at which crossovers occur in a high-throughput manner [9].

There are two approaches for analyzing meiotic products according to which genetic markers can be observed [9]. The first method uses single nucleotide polymorphisms (SNPs) as genetic markers in natural accessions or varieties in plants, which can be identified by PCR or whole-genome sequencing [9]. SNPs provide the exact number and location of crossovers on a genome-wide scale [14]. However, it is necessary to use hybrid plants to achieve a genome-wide SNP density. The second method uses fluorescence-tagged lines (FTLs) that contain T-DNAs expressing different fluorescent proteins in the seed coat or pollen. Segregation analysis of fluorescent seeds or pollen grains from FTLs allow the measurement of genetic distance between hemizygous fluorescent markers in individual plants in a high-throughput manner [15]. Seed fluorescent markers express *eGFP* and *dsRED* reporter genes in the seed coat using the seed-specific *napA* storage protein (*NAP)* gene promoter [16]. Alternatively, fluorescent pollen markers express three different fluorescent colors (dsRed, eYFP, and eCFP) in mature pollen using the pollen-specific *LAT52* promoter

[17]. Pollen FTLs were created in *a quartet1* mutant background, making four daughter cells from one male meiotic progenitor cell physically attached to one another. Thus, the three-color pollen FTL system enables visual tetrad analysis that measures the crossover frequency and interference in *Arabidopsis* [15, 18]. Seed and pollen FTL systems allow measuring crossover frequency in individual plants by analyzing the segregation of fluorescence reporters in a high-throughput manner [15, 19]. In particular, the seed FTL system has been used for natural variation studies and genetic screening of altered crossover frequency mutants using the CellProfiler tool for automatic scoring [20, 21].

Microscopy is the first and most straightforward method for *Arabidopsis* fluorescent pollen tetrad analysis [15, 19]. However, general image analysis tools such as CellProfiler and ImageJ cannot detect pollen tetrads efficiently for tetrad analysis. It takes time to manually analyze and classify 12 tetrad types for tetrad analysis. Flow cytometry has been used in pollen FTLs in *qrt/+* and wild-type plants for high-throughput pollen analysis. However, flow cytometry requires specific equipment and additional experimental steps to obtain a high-purity of pollen grains [9, 22]. In addition, flow cytometric analysis of crossover is not able to provide substantial information regarding double crossovers in a single interval [9]. Recently, deep learning-based object recognition has been applied to various scientific and industrial fields, including bioimaging [23, 24]. We developed a deep leaning-based pollen image analysis tool, DeepTetrad, for fluorescent pollen tetrad analysis. DeepTetrad precisely recognizes pollen tetrads and single pollen grains with a pretrained recognition model and then combines the fluorescent intensity of pollen for tetrad analysis [22]. DeepTetrad rapidly classifies tetrads into 12 different types from three color-FTL hemizygous plants according to the segregation pattern of three fluorescent genes of FTL to calculate the crossover rate and interference ratio [15, 22]. DeepTetrad has improved the efficiency and accuracy of crossover measurements, which accelerates the genetic dissection of meiotic recombination in *Arabidopsis.* We describe the procedures for establishing the DeepTetrad package, sampling pollen tetrads, and fluorescent pollen tetrad analysis using DeepTetrad.

2 Materials

2.1 Pollen Tetrad Imaging

1. *Arabidopsis thaliana qrt1* (Col) seeds (Arabidopsis Biological Resource Center, CS8846).
2. *Arabidopsis thaliana FTL* seeds.
3. PGM solution: 17% w/v sucrose, 2 mM $CaCl_2$, 1.625 mM boric acid, 0.1% v/v Triton-X-100, pH 7.5, titration with NaOH. The complete solution is filtered with 0.2 μm filter and stored at 4 °C.

4. 1.5 mL e-tube.
5. Fine-point stainless-steel tweezers.
6. 80 μm nylon mesh (cut into 5 × 5 cm).
7. 1 mL, 200 μL, 10 μL pipette set and pipette tips.
8. Slide glass (76 × 26 × 1 mm, flat).
9. Cover glass (9 × 9 mm, #1 thickness).
10. Rotator (suitable for 1.5 mL e-tube).
11. Centrifuge.
12. Stereo fluorescent microscope and connected PC with imaging program.
13. ET RFP filter for stereo fluorescent microscope.
14. ET YFP filter for stereo fluorescent microscope (*see* **Note 1**).
15. ET CFP filter for stereo fluorescent microscope.
16. Linux machine with Ubuntu 18.04.05 LTS 64 bit and CUDA-enabled GPU with at least 12 GB graphic memory.

2.2 Install DeepTetrad

1. The DeepTetrad pipeline requires a CUDA-enabled GPU with at least 12 GB graphic memory for running and enables the measurement of crossover frequency and interference by precisely classifying 12 types of fluorescent pollen tetrads (Fig. 1).
2. Install DeepTetrad with the given script
   (a) Install Anaconda (https://www.anaconda.com/products/individual).
   (b) Create a Conda environment

```
conda create -n pac
```

   (c) Install DeepTetrad in Linux system.

```
conda activate pac
conda install -c abysslover deeptetrad
```

   (d) Install DeepTetrad in Windows system.

```
conda activate pac
conda install -c abysslover deeptetrad
pip install deeptetrad
```

   (e) Run DeepTetrad.

```
conda activate pac
deeptetrad --physical_loc=T_DNA.txt --path=./test
```

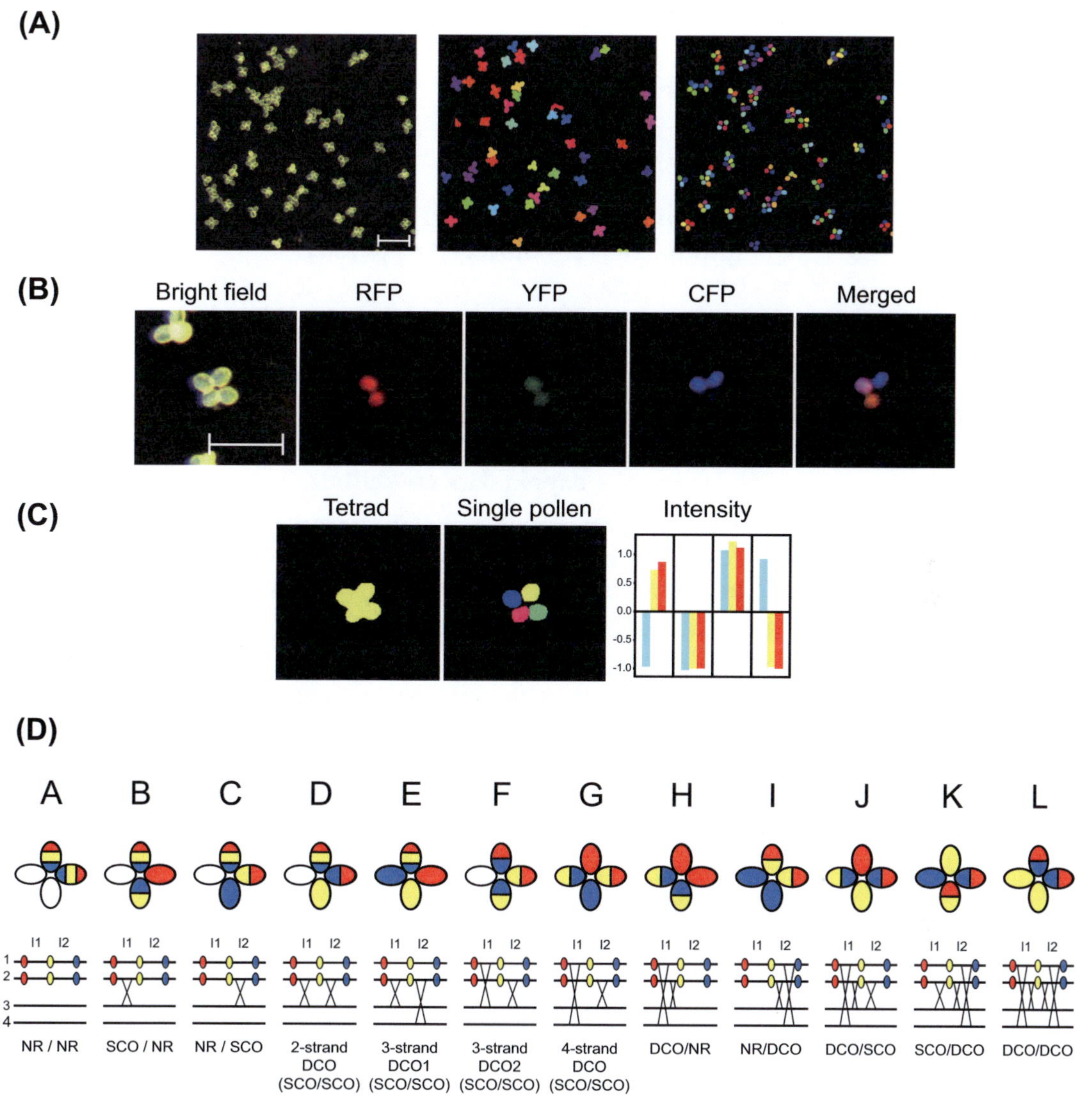

Fig. 1 Principal of pollen type classification in DeepTetrad. (**a**) Tetrad and single pollen recognition by DeepTetrad. DeepTetrad generates screens of segmented tetrads and single pollen grains and combines two screens to obtain sets of pollens within the same tetrad. (**b**) Bright field, RFP, YFP, CFP filter, and merged fluorescent image of a pollen tetrad (**c**) Tetrad and single pollen screen of a pollen tetrad from (**b**) and measured fluorescent intensity (RFP, YFP, CFP) of each pollen grain. (**d**) Diagram of twelve tetrad types with crossover events from three-color FTL (*RYC/+++*). The three transgenes are marked as a colored dot and four chromatid strands are numbered (1–4). Three transgenes define two adjacent intervals (I1 and I2) and crossover events are described as lines between chromatid strands. The twelve tetrad types are no recombination (**a**), a single crossover of interval 1 (**b**; SCO/NR), single crossover of interval 2 (**c**; NR/SCO), two-strand double crossover (**d**; 2-strand DCO), three-strand double crossover 1 (**e**; 3-strand DCO1), three-strand double crossover 2 (**f**; 3-strand DCO2), four-strand double crossover (**g**; 4-strand DCO), double crossover interval 1, noncrossover interval 2 (**h**; DCO/NR), noncrossover interval 1, double crossover interval 2(**i**; NR/DCO), double crossover interval 1, single crossover interval 2 (**j**; DCO/SCO), single crossover interval 1, double crossover interval 2 (**k**; SCO/DCO), and double crossover interval 1, double crossover interval 2 (**l**; DCO/DCO)

(f) Update DeepTetrad.

```
conda activate pac
conda update -c abysslover deeptetrad
deeptetrad -r
```

(A more detailed installation guide is available at https://github.com/abysslover/deeptetrad).

3 Methods

3.1 FTL Plant Growth

1. Sow the seeds of homozygote FTLs (Fluorescence Tagged Lines, e.g., FTL *I3bc YRC/YRC qrt1*) in soil and incubate at 4 °C for 3–4 days in dark conditions.
2. Transfer and grow the homozygote FTLs (Fluorescence Tagged Lines) in a grow room or chamber at 20 °C under long-day conditions (16 h light/8 h dark).
3. Cross the FTLs with mutants of interest to produce F_1 plants (*I3bc YRC/+++*, *mutant/+*, *qrt1/+*). Simultaneously, cross these mutants with *qrt1* to generate F_1 plants (*mutant/+*, *qrt1/+*) if the T-DNA of FTLs, are located across *the qrt1* mutation site of chromosome 5 (e.g., *I5ab*).
4. Cross FTLs with *qrt1* to produce F_1 plants as the control. The control F1 plants are *qrt1* homozygotes and scorable for tetrad analysis.
5. Grow the F_1 plants and self-fertilize them (*see* **Note 2**).
6. Once F_1 plants flower, take a flower sample using a tweezers and spread the pollen grains on a glass slide.
7. Determine the phenotype of the fluorescence of pollens from F_1 plants under a fluorescent microscope and confirm whether separate fluorescent pollen grains are linked or segregated (*see* **Note 3**).
8. Harvest seeds of F_1 plants.

3.2 Phenotyping for Measurable FTLs

1. Grow the F_2 seeds of self-fertilized F_1 plants (*I3bc YRC/+++*, *mutant/+*, *qrt1/+*) (*see* **Note 4**).
2. Determine the phenotype of each F_2 plant by observing the fluorescent pollen tetrad pattern under a fluorescent microscope.
3. Identify the plants with scorable tetrads (*YRC/+++; qrt1*) for tetrad analysis (Fig. 2). *YRC/YRC; qrt1* plants can be harvested or crossed with *+++/+++; qrt1* to produce F_1 plants.

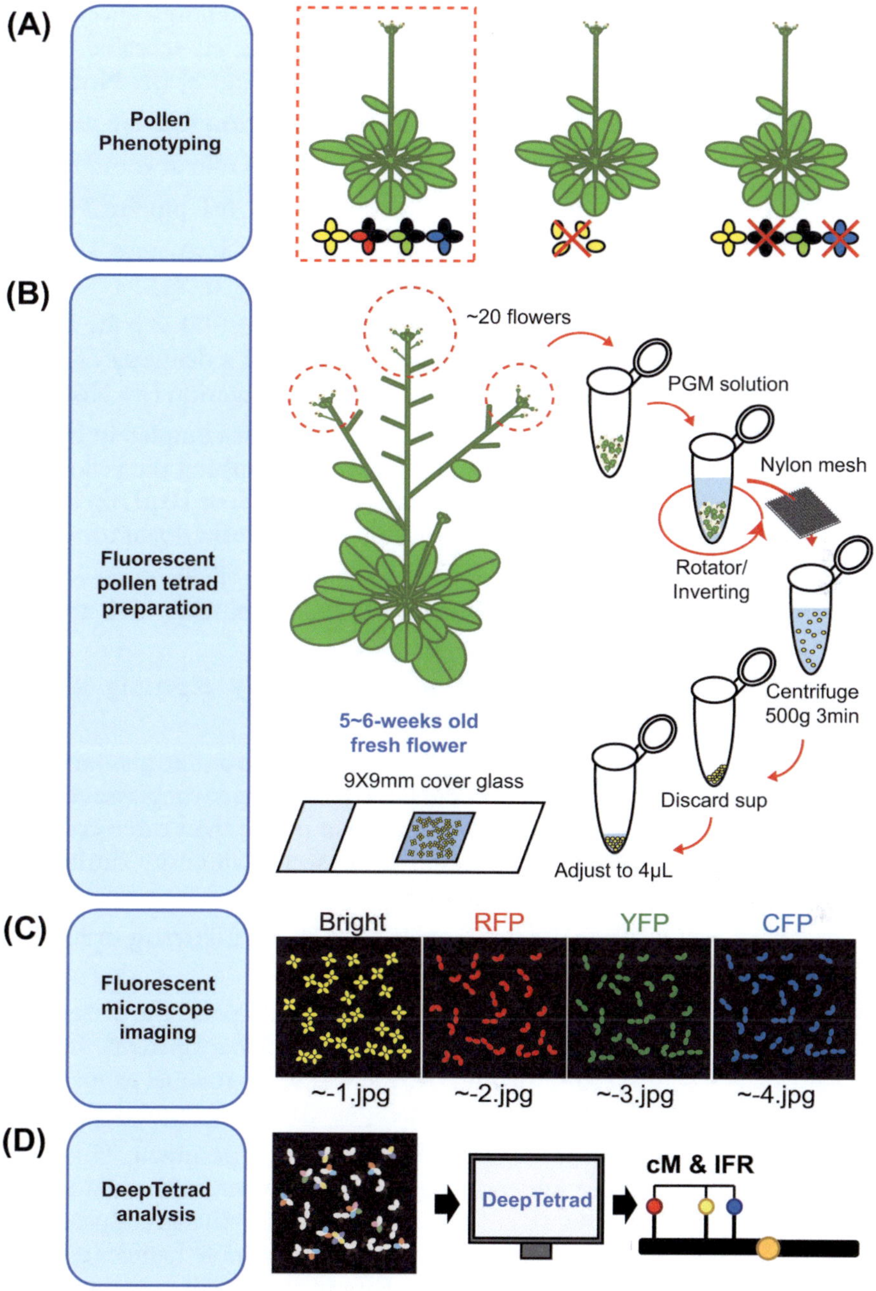

Fig. 2 Workflow of high-throughput fluorescent pollen tetrad analysis by DeepTetrad. Pollen tetrad analysis using DeepTetrad has four sample preparation steps. (**a**) Pollen phenotyping. Perform pollen phenotyping of the *Arabidopsis* FTL plants and select *Arabidopsis* plants hemizygous for all three fluorescent reporters and in *quartet* background. (**b**) Fluorescent pollen tetrad preparation. Sample flowers and prepare slides for fluorescent pollen imaging. (**c**) Fluorescent microscope imaging. Take photos of fluorescent pollen tetrads under a microscope, using bright, RFP, YFP and CFP filters. (**d**) Run DeepTetrad to analyze the image files and calculate crossover frequency and interference ratio

3.3 Pollen Tetrad Imaging

1. Collect ~20 open flowers or ~6 inflorescences from primary and secondary shoots of preselected scorable FTL plants in 1.5 mL e-tubes using tweezers (Fig. 2) (*see* **Note 5**).
2. Add 1 mL of ice-cold PGM solution to a 1.5 mL e-tube.
3. Gently mix with the rotator for 3 min at room temperature.
4. Gently mix the flowers within a 1 mL pipette 2–3 times.
5. Pipet all the solution and filter into new 1.5 mL e-tubes through a 80 μm nylon mesh (5 × 5 cm).
6. Centrifuge the filtered solution at 500 × *g* for 3 min at room temperature. A yellow pellet with a diameter of approximately 2–3 mm will appear after centrifugation (*see* **Note 6**).
7. Carefully remove the supernatant completely by pipetting or vacuum aspiration. To avoid disturbing the yellow pellet, place the pellet upward and use a 200 μL or 10 μL tip for pipetting or vacuum aspiration. If the pellet breaks down too quickly, repeat **step 6** to obtain the pollen pellet again.
8. Add ~2–3 μL of fresh PGM solution and adjust the total solution volume to 4 μL.
9. Resuspend the pellet by gently pipetting 4–5 times (*see* **Note 7**).
10. Transfer 4 μL of pollen solution to a slide glass and cover it with 9 × 9 mm cover glass. Up to three cover glasses can placed on a single slide glass. However, more than three cover glasses will result in decreased fluorescence intensity during microscope imaging (*see* **Note 8**).
11. Place the fluorescent filter in the following order: bright field, ET RFP, ET YFP, and ET CFP.
12. Set the filter to a bright field and place the prepared slide glass on the mount with a solid black background. In the reference system, the microscope was set to manual exposure mode with 50 ms exposure time, gain 10.1×, gamma 1.5, saturation 1.0.
13. Set experiment folder for each experiment. Under the experiment folder, set grandparent folders with used FTL traffic line names (e.g., *I1bc, I3bc*). Under the grandparent folder, set parent folders with individual sample names (e.g., *I1bc WT-1, I3bc fancm-3*). The Parent folder will contain pictures from a single cover glass.

    Example of Folder structures:

Experiment folder:	D:/DeepTetrad/I1bc fancm F2
Grandparent folder:	D:/DeepTetrad/I1bc fancm F2/I1bc

(continued)

Parent folder:	D:/DeepTetrad/I1bc fancm F2/I1bc/I1bc WT-1
Individual images:	D:/DeepTetrad/I1bc fancm F2/I1bc/I1bc WT-1/1-1.jpg

14. Adjust the focus and set the magnification at 50×, bright field images, ET RFP, ET YFP, and ET CFP.
15. Save four images from the same pollen with each different filter with same sample name and different suffix according to each used filter (-1.jpg for the bright field, -2.jpg for ET RFP, -3.jpg for ET YFP, and -4.jpg for ET CFP). The resolution of the fluorescent image should be 2560 × 1920. A representative image for each fluorescent filter can be seen in Fig. 3.
16. Repeat **steps 14** and **15** with different sections of the cover glass and twelve sets (12 × 4 images) of images will be generated from a single cover glass with 50× magnification.

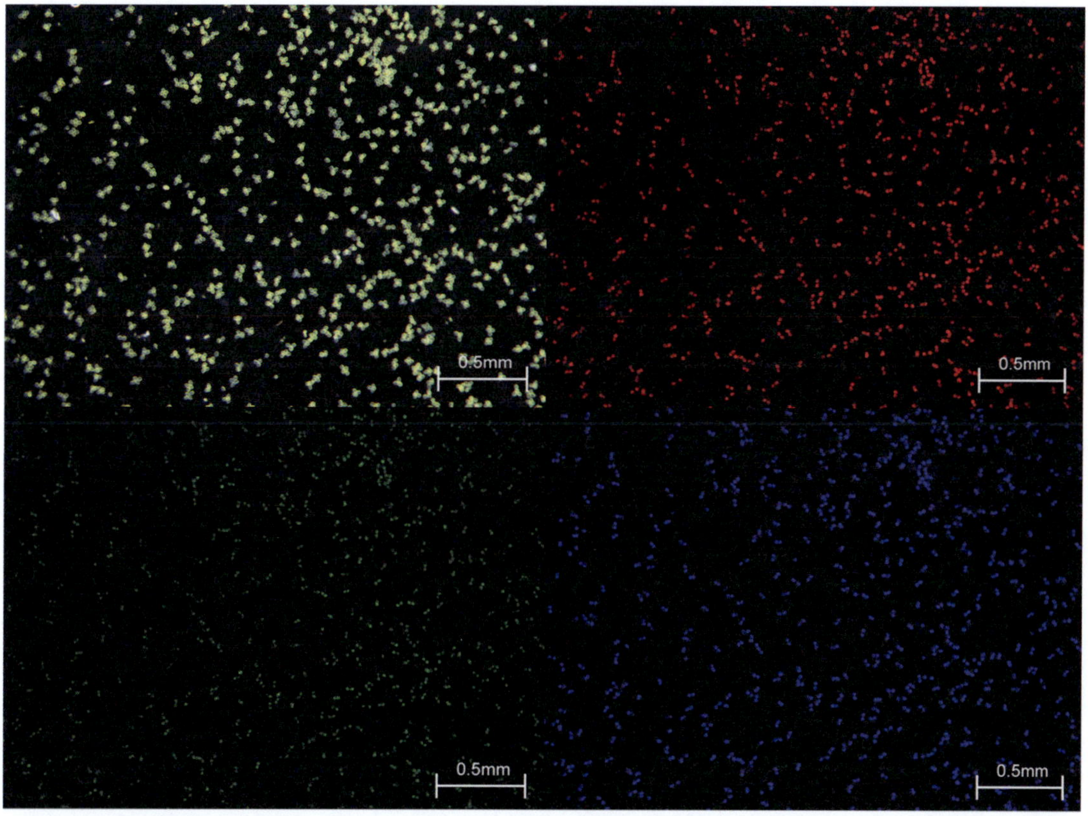

Fig. 3 Representative fluorescent pollen tetrad image for DeepTetrad analysis. Images of fluorescent pollen tetrads from the bright field, ET RFP, ET YFP, and ET CFP filter with 50× magnification. DeepTetrad requires a suitable density of tetrads for accurate analysis. In the bright field image, pollens should form tetrads and each pollen grains of tetrad should be distinguished from one another. In the three fluorescent images, pollens with fluorescent should be distinguished from pollens without fluorescent

3.4 Running DeepTetrad and Data Analysis

1. Activate conda environment which contains DeepTetrad package (pac for example).

```
conda activate pac
```

2. Move to experiment folder location.

```
cd (experiment folder location)
```

3. Run DeepTetrad analysis.

```
deeptetrad --path=./
```

4. DeepTetrad running takes approximately an hour per ~7000 tetrads in the reference system. Therefore, running time will alter the system requirements and image quality. Seven CSV files will be the final outputs of the three-color FTLs and four CSV files for the two-color FTLs. Tables 1 and 2 provide a brief explanation of the contents of each file (file name and information) (*see* **Note 9**).
5. DeepTetrad provides the pollen number, interval, and interference ratio of each individual (Table 3). In addition, the interval and interference ratio can be calculated using tetrad type data (*see* **Note 10**). Data given by DeepMonad is calculated using only a single pollen type classification.

Table 1
Final output of DeepTetrad for three-color analysis (*I1bc, I1fg, I2ab, I2fg, I3bc, I5ab*). DeepTetrad pipeline analyzes fluorescent pollen tetrad images of three-color FTL and provides 7 csv files with names and information

File name	Information
DeepTetrad_each_file_data_three_channels.csv	Summary of data from each file (image sets)
DeepTetrad_individual_interferences_three_channels.csv	Interference ratio of each individual (parent folders)
DeepTetrad_individual_intervals_three_channels.csv	Interval of each individual (parent folders)
DeepTetrad_individual_types_three_channels.deepmonad.csv	Pollen type of each individual (parent folders)
DeepTetrad_individual_types_three_channels.deeptetrad.csv	Tetrad type of each individual (parent folders)
DeepTetrad_line_intervals_three_channels.deepmonad.csv	Interval of each line (grandparent folders) analyzed with pollen type
DeepTetrad_line_intervals_three_channels.deeptetrad.csv	Interval of each line (grandparent folders) analyzed with tetrad type

Table 2
Final output of DeepTetrad for two-color analysis (*CEN3*). DeepTetrad pipeline analyzes fluorescent pollen tetrad images of two-color FTL and provides 4 csv files with names and information

File name	Information
DeepTetrad_each_file_data_two_channels.csv	Summary of data from each file (image sets)
DeepTetrad_individual_intervals_two_channels.csv	Interval of each individual (parent folders)
DeepTetrad_individual_types_two_channels.deepmonad.csv	Pollen type of each individual (parent folders)
DeepTetrad_individual_types_two_channels.deeptetrad.csv	Tetrad type of each individual (parent folders)

DeepTetrad calculates the crossover frequency of each interval and the IFR in a three-color assay with the given equations [25].

$$\chi_1\ (\text{Interval 1}) = \frac{0.5\text{T}_1 + 3\text{NPD}_1}{\text{PD}_1 + \text{T}_1 + \text{NPD}_1} = \frac{0.5(\text{B}+\text{D}+\text{E}+\text{F}+\text{G}+\text{K}) + 3(\text{H}+\text{J}+\text{L})}{\textit{Total tetrad}} \tag{1}$$

$$\chi_2\ (\text{Interval 2}) = \frac{0.5\text{T}_2 + 3\text{NPD}_2}{\text{PD}_2 + \text{T}_2 + \text{NPD}_2} = \frac{0.5(\text{C}+\text{D}+\text{E}+\text{F}+\text{G}+\text{J}) + 3(\text{I}+\text{K}+\text{L})}{\textit{Total tetrad}} \tag{2}$$

$$\chi_3\ (\text{Interval 3}) = \frac{0.5\text{T}_3 + 3\text{NPD}_3}{\text{PD}_3 + \text{T}_3 + \text{NPD}_3} = \frac{0.5(\text{B}+\text{C}+\text{E}+\text{F}+\text{J}+\text{K}) + 3(\text{G}+\text{H}+\text{I})}{\textit{Total tetrad}} \tag{3}$$

$$\chi_\delta\ (\text{w/o adj CO}) = \frac{0.5\text{T}_\delta + 3\text{NPD}_\delta}{\text{PD}_\delta + \text{T}_\delta + \text{NPD}_\delta} = \frac{0.5(\text{B}) + 3(\text{H})}{(\text{A}) + (\text{B}) + (\text{H})} \tag{4}$$

$$\chi_\gamma\ (\text{w adj CO}) = \frac{0.5\text{T}_\gamma + 3\text{NPD}_\gamma}{\text{PD}_\gamma + \text{T}_\gamma + \text{NPD}_\gamma} = \frac{0.5(\text{D}+\text{E}+\text{F}+\text{G}+\text{K}) + 3(\text{J}+\text{L})}{(\text{C}+\text{I}) + (\text{D}+\text{E}+\text{F}+\text{G}+\text{K}) + (\text{J}+\text{L})} \tag{5}$$

$$\sigma\ (\text{IFR}) = \frac{\chi_\gamma}{\chi_\delta} \tag{6}$$

DeepTetrad calculates crossover frequency in two-color assay with Perkin's equation [26].

$$\text{cM} = \frac{0.5T + 3\text{NPD}}{(\text{PD} + T + \text{NPD})} \times 100$$

Table 3

Measurement of crossover frequency and interference ratio of FTLs by DeepTetrad. DeepTetrad counts twelve tetrad types from three-color FTLs and calculates the crossover frequency of intervals and interference ratio. The crossover pattern of each letter code is described in Fig. 1d. The crossover frequency of interval 1 is calculated from Eq. (1), interval two from Eq. (2), and interval three from Eq. (3). Interval 3 indicates an interval of longest interval combination from three T-DNAs. The crossover interference ratio (IFR, Eq. (6)) is the ratio of the map distance with adjacent crossover (w adj CO, Eq. (5)) to the map distance without adjacent crossover (w/o adj CO, Eq. (4))

FTL	A	B	C	D	E	F	G	H	I	J	K	L	total	interval 1	interval 2	interval 3	w/o adj CO	w adj CO	IFR
I1bc	9457	2397	6144	123	97	115	104	24	108	3	12	0	18,584	8.10	19.66	27.40	0.11	0.03	0.33
I1fg	11,869	1732	474	17	2	4	5	8	0	0	1	0	14,112	6.41	1.80	8.12	0.07	0.03	0.44
I2ab	1430	57	108	0	0	1	1	0	1	0	0	0	1596	1.85	3.63	5.57	0.02	0.01	0.47
I2fg	8389	1410	909	24	9	7	8	2	3	0	0	0	10,761	6.83	4.53	11.21	0.07	0.03	0.34
I3bc	6292	3377	811	34	28	32	40	47	6	2	0	0	10,669	17.83	4.61	22.53	0.19	0.08	0.41
I5ab	3006	2745	1590	89	88	91	84	51	14	8	6	0	7772	22.24	13.32	34.88	0.26	0.10	0.39

4 Notes

1. It is essential not to use ET GFP filter because the GFP spectrum overlaps to CFP.
2. Beware not to give too much stress, such as drought or heat. Under such stress conditions, fluorescent reporters can be silenced, and pollen number can decrease.
3. If F_1 plants are heterozygous for *qrt1*, they only produce single pollen grains.
4. Scorable mutant FTL plant (*I3bc YRC/+++; mutant qrt1)* is one per 32–40 F_2 plants so it is recommended to sow at least 50–120 F_2 seeds.
5. Approximately 500–600 tetrads from ~20 flowers can be collected and recognized by DeepTetrad. More flowers will result in more tetrads for analysis but result in improper tetrad recognition because of high pollen density.
6. If more or fewer flowers are used, the pellet diameter will change with the number of pollen grains.
7. If the prepared tetrad breaks down into a single pollen grain, lower the acceleration and deceleration of the centrifuge.
8. After sample preparation, pollens can hold for approximately 2 h at 4 °C or 30 min under room temperature conditions. After that, the fluorescent intensity decreases and fails to obtain a good image for analysis.
9. Suppose you want to collect multiple individual data in a single output. In that case, you can add the tetrad type of each individual and calculate the interval and interference ratio with the sum of the tetrad type.
10. Approximately 500 tetrads are sufficient to calculate the interval of an individual, and ~2000 tetrads are sufficient to calculate the interference ratio of the individual.

References

1. Mercier R, Mézard C, Jenczewski E et al (2015) The molecular biology of meiosis in plants. Annu Rev Plant Biol 66:297–327
2. Villeneuve AM, Hillers KJ (2001) Whence meiosis? Cell 106:647–650
3. Grelon M, Vezon D, Gendrot G et al (2001) AtSPO11-1 is necessary for efficient meiotic recombination in plants. EMBO J 20:589–600
4. Ferdous M, Higgins JD, Osman K et al (2012) Inter-homolog crossing-over and synapsis in Arabidopsis meiosis are dependent on the chromosome Axis protein AtASY3. PLoS Genet 8:e1002507
5. Li Y, Qin B, Shen Y et al (2018) HEIP1 regulates crossover formation during meiosis in rice. Proc Natl Acad Sci U S A 115: 10810–10815
6. Mercier R, Jolivet S, Vezon D et al (2005) Two meiotic crossover classes cohabit in Arabidopsis: one is dependent on MER3, whereas the other one is not. Curr Biol 15:692–701

7. Copenhaver GP, Housworth EA, Stahl FW (2002) Crossover interference in Arabidopsis. Genetics 160:1631–1639
8. Hollingsworth NM, Brill SJ (2004) The Mus81 solution to resolution: generating meiotic crossovers without Holliday junctions. Genes Dev 18:117–125
9. Yelina NE, Ziolkowski PA, Miller N et al (2013) High-throughput analysis of meiotic crossover frequency and interference via flow cytometry of fluorescent pollen in Arabidopsis thaliana. Nat Protoc 8:2119–2134
10. López E, Pradillo M, Oliver C et al (2012) Looking for natural variation in chiasma frequency in Arabidopsis thaliana. J Exp Bot 63: 887–894
11. Sanchez-Moran E, Armstrong SJ, Santos JL et al (2002) Variation in chiasma frequency among eight accessions of Arabidopsis thaliana. Genetics 162:1415–1422
12. Chelysheva L, Grandont L, Vrielynck N et al (2010) An easy protocol for studying chromatin and recombination protein dynamics during Arabidopsis thaliana meiosis: Immunodetection of Cohesins, histones and MLH1. Cytogenet Genome Res 129:143–153
13. Lhuissier FGP, Offenberg HH, Wittich PE et al (2007) The mismatch repair protein MLH1 marks a subset of strongly interfering crossovers in tomato. Plant Cell 19:862–876
14. Rowan BA, Heavens D, Feuerborn TR et al (2019) An ultra high-density Arabidopsis thaliana crossover map that refines the influences of structural variation and epigenetic features. Genetics 213:771–787
15. Francis KE, Lam SY, Harrison BD et al (2007) Pollen tetrad-based visual assay for meiotic recombination in Arabidopsis. Proc Natl Acad Sci U S A 104:3913–3918
16. Melamed-Bessudo C, Yehuda E, Stuitje AR et al (2005) A new seed-based assay for meiotic recombination in Arabidopsis thaliana. Plant J 43:458–466
17. van Tol N, Rolloos M, van Loon P et al (2018) MeioSeed: A CellProfiler-based program to count fluorescent seeds for crossover frequency analysis in Arabidopsis thaliana. Plant Methods 14:32
18. Rhee SY, Somerville CR (1998) Tetrad pollen formation in quartet mutants of Arabidopsis thaliana is associated with persistence of pectic polysaccharides of the pollen mother cell wall. Plant J 15:79–88
19. Berchowitz LE, Copenhaver GP (2008) Fluorescent Arabidopsis tetrads: a visual assay for quickly developing large crossover and crossover interference data sets. Nat Protoc 3:41–50
20. Ziolkowski PA, Underwood CJ, Lambing C et al (2017) Natural variation and dosage of the HEI10 meiotic E3 ligase control Arabidopsis crossover recombination. Genes Dev 31: 306–317
21. Lawrence EJ, Gao H, Tock AJ et al (2019) Natural variation in TBP-ASSOCIATED FACTOR 4b controls meiotic crossover and germline transcription in Arabidopsis. Curr Biol 29: 2676–2686.e3
22. Lim EC, Kim J, Park J et al (2019) DeepTetrad: high-throughput image analysis of meiotic tetrads by deep learning in Arabidopsis thaliana. Plant J 1:1–11
23. Moen E, Bannon D, Kudo T et al (2019) Deep learning for cellular image analysis. Nat Methods 16:1233–1246
24. Eraslan G, Avsec Ž, Gagneur J et al (2019) Deep learning: new computational modelling techniques for genomics. Nat Rev Genet 20: 389–403
25. Perkins DD (1962) Crossing-over and interference in a multiply marked chromosome arm of Neurospora. Genetics 47:1253–1274
26. Perkins DD (1949) Biochemical mutants in the smut fungus Ustilago maydis. Genetics 34: 607–626

Chapter 20

Barley (*Hordeum Vulgare*) Anther and Meiocyte RNA Sequencing: Mapping Sequencing Reads and Downstream Data Analyses

Miriam Schreiber, Jamie Orr, Abdellah Barakate, and Robbie Waugh

Abstract

RNA sequencing (RNA-seq) data is by now the most common method to study differential gene expression. Here we present a pipeline from RNA-seq generation to analysis with examples based on our own barley anther and meiocyte transcriptome. The bioinformatics pipeline will give everyone, from a beginner to a more experienced user, the possibility to analyze their datasets and identify significantly differentially expressed genes. It also allows differential alternative splicing analysis which will become increasingly common due to the high regulatory impact on the gene expression. We describe use of the Galaxy interface for RNA-seq read quantification and the 3D RNA-seq app for the downstream data analysis.

Key words RNA sequencing, Differential gene expression, Alternative splicing, 3D-RNA-seq, Barley, Meiosis

1 Introduction

Analyzing RNA-seq data has become the norm for studying differential gene expression. It enables the identification of alternatively spliced transcripts and new genes or transcripts in the tissues studied. Reads can be mapped to either a genome or a transcriptome reference. In this methods chapter we will focus on differential expression and alternative splicing analysis which is based on mapping the reads to an available reference transcriptome. Several plant transcriptomes are available through EnsemblPlants (http://plants.ensembl.org/index.html). If a de novo transcriptome is required, the following papers are good starting points to understand their construction: [1, 2]. The appropriate number of replicates and required sequencing depth for RNA-seq experiments is a fiercely debated topic [3], which makes providing specific recommendations very difficult. In general, more replicates and greater sequencing depth are likely to increase reproducibility and

Christophe Lambing (ed.), *Plant Gametogenesis: Methods and Protocols*, Methods in Molecular Biology, vol. 2484, https://doi.org/10.1007/978-1-0716-2253-7_20,

resolution, respectively. This should be given considerable thought during the experimental design stage to be able to achieve meaningful results.

In plants around 60–70% of genes undergo alternative splicing [4, 5]. Including alternative splicing analysis instead of only focusing on differential gene expression is important as genes might show a stable overall transcript abundance but the proportion of a particular isoform contributing to the total transcript abundance might have switched. For our analysis we are using a so-called padded transcriptome where all isoforms for the corresponding gene have been modified to contain the same 3′ and 5′ positions as this has been shown to improve quantification in previous experiments in *Arabidopsis* [5].

In our anther and meiocyte transcriptome analysis [6] we were able to show that while transcription appears stable throughout meiotic prophase I, the major transcriptional regulation happens at the onset of meiosis. The change from premeiosis to the leptotene–zygotene stage showed a significant downregulation of long noncoding RNAs. Studying those transcriptional changes in other plant species will be of great interest.

To make the bioinformatics pipeline for RNA-seq analysis as easy to use as possible for everyone here we use our previously published samples [6]. The experiment was set up with four sample groups covering a combination of anthers that contain meiocytes at different developmental stages embedded within the biological tissue, and two sample groups of isolated meiocytes themselves. We had three biological replicates of each sample group (Table 1). The meiotic stages cover barley anthers and meiocytes before and during meiotic prophase I.

2 Materials

2.1 Materials, Consumables, and Chemicals

1. 9 cm disposable petri dishes.
2. 9 cm Whatman filter paper discs of grade 1.
3. Insulin-needle attached to 1 ml syringe to detach and transfer anthers.
4. 1.5 ml Safe-Lock Eppendorf LoBind microcentrifuge tubes.
5. Microtube Pestles for grinding anthers and meiocytes in 1.5 ml Eppendorf LoBind microcentrifuge tubes.
6. 2% acetocarmine: 2 g of carmine in 100 ml of 45% glacial acetic acid.
7. Vectashield containing DAPI.
8. RNA*later*.

Table 1
SRA accessions of example RNA-seq. Data and sample information

NCBI Sequence read archive	Barley Cultivar	Anther size	Tissue type	Replicate	Anther stage
SRR9890003	Golden promise	0.3–0.4 mm anthers	Anther	1	Premeiosis
SRR9889994	Golden promise	0.3–0.4 mm anthers	Anther	2	Premeiosis
SRR9889993	Golden promise	0.3–0.4 mm anthers	Anther	3	Premeiosis
SRR9890004	Golden promise	0.5–0.9 mm anthers	Anther	1	Leptotene/Zygotene
SRR9889992	Golden promise	0.5–0.9 mm anthers	Anther	2	Leptotene/Zygotene
SRR9889991	Golden promise	0.5–0.9 mm anthers	Anther	3	Leptotene/Zygotene
SRR9889990	Golden promise	1.0–1.2 mm anthers	Anther	1	Pachytene/ Diplotene
SRR9889989	Golden promise	1.0–1.2 mm anthers	Anther	2	Pachytene/ Diplotene
SRR9889996	Golden promise	1.0–1.2 mm anthers	Anther	3	Pachytene/ Diplotene
SRR9889995	Golden promise	1.3–1.4 mm anthers	Anther	1	Metaphase1ToTetrad
SRR9890001	Golden promise	1.3–1.4 mm anthers	Anther	2	Metaphase1ToTetrad
SRR9890002	Golden promise	1.3–1.4 mm anthers	Anther	3	Metaphase1ToTetrad
SRR9890006	Golden promise	0.5–0.9 mm anthers	Meiocyte	1	Leptotene/Zygotene
SRR9890005	Golden promise	0.5–0.9 mm anthers	Meiocyte	2	Leptotene/Zygotene
SRR9889998	Golden promise	0.5–0.9 mm anthers	Meiocyte	3	Leptotene/Zygotene
SRR9889997	Golden promise	1.0–1.2 mm anthers	Meiocyte	1	Pachytene/ Diplotene
SRR9890000	Golden promise	1.0–1.2 mm anthers	Meiocyte	2	Pachytene/ Diplotene
SRR9889999	Golden promise	1.0–1.2 mm anthers	Meiocyte	3	Pachytene/ Diplotene

9. 0.01 M citrate buffer pH 4.5: Prepare 100 ml of 0.1 M citrate buffer by mixing 44.5 ml of 0.1 M citric acid monohydrate and 55.5 ml of 0.1 M trisodium citrate dihydrate to obtain a pH 4.5 buffer.
10. 1× PBS buffer.
11. TRIzol Plus RNA Purification Kit.
12. Full-length cDNA synthesis and amplification kit.

2.2 Equipment

1. Stereomicroscope with eyepiece graticule for measuring anther length.
2. Two fine tweezers for holding spikes and removing awn under stereomicroscope.
3. Cavity slide for releasing meiocytes from anthers.
4. Bioanalyzer 2100 for RNA quality checks.
5. qPCR and Qubit fluorometer for RNA quantification.

2.3 Datasets

A barley reference transcriptome (AntherTranscriptomeBAnTr-Padded.fasta) is available from figshare: https://doi.org/10.6084/m9.figshare.12136773.v1

The corresponding RNA-seq sequence datasets are available through NCBI [7] (sequence read archive numbers in Table 1).

2.4 Software and Tools

The proposed bioinformatics pipeline has the advantage of being highly flexible and adjustable. The tools for RNA-seq processing can be used on the public Galaxy server [8] (as shown in this example) or used as command line tools on a high-performance Linux cluster. Differential gene expression analysis can be done through a web-based user interface (3D RNA-seq app) [9] as shown in the example or using the individual R packages wrapped in the 3D RNA-seq app. Therefore, this pipeline is suitable for people confident in bioinformatics who want the flexibility of changing parameters, but also for people with limited prior experience of RNA-seq analysis.

2.4.1 RNA-Seq Data Processing

For RNA-seq processing we used the Galaxy web platform on the public server (usegalaxy.org).

1. Faster Download and Extract Reads in FASTQ [7] is a tool based on fastqer-dump from the SRA toolkit (http://ncbi.github.io/sra-tools/). It downloads the data from NCBI SRA archive onto the galaxy server in fastq format.
2. Trimmomatic [10] is a tool for removing adapter sequences and gently quality trimming at the ends.
3. Salmon [11, 12] is a tool for fast transcript quantification. It is first used to index the transcript reference and then for quantification of the RNA-seq data.

2.4.2 Differential Gene Expression and Alternative Splicing Analysis

We used the public user interface of the 3D-RNAseq app (https://3drnaseq.hutton.ac.uk/). It integrates multiple R [13] packages from Bioconductor (https://bioconductor.org/).

1. edgeR [14] is used for transcript normalization across samples.
2. RUVseq [15] stands for "remove unwanted variation" and is used to remove batch effects.
3. Limma [16] analyses differential expression.
4. TSIS [17] is available through GitHub (https://github.com/wyguo/TSIS) and is used to identify significant isoform switches between two conditions.

2.4.3 Further Downstream Analysis and Visualization

The last part of the downstream analysis is performed in R with scripts available from: https://github.com/SchreiberM/Barley-anther-and-meiocyte-transcriptome-analysis

1. The Gene Ontology (GO) analysis uses two different R packages: topGO [18] for the GO enrichment and GOPlot [19] for the visualization.
2. UpSet, Venn, and Euler diagram plotting scripts use tidyR [20] to reformat 3D RNA-seq. Data. UpSet and alluvial plots are created using UpSetR [21] and ggalluvial [22], respectively. A number of alternative programs to make Venn [16, 23–25] and Euler diagrams [26, 27] are presented.

3 Methods

3.1 Anther and Meiocyte Collection

1. Barley cv. Golden Promise plants are grown for 5–7 weeks (*see* **Note 1**) in cereal compost in a controlled growth room at 18 °C for 16 h light and 14 °C for 8 h dark and 70% humidity (*see* **Note 2**).
2. Collect one spike at a time (*see* **Note 3**) and place it in a petri dish with wet filter paper using RNase-free milli-Q water.
3. Under a stereomicroscope, use fine tweezers to remove awns and collect anthers of one floret from the middle of the spike. Measure the length of one anther and stage a second one using acetocarmine staining [28] (*see* **Note 4**).
4. Collect the remaining anthers of all florets between the fifth and fifteenth position from the bottom resulting in 60 anthers on average. Transfer isolated anthers into 1.5 ml Safe-Lock Eppendorf LoBind microcentrifuge tube containing 100 ml of RNA*later*. Set aside 10 anthers in another tube containing 1× PBS buffer for stage validation with immunostaining using meiosis specific antibodies [29] (*see* **Note 5**). Anthers in RNA*later* can be stored in the fridge until all samples are ready for RNA extraction.

5. To isolate meiocytes, anthers are collected in 1.5 ml Eppendorf tube containing 0.01 M citrate buffer pH 4.5 and kept on ice setting aside 10 anthers in another tube with 1x PBS for immunostaining as above.
6. A maximum of 5 anthers are transferred onto a cavity slide with cold 5–10 μl of citrate buffer. The slide is placed on a cooling pack under the stereomicroscope to squeeze out meiocyte columns, which are then transferred into a new 0.5 ml Eppendorf tube containing 10 μl of citrate buffer with RNase Inhibitor (*see* **Note 6**).
7. When all the anthers are processed, an aliquot of 10 μl is transferred onto a slide and the sample mounted in Vectashield® containing DAPI to check the purity of the meiocyte columns. The remaining meiocytes are topped with 50 μl TRIzol™ and stored in the fridge until all samples are ready for RNA extraction.

3.2 RNA Extraction, Quality Control and RNA-Seq Libraries

1. Extract total RNA from anthers and meiocytes using a TRIzol based kit (*see* **Note 7**).
2. Check RNA quality and integrity using a Bioanalyzer 2100 (*see* **Note 8**).
3. Quantify RNA concentration using both qPCR and Qubit fluorometer methods.
4. Perform RNA-seq on a NextSeq 550 or appropriate alternative using recommended procedures (*see* **Note 9**).

3.3 Bioinformatics Analysis

The RNA-seq analysis is split into two parts. In the first part, we will load the data onto the Galaxy web platform, perform quality control, and quantify the RNA-seq reads. In the second part, we will load the data into the 3D RNA-seq pipeline web interface then determine differential expression, differential alternative splicing, and identify genes of interest.

3.3.1 Mapping Sequencing Reads

Register an account with usegalaxy.org. After the first files have been added, the user should rename their galaxy history with a reasonable name. This will be useful for revisiting the analysis, checking parameters, and it enables the export of necessary citations.

1. Upload File.

   The Upload File tool on the galaxy web interface allows you to download a file from an online repository to the galaxy server or upload a file from your computer. Clicking the start button will add the file to the queue and upload it automatically in the background. We use the same Galaxy interface to provide a list of the needed SRR numbers by preparing a simple text file (named "samples_SRR.txt") with one SRR number per line and uploading it to the history (*see* **Note 10**).

2. Upload sequence data.

To upload the RNA-seq data onto the server we use the "Faster download and extract reads in fastq" tab. As we will be working with multiple datasets, we select as the input type "List of SRA accession, one per line" and select as the list the above prepared text file "samples_SRR.txt". After clicking "Execute" the SRR data will be uploaded to the server.

3. Perform quality control.

As RNA-seq reads can be contaminated with adapters still attached to the reads and the quality of the reads can fall off at the end, some quality control should be performed to improve the following mapping step. With Trimmomatic the reads can be trimmed to set quality scoring cutoffs using a sliding window approach. In Galaxy, the first step is to select if the data is from single-end or paired-end reads (*see* **Note 11**). It can be selected if adapter and Illumina-specific sequences should be cut. We use four different Trimmomatic operations:

(a) Drop reads if they are below a minimum length (MINLEN) of 60 nucleotides (*see* **Note 12**).

(b) Select TRAILING which removes nucleotides at the end of each read if they are below a quality threshold of 3.

(c) Select LEADING which removes nucleotides at the start of each read if they are below a quality threshold of 3.

(d) Sliding window approach which looks at the average quality score of four base calls and cuts when it falls below 20.

This produces four output files: two files are needed for the mapping approach, which are forward.paired and reverse.paired; and two files contain the corresponding unpaired reads.

4. Use Salmon tool.

(a) Index the reference transcriptome (Salmon index).

(b) Select "Paired-end Dataset Collection" to automatically select the corresponding forward and reverse reads for quantification. The relative orientation and strandedness of the reads need to be set based on the experimental design (*see* **Note 13**).

5. Once quantification is complete the data can be viewed in the galaxy browser (Fig. 1). The whole dataset can be downloaded as a collection (Fig. 1) and will be saved as a zipped folder on your computer.

3.3.2 Downstream Analysis

The next part is done on the 3D RNA-seq analysis website (https://3drnaseq.hutton.ac.uk/). For the barley anther dataset all necessary input files can be downloaded from (https://figshare.com/articles/journal_contribution/3D_RNA_seq_BAnTr_Input/11974182). Figures and tables produced when

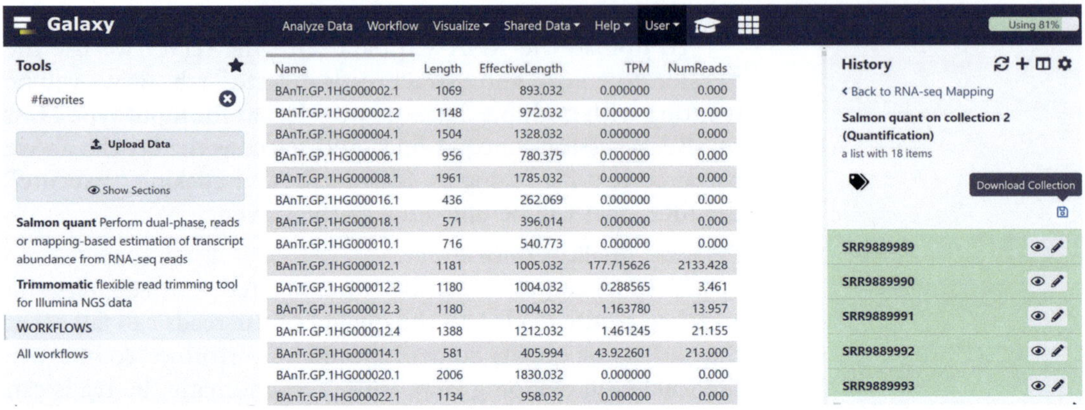

Fig. 1 Screenshot of the Galaxy user interface. On the left side is the "Tools" tab. The middle shows the first lines of the gene expression from the Salmon quant output file. The right tab shows the Galaxy history with the "Download Collection" button highlighted

running through the pipeline can be saved and will be added to a temporary working folder. All data, figures, and tables will be exported as part of the last report generation step.

1. Data preparation and generation.

   Opening the "Data generation" tab on the 3D RNA-seq app will lead you through a three-step setup.

2. Upload three different input files.
   (a) The zipped quantification file prepared and downloaded as described above.
   (b) A meta-data csv file which for our example dataset looks like Table 2.
   (c) A transcript mapping file which is a table mapping the transcript ID to the corresponding gene ID. The first 13 lines of such a file are shown in Table 3.

3. Provide the experimental design (*see* **Note 14**). After pressing "Add selected information to analysis" a visual overview is generated and should be checked to ensure that all files have been loaded and the appropriate parameters have been selected.

4. Generate the read counts and TPMs by selecting "lengthScaledTPM".

5. The next tab is "Data preprocessing" in which samples are filtered, checked for outliers, batch effects calculated, and reads normalized across the different libraries.

6. Low-abundance transcripts often prove problematic for differential gene expression and should therefore be excluded. Noise due to wrongly mapped reads and real low expressed genes are hard to distinguish which necessitates their removal. Filtering is

Table 2
A metadata table example as needed as input into 3D RNA-seq app

Condition	bio_rep	quant.files
A.Pre	rep1	AB_01-40013738
A.Pre	rep2	AB_02-40025999
A.Pre	rep3	AB_03-40016663
A.Lep/Zyg	rep1	AB_04-40023203
A.Lep/Zyg	rep2	AB_05-40024213
A.Lep/Zyg	rep3	AB_06-40024999
A.Pac/Dip	rep1	AB_07-40016670
A.Pac/Dip	rep2	AB_08-40026989
A.Pac/Dip	rep3	AB_09-40026021
A.Met/Tet	rep1	AB_10-40026992
A.Met/Tet	rep2	AB_11-40016679
A.Met/Tet	rep3	AB_12-40022212
M.Lep/Zyg	rep1	2017_01_25_M_Lept-Zyg1-42525794
M.Lep/Zyg	rep2	2017_01_25_M_Lept-Zyg2-42526802
M.Lep/Zyg	rep3	2017_01_25_M_Lept-Zyg3-42528790
M.Pac/Dip	rep1	2017_01_25_M_Pach-Dipl1-42514790
M.Pac/Dip	rep2	2017_01_25_M_Pach-Dipl2-42525832
M.Pac/Dip	rep3	2017_01_25_M_Pach-Dipl3-42516796

done by choosing a count per million (CPM) reads cutoff which needs to be exceeded in a selected number of samples. This step provides a plot to help with the decision on finding the best cut-off. The plot illustrates the log transformed read counts versus the standard variation with a mean-variance trend curve indicating a drop trend with too many low expressed transcripts which should disappear after filtering.

7. A principal component analysis (PCA) is used to identify outliers in the dataset (*see* **Note 15**).
8. Carefully consider if the data contains a batch effect based on the PCA output. In most cases no batch effect can be observed and correction for batch effects can be skipped (*see* **Note 16**).
9. Transcript normalization is done to account for the different sequencing read depth per library.
10. In the 3D analysis tab, there are two important steps which need to be performed.

Table 3
Example of the top transcript to gene ID mapping

TxID	GeneID
BAnTr.GP.1HG000002.1	BAnTr.GP.1HG000002
BAnTr.GP.1HG000002.2	BAnTr.GP.1HG000002
BAnTr.GP.1HG000004.1	BAnTr.GP.1HG000004
BAnTr.GP.1HG000006.1	BAnTr.GP.1HG000006
BAnTr.GP.1HG000008.1	BAnTr.GP.1HG000008
BAnTr.GP.1HG000016.1	BAnTr.GP.1HG000016
BAnTr.GP.1HG000018.1	BAnTr.GP.1HG000018
BAnTr.GP.1HG000010.1	BAnTr.GP.1HG000010
BAnTr.GP.1HG000012.1	BAnTr.GP.1HG000012
BAnTr.GP.1HG000012.2	BAnTr.GP.1HG000012
BAnTr.GP.1HG000012.3	BAnTr.GP.1HG000012
BAnTr.GP.1HG000012.4	BAnTr.GP.1HG000012
BAnTr.GP.1HG000014.1	BAnTr.GP.1HG000014

11. Defining contrast groups. This step takes the experimental design into account and compares different tissues or time points (for our example *see* Fig. 2 showing which groups were compared).
12. Conduct 3D analysis. To start differential gene and transcript expression analysis one needs to decide multiple parameters (*see* **Note 17**).
    (a) Choose a pipeline. The default pipeline uses the R package limma with voom transformation. In case of a looser clustering of replicates in the PCA plot, voomWeights can be chosen instead which is more lenient in the analysis.

    Set differential expression and alternative splicing thresholds. Define which adjusted *p*-value should be used to determine the statistical significance cut off for differential expression between contrast groups. Increasing this value will result in a larger number of false positives while decreasing the value might exclude biologically meaningful results. In addition, log2 fold change (FC) cutoffs must be set for differential expression between contrast groups (*see* **Note 18**). The last parameter ΔPS defines how big the difference in percentage between two transcripts needs to be to determine differential alternative splicing.

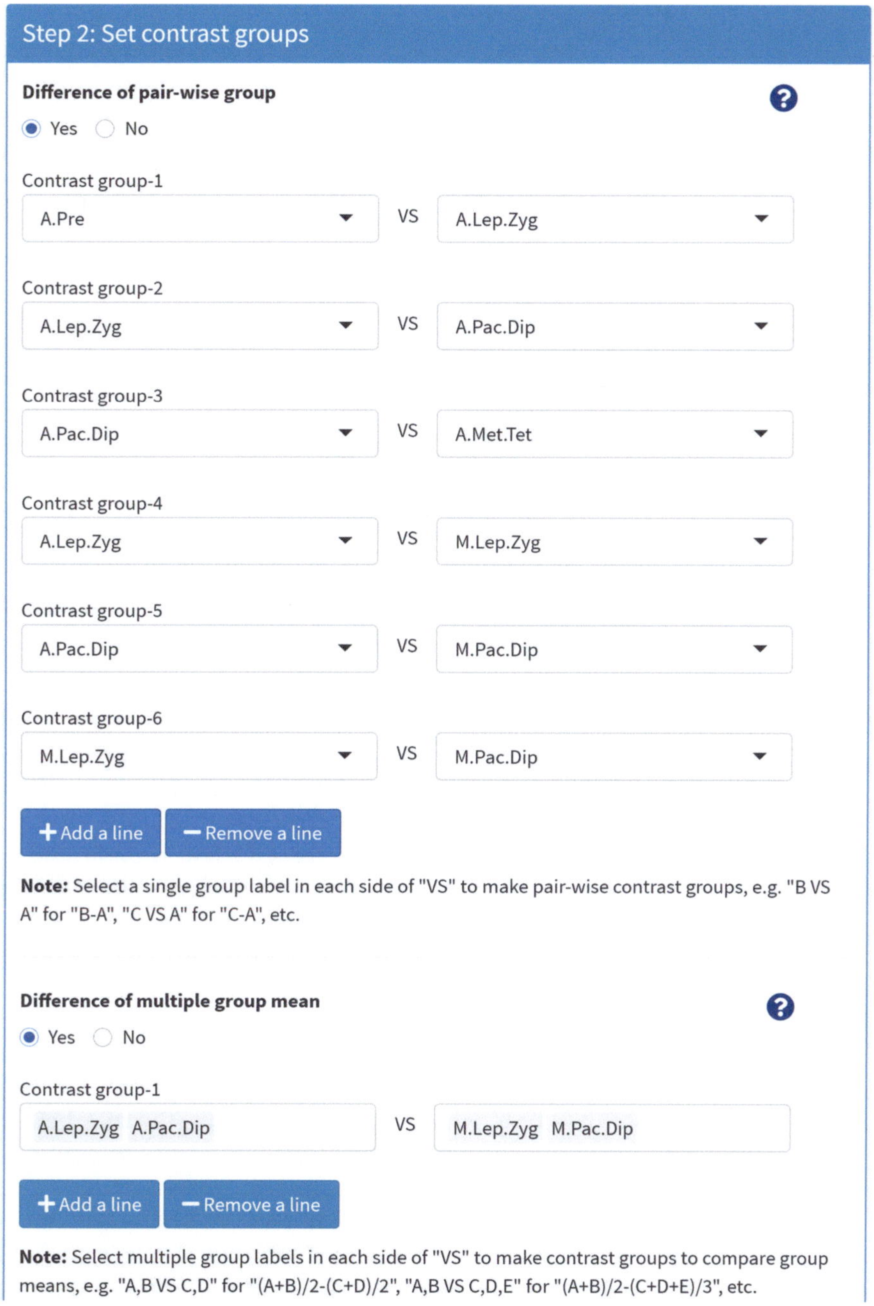

Fig. 2 Screenshot of the experimental design setup step for the 3D RNA-seq app

13. Additional analysis steps.

    For this method chapter we will skip the additional steps on further data mining. Those include different visualization options, integrating gene ontology terms, and looking at time-series trends or isoform switches. For the experienced scientist those might prove useful to come back to once familiarized with the analysis.

14. Generate Report.

    The final step is saving the results, generating a report, and downloading the results. This will provide an overview of which parameters have been set and will always allow the user to go back and repeat or revise the pipeline. This includes R files and comma-separated result files which can be imported into R and with the following R scripts transformed into publication quality figures and results.

3.4 Further Downstream Analysis and Visualization

3.4.1 GO Enrichment

One aspect of downstream analysis is the identification of enriched GO terms in the differential expressed genes. The analysis using GO_pipeline.r script (from the GitHub repository) is performed in R. A GO annotation file is needed corresponding to your transcriptome (*see* **Note 19**), an example of the first few lines of which can be seen in Table 4.

1. Read in the annotation file and the "DE gene testing statistics.csv" file from the 3D RNA-seq output.
2. Identify the experimental setup for GO enrichment analysis and set maximum adjusted *p*-value and log2FC cut-offs (*see* **Note 20**).
3. GO enrichment is conducted using topGO with the weight01 algorithm and fisher test statistics to identify enriched GO terms for molecular function (MF), biological process (BP), and cellular component (CC).
4. Plotting is performed using GOPlot Bubble and a comma separated output file with the results is produced (*see* **Note 21**).

3.4.2 Visualizing Relationships Between Contrast Groups

Often, we want to determine how differential expression in one or more of our contrast groups relates to differential expression in the other contrast groups. This can help to identify genes of interest but can be difficult to interpret in large tabular data sets. A good first step is to visualize the intersections between contrast groups. How this is best achieved is dependent on the number of contrast groups in your experimental design and the biological context of your sample data. The most common representation of this would be a Venn diagram (Fig. 3). However, Venn diagrams are limited in the number of groups you can effectively plot at once. In fact, most R packages for Venn diagram plotting don't accommodate more than four groups. In this example, we have five contrast groups to represent; ten if you wished to distinguish between differentially expressed genes (DEGs) which are upregulated and those that are downregulated in each comparison.

Euler diagrams (Fig. 4) are similar to Venn diagrams in that intersections are represented by overlapping shapes. However, the size of the shapes and overlaps is proportional to the total size of the group they represent, making this information somewhat more

Table 4
Example of the gene ontology table which is needed for GO enrichment analysis

BAnTr. GP.1HG000100	GO:0016021,GO:0016874,GO:0006464,GO:0036459,GO:0006511,GO:0004672,GO:0005524,GO:0016310
BAnTr. GP.1HG000102	GO:0009744,GO:0016301,GO:0009569,GO:0043562,GO:0031588,GO:1901982,GO:0019887,GO:0005515,GO:0046872, GO:0016310,GO:0045859,GO:0016021,GO:0030001,GO:0005975
BAnTr. GP.1HG000106	GO:0016747,GO:0015937,GO:0010287,GO:0004140,GO:0005524,GO:0016021,GO:0033306,GO:0016310,GO:0019432
BAnTr. GP.1HG000108	GO:0016021,GO:0003677,GO:0005634
BAnTr. GP.1HG000402	GO:0016021
BAnTr. GP.1HG000404	GO:0016021
BAnTr. GP.1HG000406	GO:0009451,GO:0008270,GO:0005739,GO:0000959,GO:0003678,GO:0008312,GO:0003700,GO:0016071,GO:0048500, GO:0005846,GO:0016627,GO:0003729,GO:0005634,GO:0016021,GO:0010150,GO:0009751,GO:0009753, GO:0019904,GO:0000339,GO:0009737,GO:0043565,GO:0042393,GO:0009555,GO:0031930,GO:0032508, GO:0006614,GO:0006355,GO:0006629,GO:0051028,GO:0055114,GO:0000377
BAnTr. GP.1HG000408	GO:0016874,GO:0016021
BAnTr. GP.1HG000110	GO:0015074,GO:0008270,GO:0003676
BAnTr. GP.1HG000112	GO:0008270,GO:0015074,GO:0003676
BAnTr. GP.1HG000708	GO:0045735,GO:0016021
BAnTr. GP.1HG000410	GO:0016874,GO:0016021
BAnTr. GP.1HG000412	GO:0016021

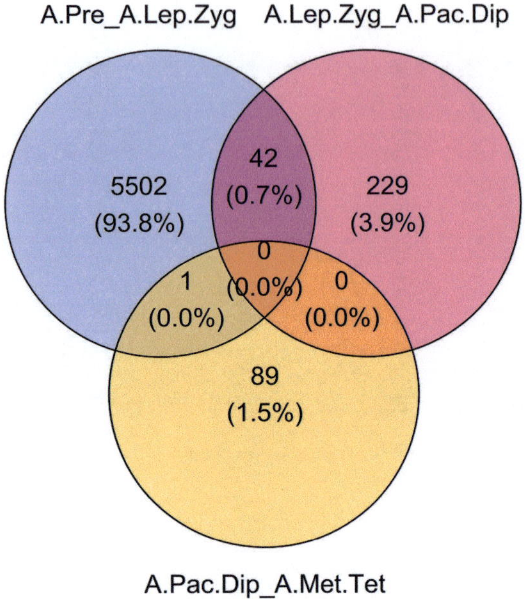

Fig. 3 Venn diagram of DEGs in anther-to-anther contrast groups created using ggvenn from 3D RNA-seq output. The comparisons are: A.Pre-A.LepZyg, premeiosis versus leptotene–zygotene; A.LepZyg- A.PacDip, leptotene–zygotene versus pachytene–diplotene; and A.PacDip-A.MetTet, pachytene–diplotene versus metaphase I–tetrad

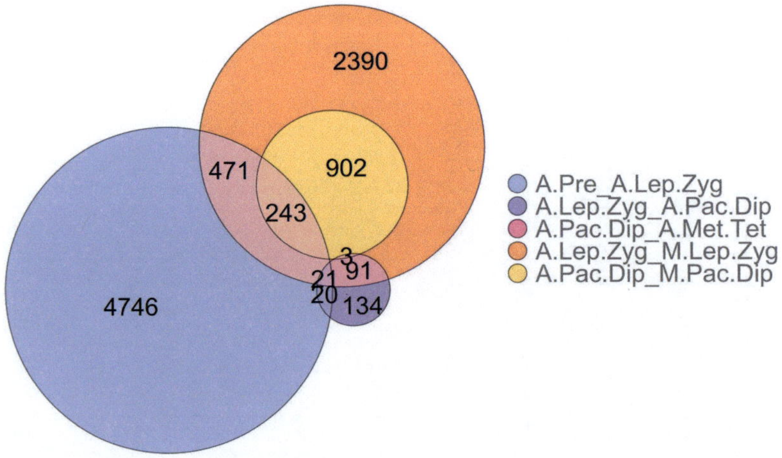

Fig. 4 Euler diagram of DEGs in all contrast groups created using Eulerr from 3D RNA-seq output. The comparisons are: A.Pre-A.LepZyg, premeiosis versus leptotene–zygotene; A.LepZyg- A.PacDip, leptotene–zygotene versus pachytene–diplotene; A.PacDip-A.MetTet, pachytene–diplotene versus metaphase I–tetrad; A.LepZyg-M.LepZyg, anthers at leptotene–zygotene versus meiocytes at the same stage; and A.PacDip-M.PacDip, anthers at pachytene–diplotene versus meiocytes at the same stage

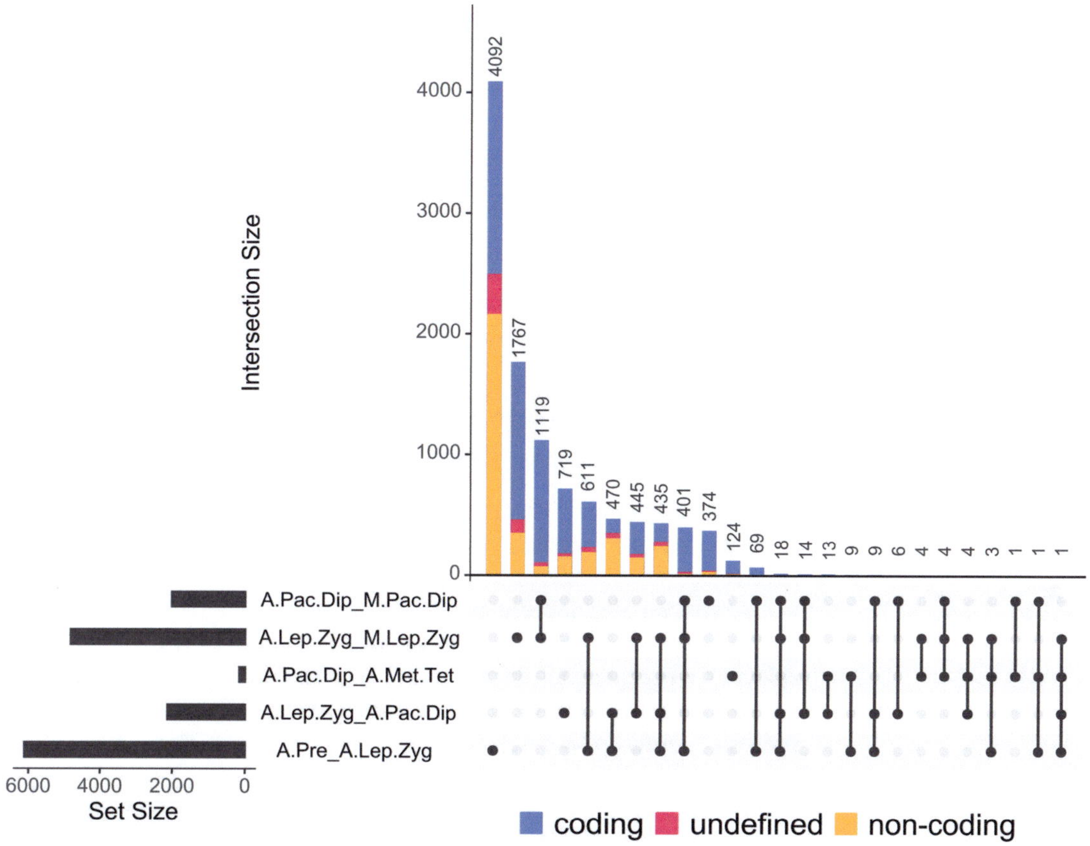

Fig. 5 UpSet plot DEGs in all contrast groups created using UpSetR from 3D RNA-seq output. The comparisons are: A.Pre-A.LepZyg, premeiosis versus leptotene–zygotene; A.LepZyg- A.PacDip, leptotene–zygotene versus pachytene–diplotene; A.PacDip-A.MetTet, pachytene–diplotene versus metaphase I–tetrad; A.LepZyg-M.LepZyg, anthers at leptotene–zygotene versus meiocytes at the same stage; and A.PacDip-M. PacDip, anthers at pachytene–diplotene versus meiocytes at the same stage

intuitive to interpret. This can remove some of the issues of crowding in Venn diagrams with a larger number of groups, but not in every circumstance, and not entirely.

A less common method of visualizing intersections is upset or intersect plots (Fig. 5). In the default form these consist of three parts. Central to the plot is an intersection matrix where the intersections between groups are represented as connected dots. To the left of this matrix a horizontal histogram represents the size of each group. The size of each intersection is represented by a vertical histogram above the intersection matrix. These are much more suitable for comparison of a larger number of groups. Additional information—such as further categorization within intersects—can be included using color or annotation. However, as these are used less often than Venn or Euler diagrams they are not always as immediately intuitive to the reader.

Using the example data, we will prepare each of these plots representing the intersection of DEGs using a number of different R packages. In all cases there are three basic steps to preparing these plots:

1. Load the appropriate packages.
2. Import and format the data.
3. Generate the plot.

The step-by-step instructions to prepare these plots from 3D RNA-seq. Output data can be found in the GitHub repository listed above.

1. Venn and Euler diagrams: venn_or_euler_diagrams.R.
2. UpSet plots: UpSet_plotting.R.

The structure of the data used to make these plots can be easily used to extract intersections of particular interest (see UpSet_plotting.R).

Lastly, an alluvial plot can be used to visualize the flow (alluvia) between groups (strata) along an axis (Fig. 6). This is particularly relevant to sequential data. As our contrast groups are sequential, we can plot the change of gene DEG status in anthers from premeiosis to metaphase–tetrad stages. For this plot, the x axis is our

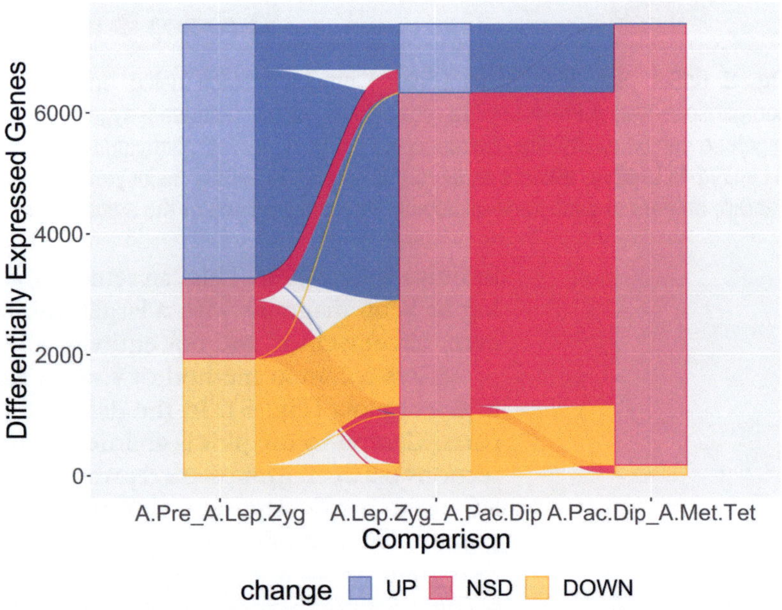

Fig. 6 Alluvial plot DEGs in anther-to-anther contrast groups created using ggalluvial from 3D RNA-seq output. The comparisons are: A.Pre-A.LepZyg, premeiosis versus leptotene–zygotene; A.LepZyg-A.PacDip, leptotene–zygotene versus pachytene–diplotene; and A.PacDip-A.MetTet, pachytene–diplotene versus metaphase I–tetrad

contrast groups, the alluvia are the genes, and our strata are whether the gene was upregulated, downregulated, or underwent no significant change in expression within contrast groups. This can be accomplished from 3D RNA-seq output by following the Alluvial_plotting.R script in the GitHub repository listed above.

3.4.3 Visualization of Differential Alternatively Spliced Transcripts Using Heatmaps or Line Graphs

In the last downstream analysis, we are taking two functional plot options from the 3D RNA-seq app and showing how to visualize them after the analysis has been completed. This enables the user to change the coloring to a color-blind friendly palette and adjust the resolution to journal requirements. In this example we are focusing on differential alternatively spliced genes. The plots can help to identify which is the major transcript isoform and where the transcript usage switches between experimental stages. We will be using DAS_plotting.R from the GitHub depository.

1. Read in multiple 3D RNA-seq output files "Transcript TPM.csv" "Significant DAS genes list and statistics.csv", "Gene TPM.csv"; plus, one of the 3D RNA-seq input file: "TranscriptToGene.csv".
2. Define the adjusted *p*-value and the maxDeltaPS value for the analysis. The maxDetltaPS value is the percentage differences between one isoform in comparison to the other. A value of 1 or −1 corresponds a 100% change of the main alternatively spliced isoform.

The heatmap and line graph show two possible ways to represent the same result (example for one differential alternatively spliced gene in Fig. 7). For expression changes as shown in the example where we have two main isoforms expressed combined with the overall gene expression, the line graph is easier to interpret. With more isoforms the plot will get harder to read and heatmaps are better at representing the data.

The same script can easily be adjusted to visualize for example the top 10 differentially expressed genes in heatmap format which is explained in detail in the R script itself.

4 Notes

1. The number of weeks to reach meiosis depends on the cultivar and environmental conditions. It is recommended to develop a reliable prediction system based on external morphological characteristics of the main tiller as described for the barley cultivar Optic [30]. We suggest to grow plants in batches twice a week for the duration of the experiment as only few plants can be harvested at a time to minimize sample variation and impact on health and safety of the collector.

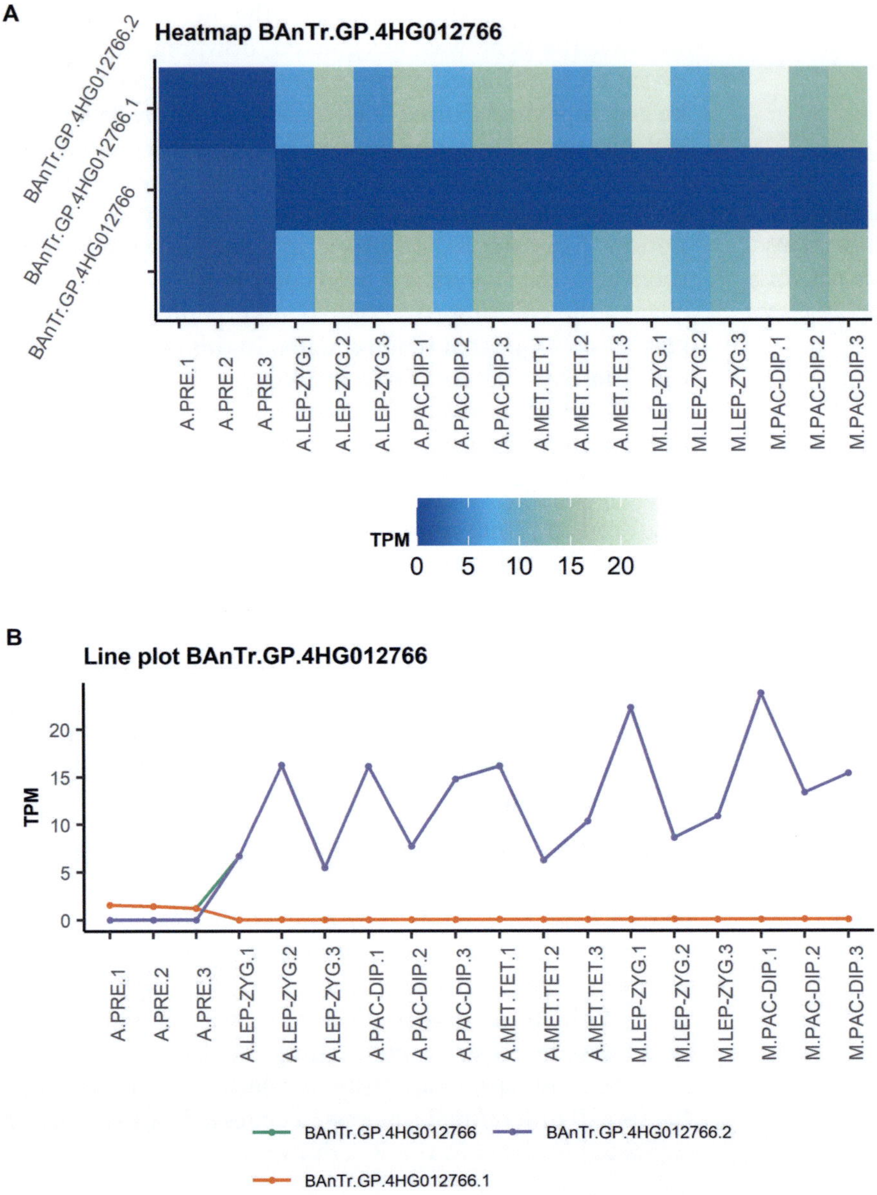

Fig. 7 Heatmap (**a**) and line plot (**b**) of the expression of the differential alternatively spliced gene BAnTr. GP.4HG012766. This represents two alternative ways to plot the same information regarding the expression of particular genes or transcripts across samples. Expression, in transcript per million (TPM), of both isoforms of this gene (BAnTr.GP.4HG012766.1 and BAnTr.GP.4HG012766.2) and of the gene as a whole (BAnTr. GP.4HG012766) is shown for each condition and each of three corresponding biological replicates . The heatmap (**a**) displays expression in each sample relative to a color scale as displayed in the legend below the *x* axis of the plot. In the line graph (**b**) expression is charted on the *Y* axis

2. It is better to grow plants in a controlled plant growth room instead of a glasshouse where the environmental conditions vary widely.
3. First, prepare a table to record sample information including plant height, number of leaves, measure emerging flag leaf and spike lengths. Collect samples for 1–2 h only at similar times during the day. Wear gloves and be quick once the spike is removed.
4. Anthers can be easily detached using insulin-needle attached to 1 ml syringe. Anther length and acetocarmine staining indicate approximate meiotic stage that need confirmation by immunocytology.
5. Top up with anthers from different spikes until desired number is collected.
6. After removing their tips with insulin needles, anthers can be tapped gently to release the meiocyte column. Using RNase-free 10 μl tips, released meiocytes can be transferred into a new 0.5 ml Eppendorf tube discarding anther tissue.
7. We find TRIzol Plus RNA Purification Kit good as it combines advantages of TRIzol and PureLink RNA Mini Kit.
8. It is important to check RNA integrity using Bioanalyzer 2100 or equivalent and use only samples with highest RNA integrity number (i.e., >8.0). It is important to prepare more replicates than needed to avoid using low quality samples.
9. It is advisable to discuss your RNA-seq experiment with your inhouse or external sequencing service provider at the start.
10. For our example data the transcriptome reference file can either be first downloaded onto your computer and then uploaded to galaxy or uploaded directly to galaxy by clicking "Paste/Fetch data" and then pasting this link: https://ndownloader.figshare.com/files/22318350.
11. In our example we are selecting "paired-end (two separate input files)". Two fastq files from the galaxy history need to be provided, the corresponding ":forward" and ":reverse" files.
12. The minimum length of reads to keep in the trimmomatic step should always be selected dependent on the total read length.
13. For the test dataset the reads were orientated toward each other (I = Inward) and not stranded (U).
14. There have been no sequencing replicates done in the example dataset. Technical replicates are especially useful if the data is sequenced across multiple sequencing runs and different sequencing machines. As column of interest, we select "condition." This column (Table 2) provides the experimental setup.

For biological replicate the "bio_rep" column is chosen and for quantification folder name the "quant.files" column is chosen.

15. In case of single obvious outliers, one would have to go back to the "Data generation" step and remove those samples before uploading the modified datasets. In case of batch clustering which reflects the experimental design, it can be corrected in the next step.
16. The example dataset does not suffer from batch effects, but it is very common that samples processed and sequenced at different times or handled by different person during the experimental steps show batch effects. Therefore, replicates should be spread out across sequencing runs and processing. This will allow to correct for those effects and still get meaningful differentially expressed genes.
17. The default parameters are a good starting point and should be used at the beginning. Depending on the results one could go back to make them more stringent. We would usually advise against more lenient parameters as this might introduce false positives.
18. A log2 fold change of 1 means that the gene is twice the expression in one tissue in comparison to the other.
19. GO annotations can be derived from annotation pipelines such as eggNOGG-mapper [31] or Pannzer2 [32]. Both are available as online servers to which batch annotations of up to 100,000 amino acid sequences in fasta format can be submitted at a time.
20. The GO_pipeline.r script on GitHub includes an example of the upregulated genes with a log2FC equal to or higher than two and an adjusted *p*-value below 0.01 in the comparison A. Pre to A.LepZyg. We advise to look at upregulated and downregulated genes separately.
21. The lowest possible *p*-value which is produced by the statistical test is <1e-30. As this is a value which cannot be plotted, it is changed to 1e-30 for visualization purposes.

References

1. Geniza M, Jaiswal P (2017) Tools for building de novo transcriptome assembly. Curr Plant Biol 11-12:41–45
2. Hölzer M, Marz M (2019) De novo transcriptome assembly: a comprehensive cross-species comparison of short-read RNA-Seq assemblers. GigaScience 8(5):giz039
3. Lamarre S, Frasse P, Zouine M et al (2018) Optimization of an RNA-Seq differential gene expression analysis depending on biological replicate number and library size. Front Plant Sci 9:108
4. Clark S, Yu F, Gu L et al (2019) Expanding alternative splicing identification by integrating multiple sources of transcription data in tomato. Front Plant Sci 10:689
5. Zhang R, Calixto CPG, Marquez Y et al (2017) A high quality Arabidopsis transcriptome for accurate transcript-level analysis of alternative splicing. Nucleic Acids Res 45(9):5061–5073

6. Barakate A, Orr J, Schreiber M et al (2021) Barley anther and Meiocyte transcriptome dynamics in meiotic prophase I. Front Plant Sci 11:619404
7. Leinonen R, Sugawara H, Shumway M et al (2011) The sequence read archive. Nucleic Acids Res 39(Database issue):D19–D21
8. Afgan E, Baker D, Batut B et al (2018) The galaxy platform for accessible, reproducible and collaborative biomedical analyses: 2018 update. Nucleic Acids Res 46(W1):W537–W544
9. Guo W, Tzioutziou NA, Stephen G et al (2020) 3D RNA-seq: a powerful and flexible tool for rapid and accurate differential expression and alternative splicing analysis of RNA-seq data for biologists. RNA Biol 18(11):1574–1587
10. Bolger AM, Lohse M, Usadel B (2014) Trimmomatic: a flexible trimmer for Illumina sequence data. Bioinformatics 30(15):2114–2120
11. Patro R, Duggal G, Love MI et al (2017) Salmon provides fast and bias-aware quantification of transcript expression. Nat Methods 14(4):417–419
12. Srivastava A, Malik L, Smith T et al (2019) Alevin efficiently estimates accurate gene abundances from dscRNA-seq data. Genome Biol 20(1):65
13. R Core Team. (2020) R: a language and environment for statistical computing. https://www.R-project.org/
14. Robinson MD, McCarthy DJ, Smyth GK (2010) edgeR: a Bioconductor package for differential expression analysis of digital gene expression data. Bioinformatics 26(1): 139–140
15. Risso D, Ngai J, Speed TP et al (2014) Normalization of RNA-seq data using factor analysis of control genes or samples. Nat Biotechnol 32(9):896–902
16. Ritchie ME, Phipson B, Wu D et al (2015) Limma powers differential expression analyses for RNA-sequencing and microarray studies. Nucleic Acids Res 43(7):e47
17. Guo W, Calixto CPG, Brown JWS et al (2017) TSIS: an R package to infer alternative splicing isoform switches for time-series data. Bioinformatics 33(20):3308–3310
18. Alexa A, Rahnenfuhrer J. (2020) topGO: Enrichment Analysis for Gene Ontology. R package version 2.46.0. https://bioconductor.org/packages/release/bioc/html/topGO.html
19. Walter W, Sanchez-Cabo F, Ricote M (2015) GOplot: an R package for visually combining expression data with functional analysis. Bioinformatics 31(17):2912–2914
20. Wickham H et al (2019) Welcome to the tidyverse. J Open Source Soft 4(43):1686. https://doi.org/10.21105/joss.01686
21. Gehlenborg N (2019). UpSetR: a more scalable alternative to Venn and Euler diagrams for visualizing intersecting sets. R package version 1.4.0. https://CRAN.R-project.org/package=UpSetR
22. Brunson JC and Read QD (2020). ggalluvial: Alluvial Plots in "ggplot2". R package version 0.12.3. http://corybrunson.github.io/ggalluvial/
23. Chen H (2018). VennDiagram: generate high-resolution Venn and Euler plots. R package version 1.6.20. https://CRAN.R-project.org/package=VennDiagram
24. Yan L (2021). ggvenn: Draw Venn Diagram by 'ggplot2'. R package version 0.1.8. https://CRAN.R-project.org/package=ggvenn
25. Gao C (2019). ggVennDiagram: A "ggplot2" implement of Venn diagram. R package version 0.3. https://CRAN.R-project.org/package=ggVennDiagram
26. Wilkinson L (2011). venneuler: Venn and Euler Diagrams. R package version 1.1-0. https://CRAN.R-project.org/package=venneuler
27. Larsson J (2020). _eulerr: area-proportional Euler and Venn diagams with ellipses. R package version 6.1.0. https://cran.r-project.org/package=eulerr
28. Chaudhari AK, Chaudhary BR (2012) Meiotic chromosome behaviour and karyomorphology of *Aloe vera* (L.) Burm. f. Chromosom Bot 7(1):23–29
29. Colas I, Darrier B, Arrieta M et al (2017) Observation of extensive chromosome Axis remodeling during the "diffuse-phase" of meiosis in large genome cereals. Front Plant Sci 8: 1235
30. Gómez JF, Wilson ZA (2012) Non-destructive staging of barley reproductive development for molecular analysis based upon external morphology. J Exp Bot 63(11):4085–4094
31. Huerta-Cepas J, Szklarczyk D, Heller D et al (2019) eggNOG 5.0: a hierarchical, functionally and phylogenetically annotated orthology resource based on 5090 organisms and 2502 viruses. Nucleic Acids Res 47(D1):D309–D314
32. Törönen P, Medlar A, Holm L (2018) PANNZER2: a rapid functional annotation web server. Nucleic Acids Res 46(W1):W84–W88

Chapter 21

Profiling m6A RNA Modifications in Low Amounts of Plant Cells Using Maize Meiocytes

Drin Shabani, Thomas Dresselhaus, and Stefanie Dukowic-Schulze

Abstract

RNA modifications can influence gene expression via multiple aspects such as RNA stability and alternative splicing. The most prominent RNA modification is m6A (N6-methyladenosine). Its profiling from low starting amounts of <100 cells is challenging. We describe here a complete workflow from cell isolation to data analysis that is based on using an RNA CUT&RUN-supported m6A-RIP (RNA immunoprecipitation) procedure and a subsequent adaptor-tagging library synthesis. Male meiocytes isolated from maize anthers were used as a test system to establish the protocol.

Key words m6A, RNA modification, Meiocytes, Maize, CUT&RUN, RIP

1 Introduction

In plants as well as animals, gene expression is regulated on multiple levels, starting with the primary sequence of the underlying DNA locus (with binding motifs recognized by, e.g., transcription factors), epigenetic marks of the genomic locus (e.g., DNA methylation or histone modifications), and then the RNA itself (e.g., its cellular localization and stability). RNA splicing and the regulation of its stability are now known to be influenced by so-called epitranscriptomic marks, of which N6-methyladenosine (m6A) is the most prominent one in terms of quantity and studies [1]. However, less than 0.1% of RNAs are assumed to be modified with m6A, which makes their detection and low-input protocols challenging.

Here, we describe an m6A-profiling workflow for low amounts of plant cells—the method was optimized by using approximately 50 cells of maize male meiocytes per sample. The whole workflow is shown in Fig. 1. After cell isolation and RNA extraction, m6A-modified RNA molecules are captured by a specific m6A antibody in a CUT&RUN (*C*leavage *U*nder *T*argets & *R*elease *U*sing *N*uclease) approach. CUT&RUN has initially been

Christophe Lambing (ed.), *Plant Gametogenesis: Methods and Protocols*, Methods in Molecular Biology, vol. 2484, https://doi.org/10.1007/978-1-0716-2253-7_21,

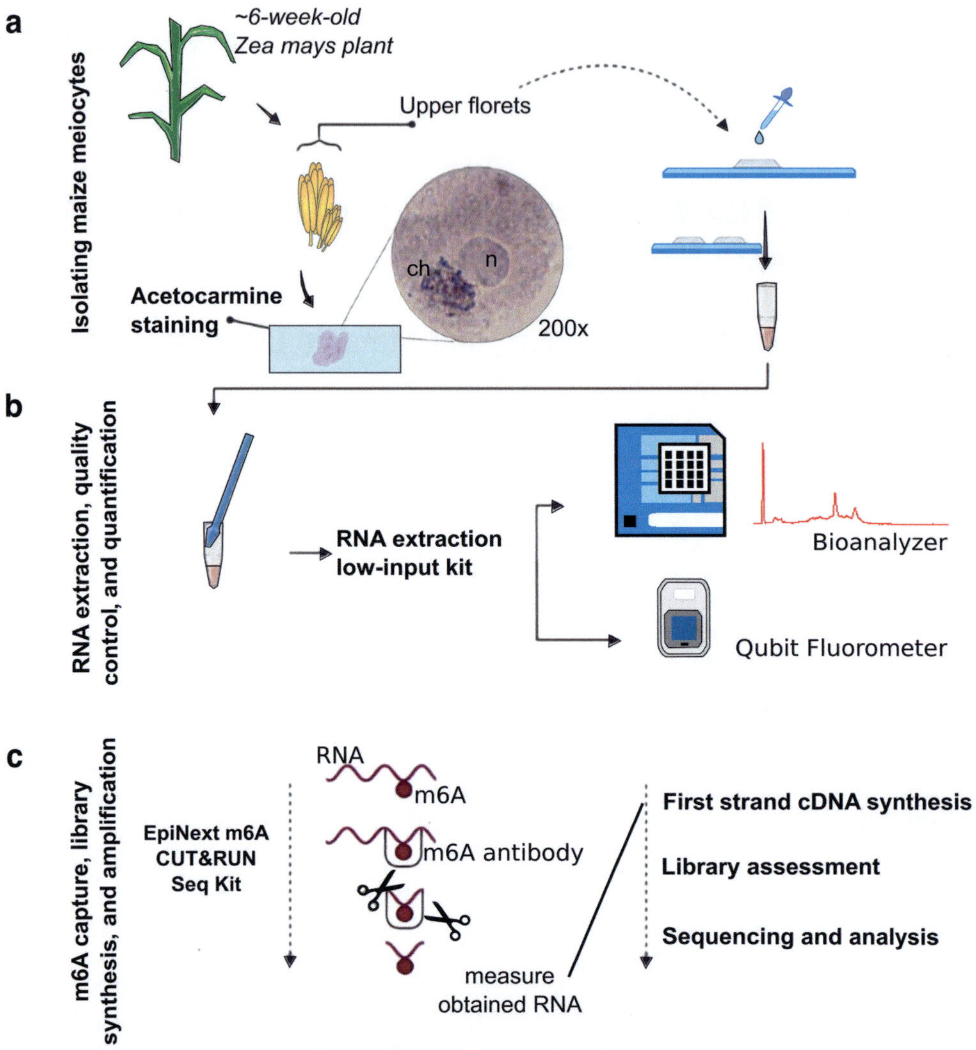

Fig. 1 Workflow overview. (**a**) Maize meiocyte isolation starts with selection of plants at the desired developmental stage and identification of the exact meiotic stage by acetocarmine staining of fixed anthers from upper florets. When the stage is confirmed, fresh upper-floret anthers are transferred into PBS, squashed by a flat-end pipette tip, more PBS is added and meiocytes are transferred by micropipetting to another PBS drop. At the end, TRIzol is added to the tube for proper storing and preventing RNA from degradation. *ch* chromosomes; *n* nucleolus. (**b**) Mechanical homogenization by plastic tube pestles is suitable for preparing samples to undergo RNA extraction. RNA extraction, here by a TRIzol-based method, is used to collect RNA released by meiocytes. Quality assessment and quantification is done by a sensitive detection device such as a Qubit Fluorometer or Bioanalyzer (*see* also Fig. 2). (**c**) An m6A antibody binds to m6A-modified RNA and protects captured RNA during a subsequent RNA fragmentation step (as implemented in the EpiNext m6A CUT&RUN Seq Kit). The m6A capture step is followed by library synthesis by first and second strand cDNA synthesis with adaptor tagging, library amplification and indexing. Bead-technology-based cleanup steps are performed. Sample libraries are sequenced on a short-read sequencer, and resulting reads are analyzed computationally

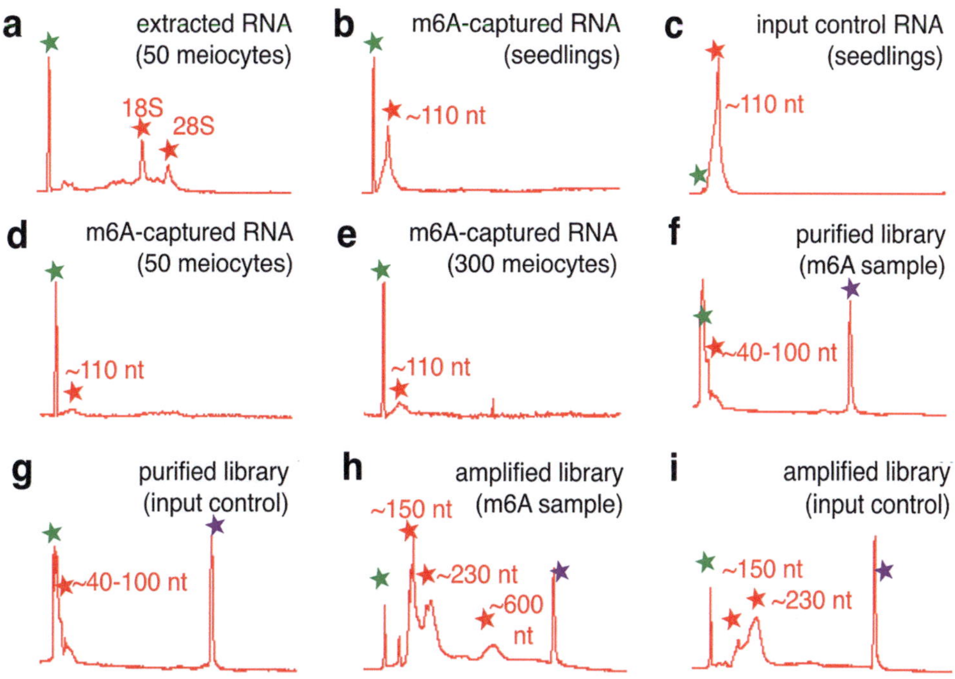

Fig. 2 Examples of quality controls (Bioanalyzer). Simplified profiles from electropherograms show size distributions of RNA and DNA at specific steps in the procedure. The *x*-axis is run-time or nucleotide size, with lower values at the left. Stars depict peaks. Green and purple stars are markers needed to identify the start and end positions, respectively. Red stars are specific peaks obtained in the analysis. Panels (**a**)–(**e**) depict RNA profiles, panels (**f**)–(**i**) depict DNA profiles from different samples using each 50 meiocytes. (**a**) Profile of extracted RNA from ~50 maize meiocytes. The RIN (RNA integrity number) is calculated from the 18S rRNA and 28S rRNA peaks. For the optimal RIN = 10, the 28S rRNA peak would be higher than the 18S rRNA peak. The RIN here was 7.1. (**b**) Profile of presumably m6A-containing RNA from seedlings captured with the m6A antibody in the CUT&RUN procedure. (**c**) Profile of fragmented RNA of the input control from seedlings without m6A capture. (**d**) As (**c**) but from 50 meiocytes. The peak is less strong. (**e**) As (**c**) and (**d**) but from 300 meiocytes. The peak is clearly present and at the same position as in (**b**)–(**d**). According to the manual, the peak is expected around 55 nts in human samples. However, the proprietary enzyme mix used in the CUT&RUN procedure results in the present peaks of ~110 nts when using maize samples, due to different cleavage frequency/efficiency in this GC-richer genome. (**f**) Profile of m6A cDNA library after first and second strand synthesis with adaptors (which add 91 nts to the fragments, resulting in maximally 146 nts in human samples, and maximally 201 nts in maize samples). Fragments are shorter than the corresponding RNA fragments because of random priming/binding of the adaptors which can bind anywhere within the captured RNA fragments. (**g**) As (**f**), but from the input control. (**h**) Profile of amplified m6A cDNA library after 20 PCR cycles (with index primers adding another 30 nts, resulting in 231 nts in maize samples). The smaller peaks ~150 nts are likely resulting from ~30 nt fragments, which is the length covered by the m6A antibody plus added adaptors and primers. The peak at ~600 nts was seen in independent samples and likely results from high density of m6A on long RNA fragments. (i) As (**h**), but from the input control. The peaks ~150 nts and ~600 nts are smaller or absent

developed for DNA as a more sensitive alternative to ChIP (chromatin immunoprecipitation) [2], and has here been adapted to remove all unbound RNA by RNA digestion. m6A capture with CUT&RUN followed by library preparation was achieved with a RNA starting amount of just ~35–60 ng. Finally, we summarize how the obtained sequencing data can be processed and analyzed.

2 Materials

2.1 Plant Material and Equipment

1. Maize plants (*Zea mays*) of the B73 inbred line are grown in the greenhouse (16 h light at 28 °C and 8 h without additional light at 22 °C; 65–75% relative humidity).
2. Scalpel (e.g., blade #10, 11, 14).
3. Tape.
4. 1.0 ml syringe with suitable needles (e.g., 0.60 × 30 mm).
5. Fine dissecting forceps (without claws; straight or bent).
6. Laboratory alcohol burner (made from glass, with wick).
7. Self-made flat-end pipette tips: Hold a 20 μl or 200 μl pipette tip and apply heat (alcohol burner or lighter) to the opening of the tip. Quickly press the melted tip opening onto a microscope slide and wait 2–3 s until it solidifies.
8. Manual microinjection system (e.g., Cell Tram or equivalent).
9. Capillary glass needles: Extend a 50 μl capillary glass tube (e.g., ringcaps or equivalent) with a micropipette puller (Sutter P-97 or equivalent) to obtain a >1 cm section with an inner diameter of ~250–500 μm (e.g., parameters: heat = 500, pull = 50, velocity = 20, time = 120), break the tube in the middle of the extended and thinned section, then attach it tightly to the manual microinjector with parafilm.
10. Plastic tube-fitting pestles.
11. Plastic tubes: 15 ml and 50 ml Falcon tubes, 1.5 ml microtubes, low-retention RNase-free 1.5 ml microtubes, 0.2 ml PCR tube stripes with attached lids.
12. Microscope slides.
13. Plastic coverslips, 22 mm × 22 mm.
14. Microscopes: Stereomicroscope, inverted microscope, compound microscope.
15. Oxidized iron nails (to supply Fe^{3+} to the staining reaction): Scrape thin long nails with sandpaper, spray them with water and wait for several days until they become rusty.
16. Micro UV spectrophotometer (e.g., Nanodrop or equivalent).
17. Dye-based fluorescence detector (e.g., Qubit or equivalent).
18. Automated microfluidic gel system (e.g., Bioanalyzer or equivalent).

2.2 Buffers, Solutions and Chemicals

All solutions are prepared with nuclease-free double-distilled water.

1. Farmer's fixative: 3:1 mix of absolute ethanol–acetic acid. For 500 ml, mix 375 ml of 100% ethanol and 125 ml of glacial acetic acid.
2. RNase decontamination solution.
3. Acetocarmine staining solution (0.5%): 0.5% of carmine (w/v), 55% of water (v/v), 45% of glacial acetic acid (v/v). For a 100 ml solution of 0.5% acetocarmine staining solution, dissolve 0.5 g of carmine in 55 ml of water (stir on heater, work in a fume hood), add 45 ml of glacial acetic acid while stirring and boil for ~30 min. Filter through Whatman filter paper and store in the dark.
4. Phosphate-buffered saline (10× PBS): 1.37 M NaCl, 0.027 M KCl, 0.1 M Na_2HPO_4, 0.018 M KH_2PO_4, pH 7.0. Dissolve 80 g of NaCl, 2 g of KCl, 14.4 g of Na_2HPO_4, and 2.5 g of KH_2PO_4 in 1 L of dH_2O. Adjust pH to 7.0 with HCl, autoclave and store at room temperature.
5. TRIzol/TRI Reagent (mixture with guanidine thiocyanate and phenol).
6. Direct-zol RNA Microprep or equivalent RNA purification kit.
7. EpiNext CUT&RUN RNA m6A-Seq Kit or equivalent kit for enrichment of RNA fragments containing m6A from low input RNA.
8. Qubit dsDNA HS Assay or equivalent accurate dsDNA quantification assay.
9. Bioanalyzer RNA/dsDNA HS Assay or equivalent RNA/DNA quality control measuring assay.

3 Methods

This protocol lists the steps for profiling m6A RNA modifications from low input amounts (<100 plant cells). We put the focus on the adaptations, optimizations, and considerations needed for these low input amounts. As a test system, we used maize meiocytes, but the method can presumably be adapted to low input samples of other cell types and tissues. For initial considerations, *see* **Notes 1–3**; also *see* Table 1.

3.1 Sample Preparation: Isolating Maize Meiocytes

Capillary collection of meiocytes (CCM) modified from [4] was applied in order to collect meiocytes of the desired specific stage, and is detailed below:

Table 1
Important points for establishing a low-input m6A profiling experiment

Step	Important aspect	Advice	Variations
Cell isolation & RNA extraction	Qualitative & quantitative assessment of cells and resulting RNA (RIN as high as possible)	– Invest time in optimizing the collection and extraction procedure – Check the quality with a Bioanalyzer High Sensitivity RNA assay	– Snap-freezing vs. TRIzol storage – Tissue homogenizer vs. pestle grinding
m6A capture	– Highest possible amount of amplifiable m6A-containing RNAs	– Use high-quality RNA only (RIN $\geq$6, better $\geq$8) – Work precisely and RNase free (*see* **Notes 2** and **3**) – Use a highly specific antibody [3] – Reduce background (e.g., by removing unbound RNA by CUT&RUN)	
Library synthesis	– High efficiency – Comparable samples without bias	– Optionally test different library preparation methods (*see* **Note 4**) – Use the same number of PCR cycles for all samples	– Adaptor-tagging library prep (included) – Other special low-input library prep methods

1. Open a plant that is assumed to have the desired stage (*see* **Note 5**) with a scalpel. Make the cut deep enough to approach the tassel without damaging it. Locate the main tassel and collect 2–3 spikelets from the mid part of the main tassel in Farmer's fixative. Carefully return the tassel and the whole area into the mid of the stem as it was prior to incision, then wrap it with tape (*see* **Note 6**). If you have multiple plants, label the tape with an ID and the date to easily distinguish samples.
2. Under a stereomicroscope, use syringe needles to dissect the upper floret anthers out of collected spikelets. Transfer the anthers to a slide containing a drop of acetocarmine solution (~35 μl). Heat the slide shortly with an alcohol burner (2–3 s) and remove it from the flame to stir it with a rusty nail in order to achieve proper staining (by leaching iron oxide). Repeat the heating and stirring until the drop becomes purple. Then use a flat-end pipette tip to squeeze anthers, allowing meiocyte cells or clusters to float freely. Add a coverslip and press it down with a finger to remove extra liquid. Use delicate task wipes to dry off extra liquid.
3. Examine the slide under a bright-field microscope at 100–400× magnification and identify the meiotic stage.

4. Verify the meiotic stage toward the top/bottom of the main tassel to identify the tassel section that consists of the same stage.
5. Bring the plant into the lab or remove the whole tassel or upper leaf whorl, and then work swiftly to reduce induction of stress response.
6. Get 30–60 stage-verified anthers from upper florets and transfer them into an ice-cold 1× PBS drop on a microscope slide or plastic coverslip (~50 μl) (*see* **Note 7**). The procedure should be done under RNase-free conditions (*see* **Note 2**), and on ice whenever possible.
7. Squeeze anthers in the PBS drop by using a flat-end pipette tip, and add extra PBS after opening up anthers (keep the drop concave).
8. Use microcapillary glass tubes to vacuum cell clusters out of the drop (*see* **Note 8**) and wash them by transfer to a new 1× PBS drop on a plastic coverslip on the same microscope slide.
9. Finally transfer the PBS drop with meiocytes onto the inner side of an RNase-free tube. Briefly centrifuge the tube to send the drop to the bottom of the tube. Meanwhile, count the cells and estimate the amount of PBS present in the tube, to use an appropriate amount of TRIzol for optimal storage (≥3–10× amount of TRIzol, *see* **Note 9**).

3.2 RNA Extraction, RNA Quality Control, and Quantification

1. For RNA extraction, follow the manual procedure of the kit of your choice. We prefer TRIzol-based protocols with columns that are suitable for small amounts of starting material (such as DirectZol RNA Microprep). For initial cell lysis *see* **Note 10**, for DNase digest *see* **Note 11**, and for recommended elution volume *see* **Note 12**.
2. Determine RNA quality and quantity by using a microfluidic system that can determine the RIN (RNA integrity number), for example, the Bioanalyzer with the RNA High Sensitivity Assay, *see* Fig. 2a. Optionally, use an appropriate fluorescent dye-based system for measuring RNA (*see* Table 2). If using a micro-UV-spectrophotometer (e.g., a NanoDrop), consider **Note 13** before dropping samples with suboptimal OD260/OD230 values.
3. Continue with samples of high quality (at least RIN >6, better >8) (*see* **Note 14**).

3.3 m6A Capture, Library Preparation and Sequencing

1. For Input Controls (which are not subjected to RNA immunoprecipitation with the m6A antibody, *see* **Note 15**), use 1/20th to 1/10th of the RNA amount used for actual m6A samples (*see* **Note 16**). Do not use an IgG negative control (*see* **Note 17**). Follow the Instruction Manual for the EpiNext

Table 2
Equipment for quantification and quality check of low amounts of DNA/RNA

Method/Equipment	Assay	Detection range	Use in this protocol
Micro UV spectrophotometer (e.g., Nanodrop or equivalent)	Direct measurement of ~1 μl solution	(dependent on brand and model) ~ 5 ng/μl to 3 μg/μl of RNA ~ 5 ng/μl to 3 μg/μl of dsDNA	– Recommended if in the expected range – For initial value for extracted RNA (to determine dilutions for, e.g., Bioanalyzer)
Dye-based fluorescence detector (e.g., Qubit or equivalent)	Measurement of 1–20 μl sample in 200 μl buffer with fluorescent dye *(measures whole RNA/DNA)*	1 μl sample with ~5–100 ng/μl of RNA (Qubit HS/ Quant-iT HS) ~0.2–100 ng/μl of dsDNA (Qubit HS/Quant-iT)	– Extracted RNA (undiluted/diluted) – Purified and amplified libraries (undiluted)
Automated microfluidic gel electrophoresis (e.g., Bioanalyzer or equivalent)	RNA/DNA High Sensitivity Assay ("Pico") (measurements can be adjusted to specific size ranges of RNA/DNA)	1 μl sample with ~0.05–5 ng/μl of RNA ~0.1–10 ng/μl of DNA libraries	– Extracted RNA (diluted) – m6A-captured RNA (undiluted) – Purified and amplified libraries (DNA) (undiluted)

m6A CUT&RUN Seq Kit for section 1 "Immunocapture and Cleavage," then for section 2 "Enriched RNA Release and Recovery." At the end of this section determine RNA amount and size distribution by quantification assays such as Bioanalyzer HS RNA assay (*see* **Note 18**). *See* Table 2 for options and Fig. 2b–e for examples.

2. Follow the Instruction Manual for the EpiNext m6A CUT&RUN Seq Kit from section 3 ("First Strand cDNA Synthesis") to section 6 ("Cleanup of Synthesized Library") and determine DNA amount and size distribution by quantification assays such as Qubit and Bioanalyzer HS DNA assay, to approximate the amount of PCR cycles needed for library amplification (*see* **Notes 19** and **20**). For examples, *see* Fig. 2f, g).

3. Finally, after section 7 ("Library Amplification and Indexing") and section 8 ("Cleanup of Amplified Library," determine DNA amounts and size distribution of amplified libraries by quantification assays such as Bioanalyzer DNA HS assay (*see* **Note 21**). For examples, *see* Fig. 2h, i.

4. Use a short-read next-generation sequencing system of your choice with advantageous read length parameters and optimally high reads-to-cost ratio (*see* **Note 22**).

3.4 Bioinformatic Analyses of m6A-Seq Data

Analyses can by performed on the command line after installation of all required tools on a personal computer/server or on a decentralized computing system. Alternatively, and as described below, the freely available and well-maintained Galaxy platform [5] offers most computing capabilities with a GUI (Graphical User Interface), thus facilitating bioinformatic analysis for a broader audience (*see* **Note 23**).

3.4.1 Initial Quality Control

1. Go to https://usegalaxy.eu/, https://usegalaxy.org/, or another hub, create an account and log in.
2. Upload data to Galaxy: Tool `Upload Data` (left pane), then drag and drop your sequencing files (.fastq) (*see* **Note 24**).
3. To keep track of your files, label them with tags that get inherited to downstream files: Click on the dataset name in the history (right pane), then click on the tag item and add tags with a hashtag in front, for example, "#sample_1", "#reverse".
4. Check the quality of the obtained sequencing data: Tool `FastQC` (left pane) [6], select multiple datasets (in middle pane), supply a list with your adapters (optional) and run the tool. Check the .html output for sequencing quality and adapter contamination.
5. Trim your reads by removing adaptors and nucleotides of low sequencing quality. Use the tool `TrimGalore!` [7] to detect adaptors automatically (*see* **Note 25**).
6. Now check the quality of the trimmed sequencing data: Tool `FastQC` (left pane) as above in Step 4. To compare all `FastQC` outputs from raw and trimmed data, use `MultiQC` (select tool from left pane) [8] on the raw data outputs. Compare amount, length and quality of sequencing reads, plus contamination with adaptors or overrepresented sequences. Optionally, check overrepresented sequences via `NCBI-BLAST` against the whole nucleotide database or the nucleotide database of your species.
7. If necessary, repeat **step 5** (Trimming) with different parameters or tools (e.g., `Trimmomatic`) [9] to optimize quality and quantity of retained reads (*see* **Note 26**).

3.4.2 Mapping Reads to the Reference Genome, Peak Calling, and Visualization

1. Use the latest and best annotated version of your reference genome and upload it to Galaxy: Tool `Upload Data` (left pane), then drag and drop your reference genome file (.fasta). For maize, the currently latest reference genome is Zm-B73-

REFERENCE -NAM-5.0 (download via https://download.maizegdb.org/Zm-B73-REFERENCE-NAM-5.0/) [10].

2. For mapping, use `Bowtie2` [11]. Define, if your data is single-end or paired-end. Then, in further drop-down menus, select your respective trimmed files and provided genome reference file. Finally, check the option `Save the bowtie2 mapping statistics to the history`.
3. For each processed file, multiple new history file items are created. Check the log file for percentage of mapping, and repeat **step 2** with different parameters for optimization (*see* **Note 27**).
4. To check the quality of the mapping, generate .bigwig files from the .bam file outputs from Bowtie2 by the tool `bamCoverage` [12].
5. Download your mapped .bam file outputs from Bowtie2 and the .bigwig files. Display them in a genome browser like IGV (Integrative Genomics Viewer) [13] (https://software.broadinstitute.org/software/igv/). For this, supply the same .fasta reference genome file used for mapping, and additionally a gene annotation file in .gff3 or .gtf format (download via https://download.maizegdb.org/Zm-B73- REFERENCE - NAM-5.0/). Carefully check sites for mapped reads in m6A samples vs. input controls: visually browse and compare quantity, distribution, and mismatch occurrences of reads.
6. As a further check, use the tool `CollectInsertSize-Metrics`. (http://broadinstitute.github.io/picard/) on your .bam files. It will output the fragment sizes between adaptors and should roughly match your Bioanalyzer library profile (*see* Fig. 2h, i). It is especially helpful to compare the insert size distribution between input and m6A samples.
7. Note down insert size peaks, mapping statistics (i.e., percentage of mapped reads), and quality of mapping as seen by visual checks in the genome browser and rerun analysis from **step 2** with different parameters or different mappers like `RNA STAR` [14] until you are satisfied with the mapping efficiency and accuracy (*see* **Note 28**).
8. Use `MACS2 callpeak` to identify sites of read accumulation in m6A samples that are higher than in input samples. For the maize genome, the ungapped length of the reference genome is ~2.18 giga base pairs which can be used as effective genome size. Select respective paired-end .bam files from samples and corresponding inputs. Set the `Band width for picking regions to compute fragment size` to the most prominent peak from your Bioanalyzer profile of RNA fragment length (here ~110 nts), optimally reflected in insert peaks obtained by `CollectInsertSizeMetrics`. Use the resulting narrow

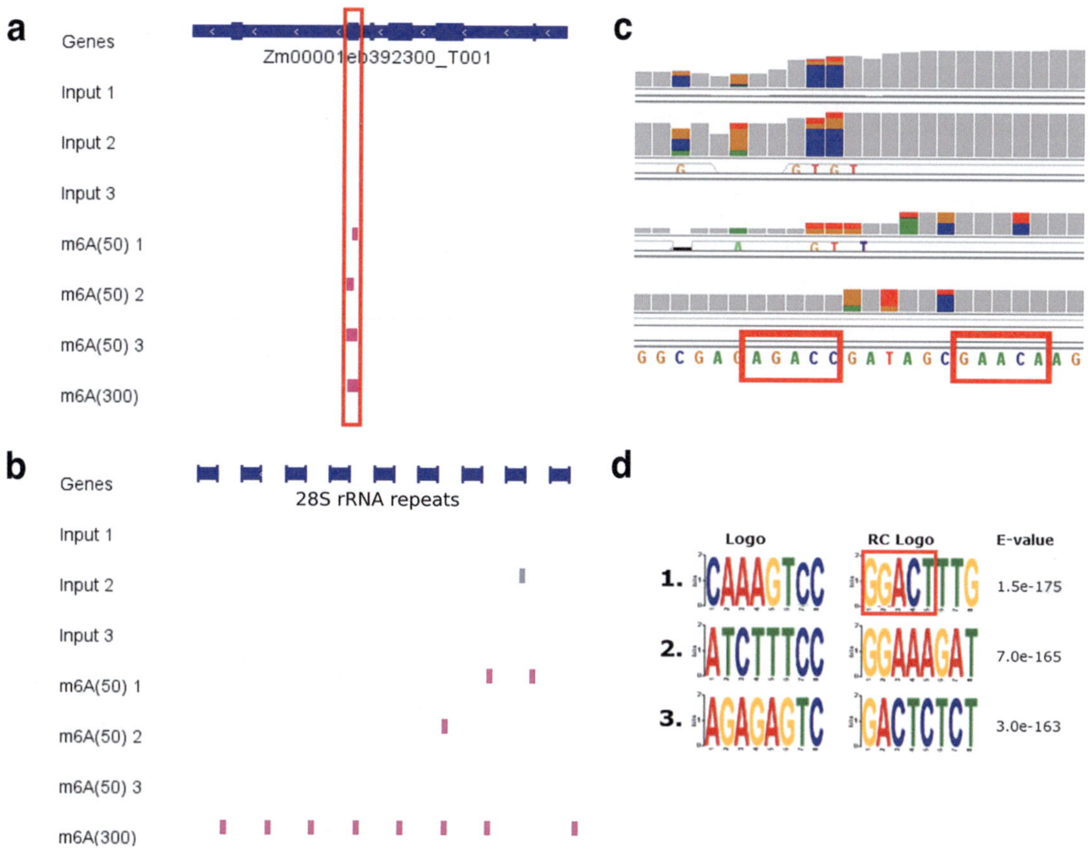

Fig. 3 Example m6A peaks and m6A motifs from meiocyte samples. (**a**) Screenshot from the IGV Genome Browser showing m6A positions identified with MACS2 after initial trimming of reads with TrimGalore! and mapping reads to the B73 maize reference genome (version 5) with Bowtie2. GFF3 annotation track displays the transcript ID (Zm00001eb392300_T001) with the gene in reverse direction, ORF in blue with bigger boxes representing CDS parts. Input samples correspond to the respective m6A samples from just 50 meiocytes ("m6A(50)"). The broad peak resulted from multiple m6A positions, was on an exon (marked by red square) and had a higher score when 300 meiocytes were used ("m6A(300)"). (**b**) Similar to (**a**) but showing a ribosomal rRNA repeat region. Identified m6A positions (MACS2-called peaks) occurred between rRNA genes and had a higher chance of detection in the sample with 300 meiocytes. (**c**) Magnified portion of an m6A peak with marked m6A motifs (AGACC and GAACA; conserved m6A RRACH (R = G/A, H = A/C/T) motif sequences; for example, *see* [15, 16]). Grey columns above nucleotides represent perfectly aligned nucleotides from reads, colored column parts represent mismatches which were detected frequently around captured m6A sites. (**d**) Topmost motifs identified in meiocyte m6A peaks with DREME; marked are 5 nucleotides of the topmost motif (in reverse complement (RC) form) which contains the conserved m6A RRACH (R = G/A, H = A/C/T) motif sequence GGACT. The high significance with which this and related motifs were detected confirms the specificity of the laboratory and analysis procedure

peak tabular files for visual analysis via IGV (*see* **Note 29**). For examples, *see* Fig. 3a, b.

9. Repeat **step 8** with different parameters or with different .bam files obtained in optimization steps before, and compare the results visually in the IGV genome browser.

3.4.3 Functional Analysis of m6A Peaks

1. As the simplest way to validate your data from both the wet lab and bioinformatic processing, use motif finder tools to identify motifs from your peak data and compare them with known m6A motifs. For this, first use the tool `bedtools GetFastaBed` [17] on your narrow peak files and the .fasta reference genome file. Then, use `DREME` [18] with default option of motif length = 8 or set to motif length = 6. A well-characterized m6A motif is RRACH (R = G/A, H = A/C/T) with the sequence GGACT being the most frequent, as also detected in our data (*see* Fig. 3c, d) (*see* **Note 30**).
2. For identifying which features (i.e., protein-coding genes, transposable elements, or noncoding RNAs) are m6A modified, use the tool `bedtools Intersect intervals` [17] on narrow peak files as "file A", with multiple "files B" supplied by .gff3's for protein-coding genes, noncoding RNAs, and transposable elements (from https://download.maizegdb.org/Zm-B73-REFERENCE-NAM-5.0/). For simplicity, do not use the original .gff3 files but .bed files created by conversion of them with `GFF-to-BED`. As `bedtools Intersect intervals` output file option, select (i) `Add data set names instead of indices`, (ii) `Single output containing intersections of any inputB lines with A`, and (iii) `Perform a left outer join. That is, for each feature in A report each overlap with B. If no overlaps are found, report a NULL feature for B (-loj).`
3. Download the output of the feature annotation for the peaks, and edit it in the following way in Excel: Remove unnecessary columns (keep peak IDs, features hit, respective locations and scores), remove duplicate values for alternative gene transcript forms by marking the feature column and deleting all _T001, _T002, _T003 (by `Find and Replace` in the `Home` tab, replace with nothing) until no more are found. If entries listing merely "chromosome" or "scaffold" are present in the feature column, replace these by "." in both the feature and the file B column to indicate no occurrence of a specific feature. Then highlight your whole table, go to the `Data` tab in Excel to select `Remove Duplicates`. Select your column containing the peak IDs and the column with the modified features, then remove duplicate rows. Next, mark the column containing the peak IDs, go to Excel's `Home` tab and select `Conditional Formatting > Highlight Cell Rules > Duplicate Values`. Check manually for remaining duplicates and how to combine them (*see* **Note 31**). In the end, the list should contain approximately as many rows as your initial peak ID list, with additional rows due to overlapping features (e.g., gene and TE hit by same m6A peak).

4. For assigning percentages to feature classes, count the instances of occurrences in the column with the peak feature origin (the different file B's or no file B) with the formula `=COUNTIF(K2:K632;B73v5_genes_bed)`. Check especially occurrences of noncoding hits for their meaning, for example, if they are due to long stretches of knobs/telomeres/centromeres. Do so via checking the chromosome locations in public databases and genome browsers. For maize, use https://jbrowse.maizegdb.org and select (i) `Official gene models for Zm-B73-REFERENCE-NAM-5.0`, (ii) `B73 non-coding gene models`, (iii) `Mikado transcriptome`, and (iv) `Knobs, centromeres and telomeres` in the left pane. After manual curation of the peak feature list, calculate the percentages of unannotated, gene/TE/noncoding RNA-overlapping peaks (*see* **Note 32**).
5. Follow up on functional analysis of genes overlapped by m6A peaks via GO analysis: Currently, we recommend to use version 3 annotations for this; thus, convert the gene IDs from v5 to v3 (*see* **Note 33**). One way to do so is to download the conversion list `B73v5_toB73v3.tsv` from https://download.maizegdb.org/Pan-genes/B73_gene_xref/. Next, use Excel's `VLOOKUP` formula to add v3 gene IDs to the list of v5 gene IDs. Then, paste the v3 gene IDs into AgriGO's web-interface (http://systemsbiology.cau.edu.cn/agriGOv2/classification_analysis.php?category=Plant&&family=Poaceae) or use another tool for GO analysis.

4 Notes

1. Experimental setup and sample selection: We recommend having at least 30–50 cells per sample, and at least 3–4 replicates. For sequencing projects more replicates with less cells and/or reads are often superior to less replicates with more cells and/or reads [19, 20]. At first, use a similar but easy to obtain cell type of which you can use 10–100× starting material for trials to identify potentially best procedures. Process all samples in the same way to avoid bias from parameters such as RNA quality or PCR amplification.
2. RNase-free working: Dedicate a workspace for RNase-free work and use best practices for RNase-free work by: handling everything only with gloves, using RNase decontamination solution and nuclease-free double-distilled water, using RNase-free filter tips, using nuclease-free double-distilled water or pretreat your water with DEPC to make it RNase free, baking flasks and bottles overnight at 180 °C before usage, using certified RNase-free enzymes (e.g., DNases).

Since RNase decontaminating cleaning agents are highly basic and can damage nucleic acids as well, we suggest rinsing cleaned surfaces with nuclease-free water and then dry them with clean delicate task wipes.

3. Use only pipettes that allow accurate and reproducible pipetting. Always verify the volume in the pipette tip by eye to avoid possible pipetting errors. Precise pipetting is crucial to determine exact concentrations and amounts, for achieving the proper sample-bead-ratios, and for uniform and diligent performance of the bead washing steps.
4. We tried an ultra-low input SMART library prep which did, however, result in rather broad peaks for m6A samples, and almost no peaks for input samples. Although more optimizations might lead to better results, we recommend to use adaptor-tagging library preparation where adaptors are tagged during the double-stranded cDNA synthesis (see EpiNext Kit Manual). Adaptor tagging is different from traditional adaptor ligation, and is used in challenging protocols such as Whole-Genome Bisulfite Sequencing (WGBS) [21] and single-cell RNA sequencing from LCM (laser-capture microdissection) samples from FFPE (formalin-fixed paraffin-embedded) tissues [22].
5. The length and age of the plant are not fully reliable to determine the exact meiotic stage of meiocytes, but the leaf collar method [23] can help to identify potential meiocyte-containing plants.
6. We recommend taping moderately tight to avoid interrupting the plant metabolism, but tight enough to reduce air contact of the inner tissues.
7. Plastic coverslips are highly recommended due to their lower adhesion properties. They can be attached to a glass microscope slide with ~5–10 μl of water.
8. Previously, we used self-made glass microcapillaries with mouth-pipetting [4]. More recently, we used a manual microinjector. It allows for smaller volumes during vacuuming, but requires both hands which prevents simultaneous movement of the slide table to move to further positions with meiocytes.
9. If the collection procedure generates high volumes, centrifuge the cells down, remove the supernatant and ensure that it does not contain cells, then add proper amounts of TRIzol to the cell pellet.
10. We could obtain higher RNA quantity and quality from 50 meiocytes when using a plastic tube-pestle instead of metal beads in a tissue homogenizer. For squashing the meiocytes in a 1.5 ml tube with a fitting pestle, we started with only

50–100 μl of TRIzol to facilitate better efficiency. In higher liquid volumes, the cells can easily evade the area between the pestle and the wall of the tube. We added 2× volumes of TRIzol after squashing, by pipetting it along the pestle to wash off any RNA still sticking on it.

11. An on-column DNase digest during RNA extraction is favored due to minimized loss and reduced timing. The digest should be carried out in its whole duration. Also, the DNase mix has to be pipetted carefully onto the middle of the column.
12. For most downstream processes, it is best to have the highest concentration possible, thus the lowest elution volume needed. However, higher elution volumes might recover more RNA, and especially two rounds of elution (e.g., 10 μl twice) might recover the most RNA. Furthermore, it has to be taken into consideration which equipment is available for quantification and RNA quality check, and how much of the RNA will be needed for this. For an overview, *see* Table 2. We would recommend the two rounds of elution with 10 μl each, then use of 1–2 μl for RNA measurements of purified or 1:10 diluted sample.
13. Suboptimal OD260/OD230 values are frequently obtained with low input samples. This is due to low amounts of contaminants that have a higher proportion to the low amount of RNA than in higher input samples where they are also present. Improvements can be achieved by higher elution volumes or sample dilution after RNA extraction though they are not necessary for the downstream application of m6A capture which dilutes the samples at least 1:10 further. We do not recommend RNA precipitation due to RNA loss by the process.
14. Select samples with comparable RINs for different tissues or conditions. Otherwise, identified m6A profile differences might be due to RNA integrity instead of sample type.
15. Input Controls can be used to establish a baseline and to filter out noise from, for example, repetitive regions or from "sticky" RNA that binds to beads by itself and is not cleaved.
16. The kit recommends 200–400 ng for the Input Controls in case of the regular range (0.5–20 μg) which constitutes ~1/100th to 4/5th of the m6A samples. We do not recommend going too low since the Input Control Library will be distributed over the whole transcriptome while the m6A samples are supposed to be on only a small proportion of the transcripts.
17. An IgG negative control should be done when material can be easily collected and acquired in larger amounts, and if budget is

available. For low input samples though, a regular Input Control is the better choice of control.

18. The obtained RNA quantity at this point is very informative. First, it indicates if the procedure worked so far or if no RNA was recovered. Second, the percentage of the yield (RNA captured divided by RNA used for experiment, in percent) allows a first estimation of m6A abundance in the samples. Our yield in seedlings (≤0.1%) was far lower than in meiocytes (0.2–0.3% in samples with 300 meiocytes; wider, less reliable range of 0.3–2.2% in samples with 50 meiocytes). Input Controls had yields of ~0.5–5%, usually one order of magnitude higher than the m6A samples. Due to the lack of other options, we used a Bioanalyzer HS RNA Assay for quantification which had the added benefit of detecting big peaks >200 nt (either environmental RNA contaminants, issues with RNA digestion, or RNA fragments with high density of m6A modification).
19. The RNA amount of specific library fragments prior amplification might be too low for proper detection. We could deduce this from the Qubit concentrations which were similar for all samples and Input Controls. Furthermore, Bioanalyzer HS DNA assays (Fig. 2f, g) showed fragments which presumably contain adaptors (33 nt), adaptor dimers, and fragments shorter than the maximal length of 176 nt (110 nt captured plus 2 adaptor with 33 nt each), due to random priming/ binding of the adaptors.
20. As stated in the previous comment, DNA quantification of pure libraries might not be reliable. We used 20 cycles to avoid creating too much bias, and used the whole volumes of the libraries, with the strategy to adjust final amplified library amounts later when loading the flow cell for sequencing. It is very important to use the same amount of PCR cycles for all libraries that will be compared. Otherwise, the differences might just be due to bias created by PCR amplification.
21. In a Bioanalyzer DNA HS assay, a specific peak ~230 bp should be apparent and relatively clean (*see* Fig. 2h, i). We sequenced the libraries containing additional lower peaks to gain information on their composition. In our case, they contained indeed specific sequences which were, however, more difficult to use for mapping. When shorter fragments are expected to be primer and adaptor molecules only, an additional cleanup step via gel electrophoresis (agarose, polyacrylamide, or automated by, for example, Blue Pippin System) can reduce the sequencing reads that are lost on unspecific sequences. Bioanalyzer outputs can be evaluated for DNA amount in a specific region of interest while Qubit assays measure only the whole DNA.

22. Single-end reads are sufficient, but paired-end reads can be used as well. Since our samples have a length of most substantially ~110 nt, and the adaptor and index sequences do not count (or not substantially) toward the read length, a 75–150 nt read length is recommended. The depth of reads needs to be adjusted to the purpose; we used only 3 million reads per sample, yielding ~50–170 m6A sites per replicate, and ~600 sites when joining all replicates (3× from 50 meiocytes each, plus 1× from 300 meiocytes vs. 3× input controls from 50 meiocytes each).

23. There are many good tutorials on how to use Galaxy (https://galaxyproject.org/learn/).

24. It is possible to create a collection for parallel processing. In the right pane, enable checkmarks, check all data files, then click `For all selected > Build list of dataset pairs` (for paired-end reads) or `> Build dataset list` (for single-end reads), then select the proper files. However, several tools do not work well with collections. It does work with the suggested tools `FastQC > Trim Galore! > FastQC > Bowtie2 > bamCoverage` and `MACS2 callpeak`, if omitting MultiQC.

25. An important parameter for trimming or later alignment is the allowed minimum size of reads. When reads are <20 nts, the chance of random alignment to the reference genome is higher, especially when allowing mismatches.

26. Instead of adjusting trimming parameters with all samples in parallel, select a sample-input pair for optimization.

27. We noticed that sequencing errors/mismatches frequently occur around m6A sites (Fig. 3c). This has been observed by others albeit with different sequencing technologies [24], and is presumably due to the structural modification of the adenosine by methylation. Thus, mismatches need to be allowed during mapping in order to identify these sites at all.

28. We obtained ~1.5% read alignment with Bowtie2 and ~20% read alignment with RNA STAR. m6A sample-specific insert size peaks that did not occur in the input controls were ~30–100 nts for Bowtie2, and ~80–100 nts for RNA STAR after trimming with Trimmomatic instead of TrimGalore! While the analysis via RNA STAR resulted in more confident peaks, analysis with Bowtie2 resulted in far higher significance levels for known m6A motifs (*see* Fig. 3d).

29. A common strategy to empower peak calling is to join the replicates before doing so. We recommend to also run peak calling on individual sample–input pairs and even samples and inputs separately to check for reproducibility and detect problematic samples.

30. Of note, motifs identified with DREME in Fig. 3d were from data processed with `Trim Galore! > Bowtie2`. Using the same data for `MEME` [25] revealed a 50 nt long motif which matches the 28S rRNA and contains the first 8 nt long motif found with DREME. We kept rRNA on purpose to obtain results for high-copy regions with presumable m6A modification as positive control. Using data processed with Trimmomatic > RNA STAR yielded also RRACH motifs enriched in the peak data with high confidence (*E*-value 5.6e-020).
31. Do not leave any duplicate values for the peaks unless they list different features. Otherwise, this will cause a distortion in subsequent GO analysis.
32. Be aware that the B73_v5 annotation is not yet foolproof. For example, rRNA repeats are still listed as protein-coding genes (*see* https://maizegdb.org/gene_center/gene/Zm00001eb261960, accessed on 07/25/2021).
33. Version 3 had a better annotation than V4 and is still actively used.

Acknowledgment

We thank the Meister lab for access to their MiSeq machine and excellent support by N. Eichner. This work was supported by the German Research Foundation DFG via Collaborative Research Center SFB960 to T.D. and a FAS grant (Financial Support for Equality) of the University of Regensburg to S.D..

References

1. Lee M, Kim B, Kim VN (2014) Emerging roles of RNA modification: m6A and U-tail. Cell 158:980–987
2. Skene P, Henikoff S, Reinberg D (2017) An efficient targeted nuclease strategy for high-resolution mapping of DNA binding sites. eLife 6:e21856
3. Weichmann F, Hett R, Schepers A et al (2020) Validation strategies for antibodies targeting modified ribonucleotides. RNA 26(10): 1489–1506
4. Dukowic-Schulze S, Garcia N, Shunmugam ASK et al (2020) Isolating male meiocytes from maize and wheat for "-omics" analyses. In: Methods in Molecular Biology. Springer, New York, pp 237–258
5. Afgan E, Baker D, Batut B et al (2018) The Galaxy platform for accessible, reproducible and collaborative biomedical analyses: 2018 update. Nucleic Acids Res 46:W537–W544
6. Andrews S (2010) FastQC - A quality control tool for high throughput sequence data. https://www.bioinformatics.babraham.ac.uk/projects/fastqc/
7. Krueger F, James F, Ewels P, et al (2021). FelixKrueger/TrimGalore: v0.6.7 - DOI via Zenodo (0.6.7). Zenodo. https://doi.org/10.5281/zenodo.5127899
8. Ewels P, Magnusson M, Lundin S et al (2016) MultiQC: summarize analysis results for multiple tools and samples in a single report. Bioinformatics 32:3047–3048
9. Bolger AM, Lohse M, Usadel B (2014) Trimmomatic: a flexible trimmer for Illumina sequence data. Bioinformatics 30:2114–2120
10. Hufford MB, Seetharam AS, Woodhouse MR, et al (2021) *De novo* assembly, annotation, and comparative analysis of 26 diverse maize genomes. bioRxiv 2021.01.14.426684

11. Langmead B, Salzberg SL (2012) Fast gapped-read alignment with Bowtie 2. Nat Methods 9: 357–359
12. Ramírez F, Ryan DP, Grüning B et al (2016) deepTools2: a next generation web server for deep-sequencing data analysis. Nucleic Acids Res 44:W160–W165
13. Robinson JT, Thorvaldsdóttir H, Winckler W et al (2011) Integrative Genomics Viewer. Nat Biotechnol 29:24–26
14. Dobin A, Davis CA, Schlesinger F et al (2013) STAR: ultrafast universal RNA-seq aligner. Bioinformatics 29:15–21
15. Chen T, Hao Y-J, Zhang Y et al (2015) m6A RNA methylation is regulated by microRNAs and promotes reprogramming to pluripotency. Cell Stem Cell 16:289–301
16. Linder B, Grozhik AV, Olarerin-George AO et al (2015) Single-nucleotide resolution mapping of m6A and m6Am throughout the transcriptome. Nat Methods 12:767–772
17. Quinlan AR, Hall IM (2010) BEDTools: a flexible suite of utilities for comparing genomic features. Bioinformatics 26:841–842
18. Bailey TL (2011) DREME: motif discovery in transcription factor ChIP-seq data. Bioinformatics 27:1653–1659
19. Liu Y, Zhou J, White KP (2014) RNA-seq differential expression studies: more sequence or more replication? Bioinformatics 30:301–304
20. Baccarella A, Williams CR, Parrish JZ et al (2018) Empirical assessment of the impact of sample number and read depth on RNA-Seq analysis workflow performance. BMC Bioinformatics 19:423
21. Miura F, Enomoto Y, Dairiki R et al (2012) Amplification-free whole-genome bisulfite sequencing by post-bisulfite adaptor tagging. Nucleic Acids Res 40:e136
22. Foley JW, Zhu C, Jolivet P et al (2019) Gene expression profiling of single cells from archival tissue with laser-capture microdissection and Smart-3SEQ. Genome Res 29:1816–1825
23. Begcy K, Dresselhaus T (2017) Tracking maize pollen development by the Leaf Collar Method. Plant Reprod 30:171–178
24. Liu H, Bekig O, Lucas MC et al (2019) Accurate detection of m6A RNA modifications in native RNA sequences. Nat Commun 10:4079
25. Bailey TL, Boden M, Buske FA et al (2009) MEME SUITE: tools for motif discovery and searching. Nucleic Acids Res 37:W202–W208

Chapter 22

Assembling Plant Genomes with Long-Read Sequencing

Quentin Gouil

Abstract

Continuous improvements in long-read sequencing allow us to tackle increasingly big and complex genomes. Here we present the principles of long-read genome assembly, taking *Solanum pennellii* nanopore sequencing as an example.

Key words Polishing, Genome, De novo assembly, Long-read sequencing, Nanopore, Solanum pennellii

1 Introduction

The assembly of the Arabidopsis genome made the humble thale cress the workhorse of plant genetics [1]. Although the Arabidopsis genome is compact (125 Mb) and repeat-poor, its initial assembly required a large effort and gaps still remain. Nowadays long-read sequencing with Pacific Biosciences (Pacbio) or Oxford Nanopore Technologies (ONT) make genome assembly far more efficient: for example, a good quality Arabidopsis genome can be assembled from 1 MinION flow cell in less than a day [2]. Repetitive or highly rearranged regions can be resolved by the long reads [2, 3], so we are no longer oblivious to large parts of the genome and their transcripts [4, 5]. Perhaps even more importantly, genomic research is no longer constrained to a small set of model species and a single reference genome; long reads are already being used to assemble many new plant genomes and capture all the genetic variation in pan-genomes [3, 6–8]. This is only the beginning of a genomic surge spearheaded by large-scale projects such as the Vertebrate Genome Project (https://vertebrategenomesproject.org/) or the Earth Biogenome Project [9].

Supplementary Information The online version of this chapter (https://doi.org/10.1007/978-1-0716-2253-7_22) contains supplementary material, which is available to authorized users.

Christophe Lambing (ed.), *Plant Gametogenesis: Methods and Protocols*, Methods in Molecular Biology, vol. 2484, https://doi.org/10.1007/978-1-0716-2253-7_22,

Sample preparation protocols, sequencing technologies and analysis tools are all rapidly evolving. As a consequence, best practices are in flux and this chapter will not provide the definitive pipeline for genome assembly. Rather, it will highlight the typical steps and factors to consider when assembling a plant genome from long-read data.

Until recently there was a severe trade-off between read length and accuracy, which was often mitigated by combining multiple technologies: long-read sequencing to obtain a draft contiguous assembly, and short-read sequencing to correct errors. Recently, PacBio HiFi reads with lengths over 20 kb and accuracy $\geq$99.5% made long-read–only assemblies realistic. Meanwhile, ONT accuracy continues to progress (currently 98% modal single-read accuracy, with a 99% accurate chemistry under testing) and "ultralong" protocols enable routine high-throughput sequencing with most of the data in reads 100 kb or longer. Long-read–only genomes are likely to become increasingly common. In this chapter we will present a nanopore-only assembly of *Solanum pennellii* LA5240, a wild relative of tomato, using data from Schmidt et al. [6]. The *S. pennellii* genome measures just over 1 Gbp, making it a realistic example of plant genome assembly while avoiding the complications of high-heterozygosity or polyploidy.

2 Materials

We require a number of specialised tools for each step of the assembly pipeline. Python-based tools can be conveniently installed within a conda environment. Installation is reasonably straightforward following the respective manuals. Many of these tools are being actively developed and we recommend to ensure the latest versions are used. The long-read-tools.org database [10] facilitates exploration of the tools and their alternatives. In this chapter we use:

1. Seqtk v1.3 [11].
2. LongQC v1.2.0 [12].
3. Porechop v0.2.4 [13].
4. Filtlong v0.2.0 [14].
5. Flye v2.8.2 [15].
6. Minimap2 v2.17 [16].
7. Medaka v1.2.1 [17].
8. Quast v5.0.2 [18].
9. Busco v5.0.0 [19].

3 Methods

3.1 Pipeline Overview

A nanopore-only genome assembly pipeline is presented in Fig. 1, beginning with sequencing and basecalling (not covered here), data preprocessing, constructing a draft assembly, polishing the draft into a consensus assembly, and evaluating the result.

3.2 Data Collection

The *S. pennellii* LA5240 nanopore sequencing by Schmidt et al. [6] was performed across 31 MinION flow cells for a total of 134 Gbases. It is worth noting that this throughput can now be achieved with 1 to 2 PromethION flow cells at a much reduced cost and effort.

Raw nanopore data are stored in a fast5 format, containing the values of the electrical current measured as a strand of DNA is translocated through the pore. This information is decoded into DNA bases and saved in a familiar fastq format during basecalling, typically using an artificial neural network. ONT's basecalling algorithms are constantly improving; higher per-read accuracy directly translates into improved quality and contiguity of the assembly [20, 21], therefore it is recommended to start the analysis by re-basecalling with the latest iteration of the ONT basecaller. However, fast5 files are very large and numerous (early on each read even had its own fast5 file, causing trouble with filesystems), so here we will not attempt to download and rebasecall the 21 TB of raw fast5 data, and instead will rely on the existing basecalling (done with Albacore 2.1.1).

The data is hosted on https://plabipd.de/portal/solanum-pennellii. We will download the fastq from each run, whose names are saved in Spenn_fastq.txt (Supp file 1):

Fastq download

```
ADDRESS=http://www.plabipd.de/projects/SpennLost/minion_alb
acore_2_1_1/; while read libname
do curl -L $ADDRESS"$libname"_all.fastq.gz >> Spenn.fq.gz
done < Spenn_fastq.txt
```

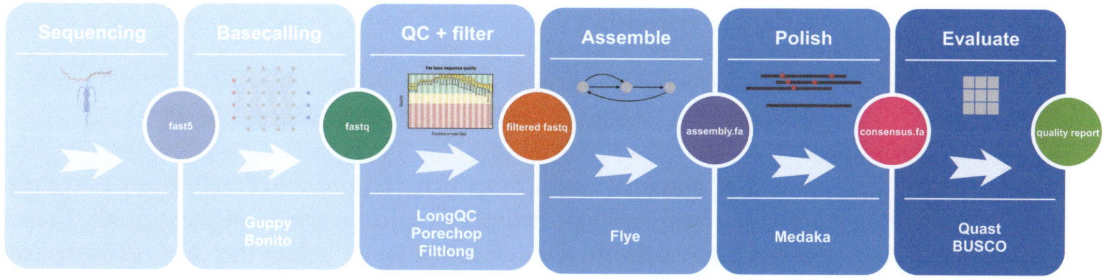

Fig. 1 Assembly pipeline for *Solanum pennellii* LA5240. We start from the basecalled reads in fastq format on **step 3**. Flowchart created with Draw.io and BioRender

We now have a large (120 GB, 134 Gbases) fastq file containing all the nanopore read sequences and quality information.

3.3 QC and Filtering

Some of the nanopore QC tools such as pycoQC [22] can use the basecalling summary text file to provide quick QC statistics for the experiment. Here, we only have the fastq data so we need to evaluate quality from the data itself, which takes longer. LonqQC [12] offers this functionality but is slow and consumes a lot of memory on a dataset of this size. To speed up the process we can select a subset of the data with seqtk [11], for example selecting 10,000 reads at random.

Initial QC

```
#subsample the data, setting the seed for reproducibility
seqtk sample -s100 Spenn.fq.gz 10000 > Spenn_subset.fq.gz
#run QC on subsamble
python LongQC/longQC.py sampleqc -x ont-ligation \
-o QC_subset Spenn_subset.fq.gz
```

We obtain a HTML report with read statistics. Among those is the length distribution (Fig. 2ai). We observe a bimodal distribution of read lengths with a mean of 12 kb and N50 of 17 kb, meaning that 50% of the data resides in reads that are 17 kb or longer. The bimodal distribution matches our expectations as Schmidt et al. employed two protocols for library preparation: one without size selection and one enriching for longer reads [6]. Adapter sequences are detected on the 5′ and 3′ ends of reads (Fig. 2aii). Although the Flye assembler [15] can account for sequencing adapters, we will trim them as part of our data preprocessing step.

The reported read quality is around 9 (88% estimated accuracy, Fig. 2aiii), although actual accuracy should be measured against a reference sequence. Porechop [13] is a popular tool for trimming adapters from ONT reads, although it is no longer maintained. Run on the whole dataset, it consumes 400 GB of memory, taking 10 h running on 12 cores. An alternative is to perform a hard trim (e.g., 50 bp) on both the 5′ and 3′ ends of reads.

Adapter trimming

```
#trim adapter sequences
porechop -i Spenn.fq.gz -o Spenn_trimmed.fq -t 12
```

The assembly step does not require extremely high coverage: 50× is considered sufficient. The longest and highest quality reads are the most useful for assembly, so we select the best reads up to a total of 50 Gb (about 50× coverage) with Filtlong [14]. The rest of the reads is later used for polishing (correcting base errors in the assembly).

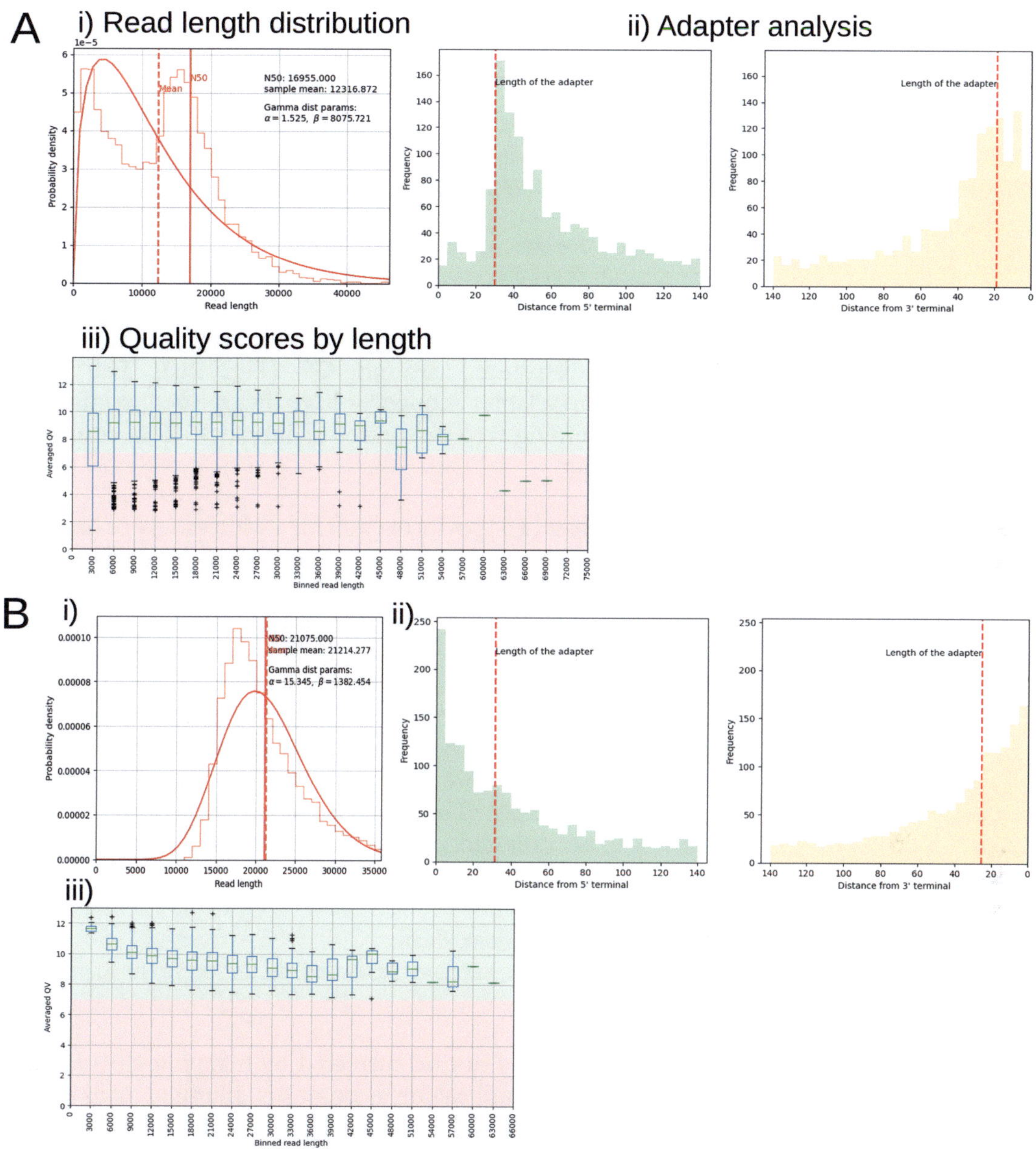

Fig. 2 LongQC read statistics. LongQC read statistics for a subset of the nanopore reads before (**a**) and after (**b**) preprocessing and filtering: (i) read lengths, (ii) presence of adapter sequences, (iii) read qualities

Filtering

```
#select most informative reads to 50× coverage
Filtlong/bin/filtlong --min_length 1000 --keep_percent 90 \
   --target_bases 50000000000 Spenn_trimmed.fq | \
gzip > Spenn_trimmed_filtered.fq.gz
#compress the input file to save space
gzip Spenn_trimmed.fq
```

We now inspect the selected data with LongQC:

Second QC

```
#subsample again
seqtk sample -s100 Spenn_trimmed_filtered.fq.gz 10000 > \
Spenn_trimmed_filt_subset.fq.gz
#run QC
python LongQC/longQC.py sampleqc -x ont-ligation \
-o QC_trimmed_filtered_subset Spenn_trimmed_filt_subset.fq.gz
```

The mean read length and N50 are now 21 kb and the quality more tightly distributed around Q9 (Fig. 2b). The peaks corresponding to adapter sequences have disappeared, confirming the adapter trimming step performed as intended. This concludes the preprocessing.

3.4 Assembly with Flye

The Flye assembler [15] performs well for a wide variety of bacterial and eukaryotic genomes, however there is an evolving collection of performant alternative tools (*see* **Note 1**). It is common practice to test multiple assemblers to find the best fit for one's data type (ONT, Pacbio) and genome characteristics (size, ploidy, heterozygosity, repeat content, etc., *see* **Note 2**). Assemblers vary widely in resource consumption, possibly becoming prohibitive on large genomes and influencing tool choice and parameter-space exploration. Run on 30 cores, Flye takes 32 h and 150 GB of memory to complete the assembly. The Canu-Smartdenovo assembly from Schmidt et al. necessitated 10 times as much cpu time [6].

Assembly

```
#run Flye on 50× longest data
flye --nano-raw Spenn_trimmed_filtered.fq.gz -t 30 -o assembly/
```

3.5 Polishing with Medaka

Errors in the assembly can be corrected by mapping all the reads back to the assembly and calling a consensus. Illumina reads can be used at this stage, but long-read–only polishing is also possible. Here, we will use Medaka to polish the assembly with the full long-read dataset. Medaka is the polisher developed by ONT and can use error correction models specific to the basecaller version. There is no model for Albacore 2.1.1 basecalling however, so we use the default (latest) model. Medaka runs minimap2 [16] in the first step to map all raw reads to the draft assembly. With 30 cores, polishing takes 16 h and 50 GB of memory.

Polishing

```
#error correction with full dataset
medaka_consensus -i Spenn_trimmed.fq.gz \
-d assembly/assembly.fasta -o medaka_consensus/ -t 30
```

Table 1
Assembly metrics. Genomic sizes are in megabases (Mb)

Assembly	N50	Total size	Largest contig	Total contigs	% complete BUSCO
Schmidt et al. raw [6]	2.45	889.92	12.32	899	29.1
Flye raw	1.79	982.82	15.10	2243	88.4
Schmidt et al. hybrid polished [6]	2.52	915.60	12.72	899	96.5
Flye long-read polished	1.79	984.56	15.14	2234	90.5
Flye hybrid-polished	1.79	982.87	15.12	2234	98.5

Polishing is often repeated multiple times, taking the output polished assembly as the input for the next round of polishing (*see* **Note 2**). The rationale is that as errors are corrected, reads can be remapped more accurately, and thus additional errors can be removed. Polishing can be run until convergence (additional rounds do not introduce any more changes) or reaching oscillation (the same positions are changed back and forth). Two to five rounds are common. Racon [23] is another popular polishing tool that can be combined with Medaka. As for assembly, trial and error remains the recommended procedure, at least for now.

If short-read data are available, further rounds of polishing with these short reads (for instance using Pilon [24]) increases base-level accuracy (*see* Table 1).

3.6 Evaluating Assembly Quality

Evaluating assembly quality is not trivial. We rely on several metrics to compare and rank the results from the various pipeline trials, and external data for validation when available. Without external data, total assembly size gives an estimate of completeness, and N50 an estimate of contiguity. The presence of a conserved set of genes in the assembly, evaluated with BUSCO [19], is another indicator of quality and completeness. If Illumina reads are available, base-level accuracy of the assembly can be estimated. The availability of a reference genome allows to detect misassemblies. Here we can use the short-read–based *S. pennellii* LA0716 reference genome [25] as a comparison, and generate QC plots and metrics for multiple assembly trials with Quast [18]. We also run BUSCO using the embryophyta_odb10 gene database for direct comparison with the Schmidt et al. assembly [6].

Assembly statistics

```
#download the reference S. pennellii LA0716 genome
mkdir -p Spenn_ref
wget ftp://ftp.solgenomics.net/genomes/Solanum_pennellii/
Spenn.fasta \ -P Spenn_ref/
#run quast on assembly pre- and post-polishing
```

```
quast --large -t 24 --circos -o quast_evaluation \
-r data/Spenn_ref/Spenn.fasta \
assembly/assembly.fasta medaka_consensus/consensus.fasta #run
BUSCO pre-polishing
busco -i assembly/assembly.fasta --out flye_assembly \
--out_path busco_flye_assembly/ -c 30 \
-l embryophyta_odb10 --mode genome
#run BUSCO post-polishing
busco -i medaka_consensus/consensus.fasta --out medaka_consensus \
--out_path busco_medaka_consensus/ -c 30 \
-l embryophyta_odb10 --mode genome
```

The results are presented in Table 1 and Fig. 3. Our Flye assembly is 11% longer than the Schmidt et al. assembly using Canu-SmartDeNovo [6] (983 Mb vs. 890 Mb, Fig. 3a) and closer to the expected genome size (1 to 1.2 Gb). Very long and repetitive regions such as centromeres are particularly complex to assemble without ultralong reads, often resulting in collapsed assemblies [2, 26]. The largest contig (continuous assembled segment) is 15 Mb, slightly longer than the published assembly (12 Mb), still a long way from the chromosome sizes (60–100 Mb). The level of fragmentation of the assemblies can also be seen from the total contigs and N50: the Flye assembly is more fragmented (over 2000 contigs) than the published long-read assembly (899 contigs [6]) and has a smaller N50 (1.79 vs. 2.45 Mb), although it is also likely more complete.

Aligning contigs from the Flye assembly to the LA0716 short-read–based reference genome (Fig. 3b) also shows fragmentation

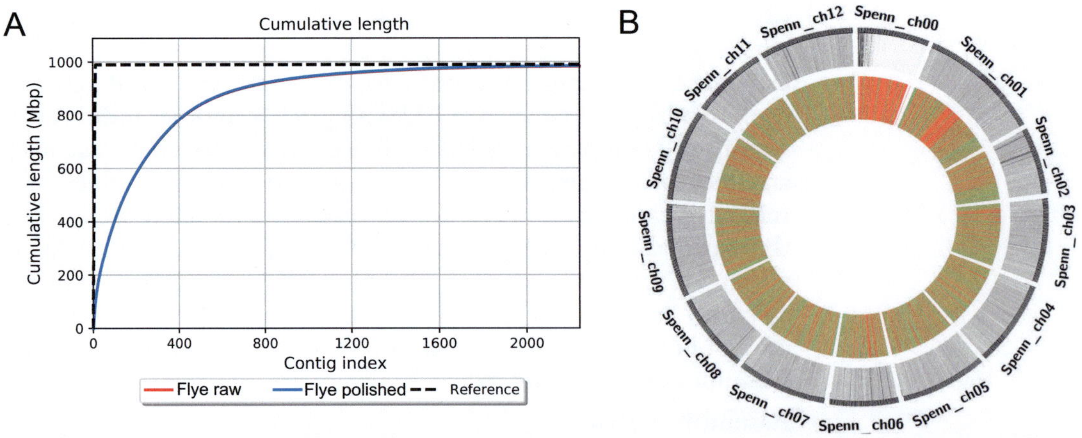

Fig. 3 Assembly statistics from Quast. (**a**) Cumulative lengths of the assemblies. They are nearly identical before and after polishing so the lines overlap. (**b**) Circos plot showing the mapping of contigs to the reference *S. pennellii* LA0716 genome [25]. The outer circle represents the LA0716 reference sequence with the GC% heatmap from 0% (white) to 62%. The inner track is the Medaka-polished Flye assembly. Aligned contigs are shown in green and "misassembled" blocks in red

and potential misassemblies. As expected, contigs from our assembly do not map to the pseudochromosome chr00 made up of unscaffolded short-read contigs, but they also fail to map to the large chromosome 1 centromere. This could be due in part to unknown sequence (Ns) in the reference genome. Gene-rich regions like the distal end of chromosome 1 are much more concordant.

Polishing can have a dramatic impact on BUSCO completeness (Table 1): in the published assembly, the BUSCO score jumped from 29% to 96.46% using a combination of short reads and long reads (hybrid polishing). The Flye unpolished assembly (using uncorrected long reads basecalled with a slightly more recent algorithm than the published assembly) already reached 88.4% BUSCO completeness, and polishing with long reads only increases the score to 90.5%. One extra round of polishing using short reads and Pilon [24] attains the highest BUSCO completeness: 98.5%. From the same raw data, updated algorithms provide a more complete and more accurate assembly. The assembly is not yet error-free and "telomere-to-telomere"; depending on the researchers' aims, additional work can be undertaken to further refine the assembly (*see* **Note 3**).

4 Notes

1. We presented a pipeline to assemble the 1 Gbp *Solanum pennellii* LA5240 genome from nanopore sequencing, current as of early 2021. Genome assembly is still an active area of research, with no one-size-fits-all approach. As sequencing technology and algorithms rapidly improve, regularly updated benchmarks are crucial to identify the current best practices (for recent examples of long-read assemblies of plant genomes *see* [21, 27]).
2. Trying out multiple assemblers and polishing rounds is recommended.
3. The most time-consuming part begins after the consensus sequence is generated: expert knowledge about the genome of the species and custom analyses are required to arrange the contigs into chromosome-length scaffolds (supported by reference genomes, Hi-C data, optical maps, or genetic maps), correct misassemblies, and annotate genes and repeats.

References

1. Provart NJ, Brady SM, Parry G et al (2021) Anno genominis XX: 20 years of Arabidopsis genomics. Plant Cell 33(4):832–845
2. Michael TP, Jupe F, Bemm F et al (2018) High contiguity Arabidopsis thaliana genome assembly with a single nanopore flow cell. Nat Commun 9(1):541

3. Michael TP, VanBuren R (2020) Building near-complete plant genomes. Curr Opin Plant Biol 54:26–33
4. Panda K, Slotkin RK (2020) Long-read cDNA sequencing enables a "gene-like" transcript annotation of transposable elements. Plant Cell 32(9):2687–2698
5. Shahid S, Slotkin RK (2020) The current revolution in transposable element biology enabled by long reads. Curr Opin Plant Biol 54:49–56
6. Schmidt MHW, Vogel A, Denton AK et al (2017) De novo assembly of a new *Solanum pennellii* accession using Nanopore sequencing. Plant Cell 29(10):2336–2348
7. Belser C, Istace B, Denis E et al (2018) Chromosome-scale assemblies of plant genomes using nanopore long reads and optical maps. Nat Plants 4(11):879–887
8. Della Coletta R, Qiu Y, Ou S et al (2021) How the pan-genome is changing crop genomics and improvement. Genome Biol 22(1):3
9. Lewin HA, Robinson GE, Kress WJ et al (2018) Earth BioGenome project: sequencing life for the future of life. Proc Natl Acad Sci U S A 115(17):4325–4333
10. Amarasinghe SL, Ritchie ME, Gouil Q (2021) long-read-tools.org: an interactive catalogue of analysis methods for long-read sequencing data. GigaScience. 10(2):giab003
11. Li H. Seqtk: toolkit for processing sequences in FASTA/Q formats. Accessed 05 Mar 2021. https://github.com/lh3/seqtk
12. Fukasawa Y, Ermini L, Wang H et al (2020) LongQC: a quality control tool for third generation sequencing long read data. G3 (Bethesda) 10(4):1193–1196
13. Wick R. Porechop: adapter trimmer for Oxford Nanopore reads. Accessed: 2021-03-05. https://github.com/rrwick/Porechop
14. Wick R. Filtlong: quality filtering tool for long reads. Accessed: 2021-03-05. https://github.com/rrwick/Filtlong
15. Kolmogorov M, Yuan J, Lin Y et al (2019) Assembly of long, error-prone reads using repeat graphs. Nat Biotechnol 37(5):540–546
16. Li H (2018) Minimap2: pairwise alignment for nucleotide sequences. Bioinformatics 34(18): 3094–3100
17. Oxford Nanopore technologies. Medaka: sequence correction. Accessed: 2021-03-05. https://github.com/nanoporetech/medaka
18. Gurevich A, Saveliev V, Vyahhi N et al (2013) QUAST: quality assessment tool for genome assemblies. Bioinformatics 29(8):1072–1075
19. Seppey M, Manni M, Zdobnov EM (2019) BUSCO: assessing genome assembly and annotation completeness. In: Kollmar M (ed) Gene prediction, Methods in Molecular Biology, vol 1962. Springer New York, New York, NY, pp 227–245
20. Wick RR, Judd LM, Holt KE (2019) Performance of neural network basecalling tools for Oxford Nanopore sequencing. Genome Biol 20(1):129
21. Murigneux V, Rai SK, Furtado A et al (2020) Comparison of long-read methods for sequencing and assembly of a plant genome. GigaScience 9(12):giaa146
22. Leger A, Leonardi T (2019) pycoQC, interactive quality control for Oxford Nanopore sequencing. J Open Source Soft 4(34):1236
23. Vaser R, Sović I, Nagarajan N et al (2017) Fast and accurate de novo genome assembly from long uncorrected reads. Genome Res 27(5): 737–746
24. Walker BJ, Abeel T, Shea T et al (2014) Pilon: an integrated tool for comprehensive microbial variant detection and genome assembly improvement. PLoS One 9(11):e112963
25. Bolger A, Scossa F, Bolger ME et al (2014) The genome of the stress tolerant wild tomato species *Solanum pennellii*. Nat Genet 46(9): 1034–1038
26. Tørresen OK, Star B, Mier P et al (2019) Tandem repeats lead to sequence assembly errors and impose multi-level challenges for genome and protein databases. Nucleic Acids Res 47(21):10994–11006
27. Lang D, Zhang S, Ren P et al (2020) Comparison of the two up-to-date sequencing technologies for genome assembly: HiFi reads of Pacific biosciences sequel II system and ultralong reads of Oxford Nanopore. GigaScience 9(12):giaa123

Chapter 23

Mapping-by-Sequencing of Point and Insertional Mutations with Easymap

Samuel Daniel Lup, David Wilson-Sánchez, and José Luis Micol

Abstract

Random mutagenesis followed by screening for phenotypes of interest is a widely used strategy for genetic dissection of biological pathways; however, identifying the causal gene traditionally required time-consuming mapping approaches based on iterative linkage analysis. Mapping-by-sequencing accelerates this process, efficiently linking the phenotype of a mutant to a narrow candidate genomic region, using next-generation sequencing (NGS) data from a mapping population segregating for the mutant phenotype. To enable researchers at any bioinformatics skill level to conduct mapping-by-sequencing, we developed the Easymap mapping software. In this protocol we break down the steps involved in mapping-by-sequencing. First, we describe different ways of obtaining a mapping population and the steps used to generate NGS data. Next, we show how to analyze the NGS data using Easymap and how to obtain a list of candidate mutations, along with comprehensive information for assessing the potential causality of each candidate. Thus, this protocol enables the user to conduct mapping-by-sequencing using Easymap, facilitating the identification of causal loci for a mutant phenotype of interest.

Key words Forward genetics, Bulk segregant analysis, Linkage analysis, High-throughput sequencing, Mapping-by-sequencing, Easymap, Allele frequency analysis, SNP (single-nucleotide polymorphism), Insertional mutation

Abbreviations

NGS	next-generation sequencing
EMS	ethyl methanesulfonate
SNP	single-nucleotide polymorphism
AF	allele frequency
RD	read depth
Gbp	giga base pairs

Christophe Lambing (ed.), *Plant Gametogenesis: Methods and Protocols*, Methods in Molecular Biology, vol. 2484, https://doi.org/10.1007/978-1-0716-2253-7_23,

1 Introduction

Forward genetic screens have long been a popular approach for the genetic dissection of biological processes. In plants, this technique typically involves an initial mutagenesis using a chemical mutagen such as ethyl methanesulfonate (EMS, which usually induces GC → AT transitions) or a biological mutagen such as insertional elements (transposons or T-DNA). Next, mutants with a phenotype of interest are isolated among the progeny of the mutagenized plants. However, identifying the causal gene for the mutant phenotype traditionally required time-consuming mapping approaches based on iterative linkage analysis. The combination of next-generation sequencing (NGS) and classical linkage analysis has greatly facilitated the identification of mutations, a strategy that has been termed "mapping-by-sequencing" ([1]; reviewed in [2]). Indeed, mapping-by-sequencing rapidly and efficiently links the phenotype of a mutant to a narrow genomic region containing a few candidate mutations, using raw NGS data from a mapping population segregating for the mutant phenotype of interest and multiple single-nucleotide polymorphisms (SNPs).

Different experimental approaches are usually taken to identify mutations induced by EMS or T-DNA using NGS. Since EMS induces multiple lesions per genome, bulked segregant analysis is performed to identify a set of mutations that cosegregate with the phenotype of interest in a mapping population. There are several options for creating such a mapping population. The outcross mapping strategy [3] involves crossing a mutant of interest to a genetically distant wild-type accession to yield F_1 progeny, which will be heterozygous for the polymorphisms created by the mutagen, as well for all the natural SNPs already existing between the parental lines. Different distantly related accessions or varieties of the same species (such as different accessions of *Arabidopsis thaliana*, or the *indica/Xian* and *japonica/Geng* subspecies of rice [*Oryza sativa*]) have large numbers of naturally occurring SNPs; these SNPs are easy to detect by NGS and thus can be used to map causal mutations. Selfing of the F_1 progeny yields an outcross F_2 mapping population, in which the causal mutation for the phenotype of interest, other mutations induced by the mutagen used, and the natural polymorphisms are segregating. In most cases the mutations under study are hypomorphic or null and recessive; therefore, the phenotypically mutant F_2 individuals will be homozygous for the causal mutation, which in turn will be in linkage disequilibrium with its neighboring SNP alleles from the parental line that was mutagenized. Then, NGS of a pool of DNA samples extracted from phenotypically mutant individuals should allow the researcher to identify the region where polymorphisms have allele frequencies

(AFs) that deviate significantly from 0.5. NGS will also simultaneously detect any natural and mutagenesis-derived base changes within this candidate interval.

For EMS-derived mutants, two other approaches are possible: the backcross mapping strategy [4] involves backcrossing the mutant of interest to its premutagenesis parental line, and the use of the mutations generated during the EMS mutagenesis as mapping markers in a backcross F_2 mapping population. This approach uses a much lower number of polymorphisms; however, this provides an advantage when working with mutants isolated in an accession that has a sequenced genome. Finally, an M_2 mapping strategy can be employed, involving the selfing of an M_1 plant, where all EMS-induced mutations are heterozygous, and isolating the new mutant of interest by using the M_2 progeny as a segregating mapping population [5].

A mapping-by-sequencing experiment requires a control sample for the discrimination of EMS-derived mutations from background polymorphisms, which is necessary for the mapping analysis and the identification of candidate causal mutations. The control sample can be one of the parents of the mapping population or a pool of phenotypically wild-type siblings from the mapping population. The choice of the mapping strategy and the control sample used depends on the specific situation.

If the mutation of interest is a base substitution or a small insertion or deletion, it can be detected along with other point mutations during mapping-by-sequencing analysis. However, mapping of large DNA insertions, such as those caused by T-DNA, usually relies on capturing the genomic sequences that flank the insertion site. For this process, algorithms that look for single reads (for NGS single-end read datasets) or read pairs (for NGS paired-end read datasets) that span the left and right insertion-genome junctions are used. Since this approach is not based on genetic mapping by linkage analysis, it cannot be called mapping-by-sequencing *stricto* sensu, but here, we use "mapping-by-sequencing" in a broad sense for large DNA insertion mapping.

In the following sections, we describe how to obtain mapping populations and how to prepare DNA samples for mapping-by-sequencing. We then describe how to use the Easymap package [6] with reads obtained from NGS of DNA, extracted from EMS-derived and insertional mutants, to eventually obtain lists of candidate causal mutations. These mutations can be validated experimentally to confirm a causal relationship between a DNA lesion and a mutant phenotype. The computer analysis section of our protocol is designed to be approachable for a wide range of researchers and does not require advanced bioinformatics skills.

2 Materials

2.1 Plants

1. Mutant plants exhibiting a monogenic recessive phenotype. For other organisms and types of mutants, *see* **Note 1**.
2. For backcross approaches, wild-type plants of the parental line: the accession that was subjected to mutagenesis, from which the mutant derives. For outcross approaches, plants of an accession genetically distant from the parental line.

2.2 Extraction of Genomic DNA

1. Equipment for the disruption and homogenization of plant tissue (TissueLyser or ceramic mortar and pestle).
2. Liquid nitrogen.
3. Centrifuge capable of reaching 3000–5000 × *g* equipped with a swing-out rotor and holders for 50-mL Falcon tubes.
4. Nanospectrophotometer that accepts 1–3 μL samples.
5. A Qubit fluorometer and a Qubit dsDNA HS Assay Kit (optional).
6. General usage laboratory equipment: pipettes and pipette tips, water bath or heating block, vortex, electrophoresis equipment, LoBind 1.5-mL Eppendorf tubes.
7. 96–100% ethanol.
8. The DNeasy Plant Maxi Kit or an equivalent plant DNA extraction kit is recommended for DNA extraction prior to massive sequencing.

2.3 Hardware and Software

1. A computer running a Linux OS or access to a server running Linux. Alternatively, a computer with Windows 10, the Windows Subsystem For Linux (WSL) enabled and an Ubuntu app (e.g., Ubuntu 18.04 LTS) installed from the Microsoft Store. For installation in Mac OS, you will need a virtual machine running a Linux OS. See the Easymap Quickstart Installation Guide (*see* **Note 2**) for further information on how to set up the environment for the bioinformatic analysis. The amount of storage required depends on the size of the genome and the size of the reads data. As a general recommendation, twice the size of all input data should be available. For Arabidopsis, this protocol requires 4 GB of RAM.
2. Easymap [6] for mapping-by-sequencing analysis. *See* **Note 2** for installation instructions.
3. fastp [7] to preprocess FASTQ reads. This is optional, *see* **Note 3** for installation instructions.

2.4 Input Files

1. The sequence of the reference strain/accession/cultivar of the organism of interest in FASTA format. If the reference

sequence is split into several text files, they must be renamed as follows: `name.1.fa`, `name.2.fa`, etc. Files with the same base name (e.g., "name") will be grouped as one genome assembly. The only file name extension allowed is `.fa`. Rename if necessary.

2. The structural annotation of the reference sequence in GFF3 format. The only file name extensions allowed are `.gff` or `.gff3`. Rename if necessary.
3. The functional annotation of the genes contained in the GFF file. This file is optional; see the Easymap documentation to learn how to format this file. The only file name extension allowed is `.txt`. Rename if necessary.
4. If you are mapping large DNA insertions, the sequence of the insertion in FASTA format is required. The only file name extension allowed is `.fa`. Rename if necessary.
5. *See* **Note 4** for information on how to obtain the required files.

3 Methods

3.1 Experimental Design

3.1.1 Linkage Analysis Mapping of Point Mutations

The following procedure assumes that you want to map a recessive mutation. You will need to obtain a test sample and a control sample. The test sample is generated from phenotypically mutant individuals of a segregating population (M_2, backcross F_2, or outcross F_2). The control sample can be obtained from different sources (see below). The following steps guide you through the process for creating these samples. Table 1 contains an overview of all the possible test and control sample combinations supported by Easymap.

1. Determine whether your mutant is in the same genetic background (accession of origin) as the available reference sequence (e.g., Col-0 in Arabidopsis) or in a divergent accession.
2. Choose the type of segregating population that will be used to create the test sample.
   (a) If the mutant is in the reference background, an M_2 or a backcross F_2 (*see* **Note 5**) is recommended, since they generally produce cleaner results than an outcross. Choose the M_2 if the priority is speed; choose a backcross F_2 if basic preliminary characterization of the mutant (heritability, fertility, inheritance mode) is desired. If an M_2 or a backcross F_2 is not desired or possible, or if it does not allow the identification of a candidate interval, an outcross F_2 can also be used to create a mapping population (*see* **Note 6**).

Table 1
Experimental designs for linkage analysis mapping supported by Easymap

Genetic background of the mutant	Source of the mapping population	Test sample	Control sample	Argument value for -ed
The same as the reference sequence provided	M_2 or backcross F_2	DNA from a pool of M_2 or F_2 phenotypically mutant individuals	Wild-type parental strain	ref_bc_parmut
			Pool of M_2 or F_2 phenotypically wild-type siblings	ref_bc_f2wt
	Outcross F_2		Wild-type parental strain	ref_oc_parmut
			Genetically distant strain crossed to the mutant	ref_oc_polstr
Different from the reference sequence provided	M_2 or backcross F_2		Pool of M_2 or F_2 phenotypically wild-type siblings	noref_bc_f2wt
	Outcross F_2		Genetically distant parental strain	noref_bc_parmut

(b) If the mutant background is different from the reference strain, an outcross to the reference strain is preferred (*see* **Note 6**). If this is not possible, M_2 or backcross F_2 approaches are also supported by Easymap (*see* **Note 7**). Choose the M_2 approach if speed is your priority; choose backcross F_2 if basic characterization of the mutant (heritability, fertility, inheritance mode) is desired.

(c) If the mutant was not generated by EMS treatment (e.g., insertional mutagenesis) but will be mapped by linkage analysis, perform an outcross to introduce markers into the mapping population.

3. Determine which sample will be used as a control. Refer to Table 1 for possible options depending on the genetic background of your mutant and the test sample used.

(a) Mutants isolated in the reference strain background: if the test sample is derived from an M_2 or a backcross F_2, use the premutagenesis parental strain or a pool of phenotypically wild-type individuals from the same M_2 or F_2 population as the control sample (*see* **Note 8**). If the test sample is derived from an outcross F_2, it is best to use the strain genetically distant to the reference genome as the control.

(b) Mutants isolated in a nonreference strain background: if the test sample is derived from an outcross F_2, use the parental strain of the mutant under study as a control sample; if it was obtained from an M_2 or backcross F_2 population, use a pool of phenotypically wild-type M_2 or F_2 individuals.

4. Determine how many mutant individuals from the mapping population will be pooled to generate the test sample. If obtaining many individuals is feasible, 100–200 individuals (200–400 chromosomes in a diploid organism) will produce a smooth and accurate mapping signal. The same applies to M_2 or F_2 control samples. If it is not possible to obtain these numbers, use as many individuals as possible. In principle, increasing the number of recombinant chromosomes in a pool will produce more accurate AFs (*see* **Note 9**).
5. Determine the read depth (RD) for the test and control samples. For M_2 or F_2 recombinant populations, a minimum RD of 25–30× is required, but we recommend using a higher RD because this will translate into more accurate AF (*see* **Note 9**). For samples that are not recombinant populations, such as a parental control sample, 25–30× is also the minimum RD required, but a higher RD (50–100×) is strongly recommended, especially for analyses that involve a nonreference strain.
6. Perform the required cross and/or self-pollination to obtain the mapping population.
7. Grow enough individuals for the segregating population (M_2 or F_2) in order to identify the desired number of plants that show an unequivocally mutant phenotype. If you are using a control sample based on M_2 or F_2 populations, isolate the desired number of plants that show an unequivocally wild-type phenotype.

3.1.2 Tagged Sequence Mapping of Large Insertions

It might be necessary to map large insertions in a genome for several reasons. If the purpose of the experiment is to identify a causal mutation, follow the directions in Subheading 3.1.1 (*see* **Note 10**). The following procedure assumes you want to identify the positions of all possible insertions in one or several lines.

1. Determine if several insertional lines will be pooled in a single NGS sample. This is possible, but it requires an additional step after Easymap analysis in which each insertion detected is assigned to one of the constituent lines of the sample by PCR genotyping of the insertions in each line.
2. Select a sufficient number of plants from each line. For example, if segregating insertions must be detected, the use of seven plants makes the probability of missing an insertion extremely low (<0.01%).

3. Determine the type of sequencing library to use. We recommend an Illumina paired-end library with fragment sizes between 400 and 1000 bp and reads of at least 100 bp. Single-end libraries can also be used if read length is at least 100 bp.
4. Determine the RD for each sample. An RD of 10× is the minimum required, but greater RDs generally translate into insertion hits with higher confidence. If you are pooling several lines in one sample, scale proportionally.
5. Grow the plants and select those that will be used for DNA extraction and sequencing.

3.2 Plant Material Collection, DNA Extraction, and Sample Preparation

1. Calculate the amount of tissue you need to collect from each individual following the recommendations of the DNA extraction kit. For the DNeasy Plant MAXI Kit, 1 g of tissue is needed per sample. For a hypothetical population of 100 plants, 10 mg of tissue should be collected from each individual. All subsamples should be equivalent to ensure a comparable contribution to the total amount of DNA. Proceed similarly for other types of samples (segregating insertional mutants, population of insertional mutants, a single control strain).
2. Clean the scissors used to collect plant material with ethanol and dry them. Repeat before collecting each new sample.
3. Collect the amount of tissue calculated in **step 1** of this section from each individual. Harvest tissue from tender leaves, and combine the subsamples in a 50-mL Falcon tube for each sample. Keep the tube on ice during this process.
4. Freeze the tube immediately in liquid nitrogen. Proceed to the next step or store the samples at −80 °C.
5. Extract the DNA with a DNeasy Plant MAXI Kit or an equivalent method. Follow the kit handbook. A typical extraction takes up to 2 h. To disrupt the plant material, use a TissueLyser or a mortar and pestle (*see* **Note 11**).
6. For the final step of the DNA extraction, elute the sample in 2 volumes of 750 μL deionized water each. The first elution is normally more concentrated and will be used for NGS. Do not combine both eluates in a single tube.
7. Check the DNA concentration and purity using a 2-μL aliquot of each sample with a nanospectrophotometer (*see* **Note 12**). The concentration must be sufficient to have a minimum of 1 μg of total DNA per sample. Purity indexes A_{260}/A_{230} and A_{260}/A_{280} must be >1.8. The DNA concentration should not be lower than 12.5 ng/μL. We typically obtain ~30 ng/μL DNA in the first elution and 15 ng/μL in the second. These

general recommendations should be suitable for most NGS services; however, check with your NGS provider for specific requirements.

8. Check DNA integrity. Load 4 μL of each sample in a 1% agarose gel and run it at 100 V for 40 min. A single, well-defined band of over 10–12 kb must be present. Check for the absence of a smear (degraded DNA) under the 10–12 kb band. Also check for the absence of ribosomal RNA bands.

3.3 Massive Sequencing

Submit your DNA samples to a commercial provider for NGS or to a sequencing facility, specifying the required RD or Gbp per sample (*see* **Note 13**). Sequence your samples using Illumina or other available technology with a similar error rate. Easymap can analyze both single-end and paired-end read data.

You will receive the reads in FastQ format. For single-end libraries, a single file per sample is expected by Easymap. For paired-end libraries, two files are expected, normally with an identical name and different suffixes (e.g., `reads_1.fq` and `reads_2.fq`). If you receive more files per sample, you may need to concatenate them (*see* **Note 14**). Your NGS provider should facilitate a quality assessment of the reads for you to evaluate whether they can be used for the following analyses and whether they require preprocessing (see next section).

3.4 Read Quality Assessment and Preprocessing

Easymap automatically checks and reports on the quality of the reads per nucleotide position. However, high-throughput raw reads may present different artifacts (adapter contamination, truncated reads, PCR duplicates) that require a preprocessing step to detect and correct them. In this protocol, this step is optional but recommended and can be performed with `fastp` [7].

1. Open a Linux terminal and navigate to the directory where your reads are stored. If your reads are not in a Linux directory, you can upload the reads to Easymap first and navigate to them with the following command:

```
cd ~/easymap/user_data/
```

2. Run `fastp;` a default execution will autodetect and trim adapter sequences, eliminate low-quality reads and trim low-quality nucleotides from reads. For single-end reads, run the following command:

```
fastp -i reads.fq -o clean_reads.fq
```

For paired-end reads, run the following command:

```
fastp -i reads_1.fq -I reads_2.fq -o clean_reads_1.fq -O
clean_reads_2.fq
```

Replace the file names in the example command above with your input and output file names. Pay attention to the information printed by the program to determine if it has run successfully.

3. You may want to delete the raw reads to optimize storage usage. To do this, run the following command and add as many file names as you want to delete:

```
rm reads_1.fq reads_2.fq
```

3.5 Mapping Analysis Using Easymap

Easymap can be used through the command line interface or through its graphical web interface. The command line interface can be more efficient for users acquainted with this type of environment and allows more flexibility (e.g., scheduling jobs). The graphical web interface increases the accessibility for other users. Data and projects produced through one interface are accessible through the other.

3.5.1 Analysis Using Easymap Through the Web Interface

The following is a step-by-step description of how to perform an analysis in which: (a) the researcher wants to map an EMS-induced mutant by linkage analysis; (b) the mutant was obtained in the reference strain background; (c) the mapping population is derived from a backcross F_2 generation; (d) the researcher wants to use a pool of wild-type F_2 individuals as the control sample.

1. Open a web browser and go to the address where you configured Easymap (e.g., http://localhost:8100 for local installation. *See* **Note 2** for Easymap installation).
2. Gather the following files: test and control sample reads in FASTQ format and with the ".fq" extension; if your files have a different extension, rename them. The reads can also be uploaded in compressed .gz format (e.g., reads.fq.gz); the genomic reference sequence in FASTA format (with a ".fa" extension); structural annotation of the reference sequence in GFF/GFF3 format; and functional annotation of the reference sequence (optional).
3. In the web interface, go to **Manage input files** and click on **Select files**. Select all the files listed in the previous step and click on **Upload files**. Wait until all files have been uploaded and appear under **Current files in Disk**. Do not refresh or close the page during the upload. Compressed reads will take additional time for extraction after the upload.
4. Click on **Run new project**. This will take you to a form that should be completed from top to bottom, as the form fields will change depending on your input.
5. **Project name**: Provide a name to identify the project.

6. Under **Mapping-by-sequencing strategy**, select **Linkage-analysis mapping**.
7. Under **Data source**, select **Use my own reads**.
8. Select a **Reference sequence** from the dropdown menu. Reference sequences consisting of multiple files sharing the same base name (as explained in Subheading 2.4) will appear as a single option for convenience. If your reference files are not listed in the dropdown menu, verify that their names have a ". fa" extension.
9. Select the **GFF3 file** with the structural annotation of your reference genome. If your file is not listed in the dropdown menu, check that its name has a ".gff" or ".gff3" extension.
10. If you have a **Gene functional annotation file**, specify it in the dropdown list. If your file is not listed in the dropdown menu, check that its name has a ".txt" extension. This file is optional.
11. Specify your experimental design. Under **Mutant background**, select **Reference**; under **Mapping cross performed**, select **Backcross**; under **Origin of the control reads**, select **F_2 wild-types**.
12. Under **Use low stringency during SNP analysis?**, we recommend that you analyze your data with this option unchecked and that you only use it if it is considered necessary, as explained in the Easymap documentation.
13. Under **Experimental reads**, select your test and control files. If you are using paired-end reads, select both files (forward and reverse reads) while holding the Ctrl or Cmd key. If your files are not listed in the dropdown menu, check that their names have a ".fq" extension.
14. Click on **Check input and run project**. Easymap will check if any of the input parameters selected need to be modified. If all checks pass, click on **Start project**. The browser will be redirected to **Manage projects**, and the newest project will appear at the top of the list and show **Project status: running**.
15. Once the project has finished running, click on **View report**.

To perform alternative analyses, modify the input accordingly. For example, if you are interested in detecting large insertions in a sample, under **Mapping-by-sequencing strategy**, choose **Tagged-sequence mapping** and select the FASTA file that contains the insertion in the **Insertion sequence file**.

3.5.2 Analysis Using Easymap Through the Command Line Interface

This section describes how to perform the analysis described in the section above through the command line interface, as well as other common analyses.

1. Place the input files listed in Subheading 2.4 in the `easymap/user_data` directory. Easymap will search for the specified files in this directory.
2. Place the read files in the `easymap/user_data` directory. For linkage analysis mapping, this consists of reads from two samples: test and control. For large insertions, only reads from a test sample are required.
3. If any file is compressed, typically with gzip (.gz extension), decompress it by running:

```
gzip -d <filename>.
```

4. If the reference sequence is split across several files, make sure that these files are named as described in Subheading 2.4.
5. Run the program. Easymap requires a single command to map the reads, perform the required mapping analysis and analyze the candidate mutations.
   (a) Example for linkage analysis of a mutant obtained in the reference strain, backcrossed to obtain a mapping population, and where the control sample consists of phenotypically wild-type F_2 individuals. Reads are paired end.

```
./easymap -n projectBC -w snp -ed ref_bc_f2wt -r TAIR10 -g
at_genes.gff -a at_annotation.txt -p f2mut_1.fq,f2mut_2.fq -c
f2wt_1.fq,f2wt_2.fq
```

   In the command above, `-n` specifies the project name for the analysis, `-w` is the Easymap workflow to use, `-ed` the experimental design, `-r` the basename of the reference sequence, `-g` the structural annotation file, `-a` the functional annotation file, `-p` the test reads, and `-c` the control reads. All fields are required except for `-a`. By default, Easymap considers only SNPs that pass certain quality checks. However, in some read datasets that are not optimal, performing more lenient filtering can help to identify a candidate interval and the causal mutation. If you analyze your reads in the default mode and obtain very few SNPs or believe that the causal mutation could have been discarded, include the flag `-ls` anywhere in the command. The Easymap Documentation describes all the arguments that Easymap can use.

   (b) Example for linkage analysis of a mutant obtained in a nonreference accession, outcrossed to the reference accession to obtain a mapping population, and where the control sample is the premutagenesis parent of the mutant. Reads are single end. All fields are required except for `-a`.

```
./easymap -n projectOC -w snp -ed noref_oc_parmut -r
TAIR10 -g at_genes.gff -a at_annotation.txt -p f2reads.fq -c
landsberg.fq
```

(c) Example for mapping the positions of a transposon in a DNA sample (*see* **Note 10**):

```
./easymap -w ins -n projectINS -r TAIR10 -i transposon.fa
-g at_genes.gff -a at_annotation.txt -p mut_1.fq,mut_2.fq
```

In the command above, `-i` specifies the file that contains the insertion sequence.

6. The execution time will vary depending on the input data and the computing power allocated for the job. To check the status of the execution, open the log file.

```
tail -f user_projects/<project_name>/2_logs/log.log
```

If the run was successful, the program will output "Easymap analysis properly completed.". Otherwise, the message "Easymap analysis failed. See log file for more information." will be printed.

3.6 Interpretation of the Results

Easymap generates a mapping report with images and tabular data. The mapping report can be displayed in the web interface by clicking on **Manage projects** and then on **View report**. If you are not using the web interface, open the file `report.html` with a web browser. This file is located in the directory `user_projects/< project_name>/ 3_workflow_output/`; the report file will look for the images in this directory to display them. To create a copy of the report, please copy the entire directory. Once opened in a web browser, the report can be printed and stored as a PDF file.

3.6.1 Linkage-Analysis Mapping Report

1. Check the RD distribution of your samples in **Input data quality assessment**. Compare the mean depth to the recommended RDs described in Subheading 3.1.1.
2. Identify a candidate interval containing the causal mutation. **Under Mapping analysis overview**, all input contigs are displayed with the AFs of the SNP used for mapping analysis. Based on the AFs, a candidate region containing the causal mutation of the phenotype is highlighted in pink. However, this prediction could miss the causal mutation. Inspect the distribution of the AFs to find a region that was selected during the phenotyping stage in the test sample (AFs above 0.5). If your control sample is a pool of phenotypically wild-type F_2 plants, the same region should be counter-selected for the control SNP (AFs below 0.5; Fig. 1).

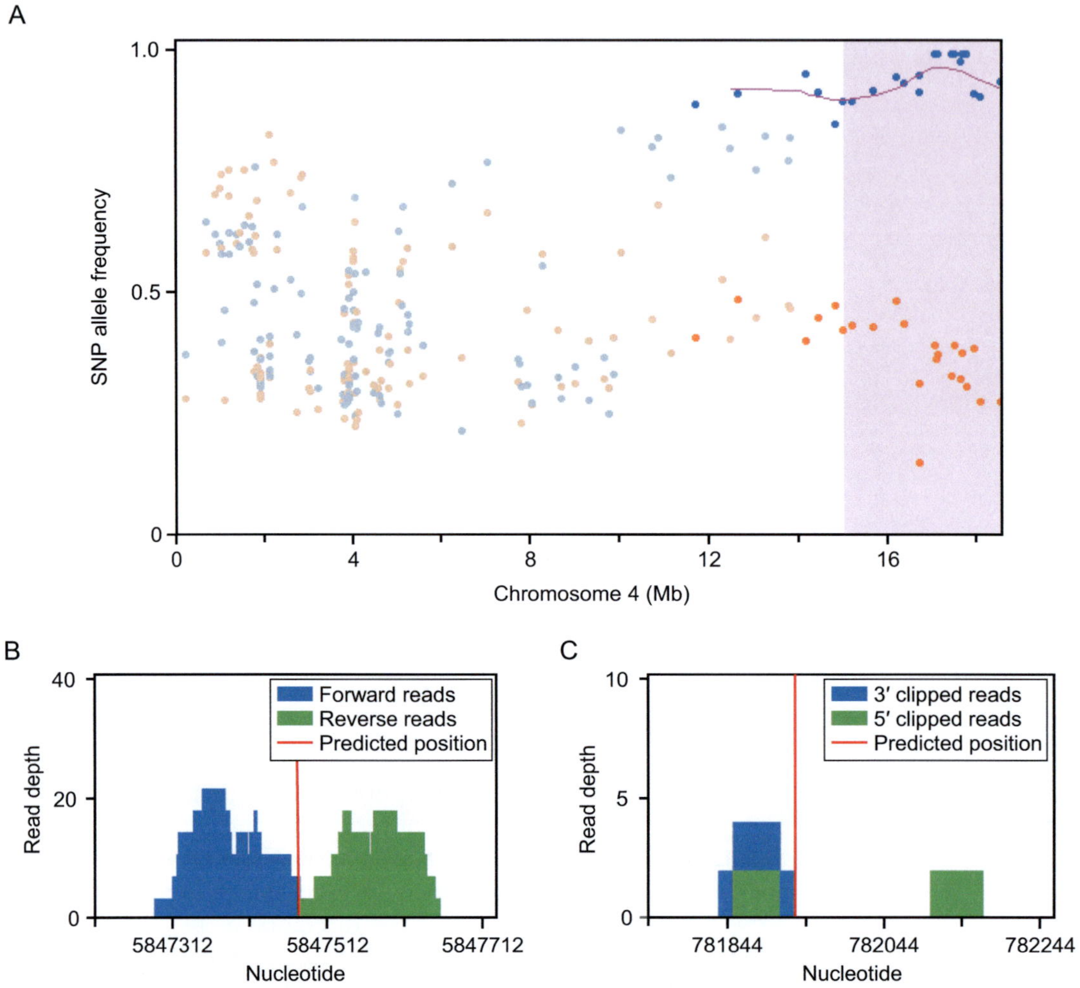

Fig. 1 Easymap output assessment. (**a**) Example of an allele frequency versus position plot for a linkage analysis experiment with a pool of phenotypically wild-type F_2 individuals as the control sample for the mapping of a loss of function allele of *SHORT-ROOT* [5]. SNPs are represented as blue and orange dots for the test and control samples, respectively, with a higher saturation for the SNP that Easymap find informative for mapping. The candidate region is highlighted in pink and the average AF in the test sample of SNP used for mapping is drawn as a purple line. Test SNPs deviate from 0.5 to 1 around the position for the causal mutation. Control SNP allele frequencies deviate to 0.33 around the same position, as this is the expected frequency of the mutation in a wild-type population of individuals for a recessive trait. (**b**, **c**) Read depth histograms for examples of paired-read analyses (**b**), and local alignment analysis (**c**). (**b**) True positive insertion, characterized by high read depths and organized data clusters. (**c**) False positive insertion, characterized by low overall read depths and disorganized data

3. **Candidate polymorphisms overview** is a magnified view of the SNPs from the test sample that fall within the candidate interval and have AFs greater than 0.8.

4. **Candidate region analysis** contains a description of the mutations within the candidate interval and their putative effects on genes. If functional annotation was provided, this will also be

displayed here. Explore the information about the candidate variants and prioritize these variants. Relevant fields for prioritizing the variants are the (a) AF (the alternative allele of a causal mutation is expected to have a frequency of 1), (b) closeness to the peak of the selected genomic region (column DTP [distance to peak] in the report) and (c) predicted effect on protein sequence or gene expression (*see* **Note 15**). To see extended information about variants in the candidate interval, or to see all of the variants in order to analyze a wider area, click on the appropriate link at the bottom of the variants table.

5. Validate the mutations of interest by Sanger sequencing. Finally, test the candidate mutations to identify the mutation that causes the observed phenotype. This can be done by phenotype complementation or by an allelism test.

3.6.2 Tagged-Sequence Mapping Report

1. Check the RD distribution of your samples in **Input data quality assessment**. Compare the mean RD to the values recommended in Subheading 3.1.2.
2. Review the detected insertions. For each insertion, check the histogram indicating the number of flanking reads that support it. If you are using paired-end reads, compare the histograms of unpaired alignments with those of local alignments: mutual agreement strongly suggests the presence of an insertion. False positives can occur, but they can normally be identified because they have much less read support than true positives (Fig. 1). If relevant to your analysis, check whether an insertion interrupts a transcript.
3. Validate the insertions of interest reported by Easymap by PCR and/or Sanger sequencing. The program will attempt to design primers to amplify the left and right junctions of each insertion. However, this is not always possible, in which case you will have to devise a strategy based on the insertion position and the sequence of the insertion.

3.7 Troubleshooting

1. In linkage-analysis mapping, if there is no clear mapping signal: (a) verify the input parameters chosen (in the report, under **Run summary**) to check for mistakes. Special attention must be paid to read input files; (b) verify that the mean RD is above the recommended threshold (Subheading 3.1.1) for both the test and control samples.
2. In linkage-analysis mapping done by backcrossing, if there are not enough SNPs to observe a mapping signal: (a) open the variants file to check all variants for candidate SNPs; (b) rerun the analysis in low stringency mode; (c) consider performing an outcross.

3. In linkage-analysis mapping involving a nonreference strain, if mutation density is very high in the candidate interval, the provided control sample may not be representative of all polymorphisms between the reference sequence and the genetically distant accession used in the experiment. High RD (50–100×) normally captures enough polymorphisms to perform effective filtering.
4. In tagged-sequence mapping, if no insertion sites are reported, please check that the file with the insertion sequence accurately represents the sequence of the expected insertion.
5. If the results are not as expected, an Easymap simulation (*see* **Note 16**) can facilitate troubleshooting.

4 Notes

1. This protocol can be used with species for which any of the mapping strategies described here is feasible. Dominant mutations or double mutants (e.g., from enhancer or suppressor screens) can also be mapped by adapting the procedures and tools described in this protocol.
2. Easymap is hosted at http://genetics.edu.umh.es/resources/easymap/. For general installation instructions, see the Quickstart Installation Guide. For detailed installation instructions, see the Easymap Documentation. To access the web interface, if your machine has Easymap installed (local installation), point your web browser to http://localhost:<port-number> (e.g., http://localhost:8100). If you are using Easymap remotely, use the IP address of the remote machine (e.g., http://11.22.33.44:8100).
3. The source code and installation instructions for fastp can be found at https://github.com/OpenGene/fastp.
4. Files with the FASTA reference sequence, structural and functional annotation information can generally be obtained from organism-specific webs (e.g., www.arabidopsis.org) or from the EMBL (http://ensemblgenomes.org).
5. An M_2 mapping population will have more segregating EMS mutations than its equivalent backcrossed F_2, which can be advantageous during mapping but could result in more candidate mutations. A mutant produced by standard EMS treatment and not previously backcrossed is expected to contain hundreds of SNPs across the genome that can be used as mapping markers and produce mapping signals. If a mutant has been backcrossed any number of times, random regions of the genome could be devoid of markers. If M_2 of backcross F_2 mapping populations do not have enough mapping markers, an outcross F_2 should be considered.

6. In some cases, a mutant phenotype cannot be easily distinguished in an outcross F_2 due to an interaction between the causal gene and other genes in the background.
7. Employing an M_2 or a backcross F_2 in a nonreference mutant background requires very high-quality sequencing data to effectively filter the hundreds of thousands of polymorphisms between the reference and mutant accessions. This can be achieved with high RD.
8. It is unknown which control sample (mutant parental line or a pool of phenotypically wild-type individuals from the same M_2 or F_2 population) is preferable in each particular mapping experiment. If you are mapping multiple mutants derived from the same parental line, sequencing information from this strain can be used as a control sample to map all the mutants. By contrast, using phenotypically wild-type plants from the mutant population will require a specific control sample to be sequenced for each mutant.
9. The number of recombinant chromosomes and the RD act as sampling steps when measuring AFs in a mapping population. As such, they can limit the power of linkage analysis mapping. To increase mapping accuracy, the sampling-limiting factor should be increased. For more guidelines based on simulated data, *see* Refs. 8, 9.
10. It is possible to perform linkage analysis and candidate identification of an insertional mutant by running Easymap twice. First, run the Easymap linkage analysis workflow with the test and control reads of a mapping population to define a candidate region; next, run the Easymap tagged-sequence workflow with the test reads to identify insertion events and check whether any of them are located within the previously determined candidate region.
11. For tissue disruption with a mortar and pestle, start by submerging the mortar and the Falcon tube containing the sample in liquid nitrogen. Grind the frozen sample until a very fine powder is obtained, keeping the sample frozen by frequently adding liquid nitrogen to the mortar. Use a clean mortar and pestle for each sample.
12. Measurement of DNA concentration using spectroscopic absorbance at 260 nm, such as that obtained on a NanoPhotometer, generally overestimates the concentration of DNA in a sample due to its sensitivity to the perturbing actions of common contaminants such as natural pigments, RNA and free nucleotides. A Qubit fluorometer can be optionally used to overcome this problem.
13. In our experience using Arabidopsis leaf DNA, achieving 50× requires around 7.2 Gbp of raw reads. This is higher than the theoretical value (6 Gbp) because chloroplast DNA takes up a

substantial fraction of the reads. It is important to take into account that RD always refers to a mean value, and that the RD distribution is bell-shaped. For this reason, increasing RD has a positive effect on polymorphism detection and AF measurement.

14. To concatenate files, open a Linux terminal and use the cat command. Example: cat readsA.fq readsB.fq > readsAB.fq. For compressed files: cat readsA.fq.gz readsB.fq.gz > readsAB.fq.gz.
15. In our experience in Arabidopsis, the vast majority of causal mutations in EMS mutants correspond to nonsynonymous mutations and mutations affecting splicing signals.
16. Simulation of a mapping-by-sequencing experiment can be helpful when designing the real experiment (type of cross, sequencing technical parameters) or when troubleshooting mapping results. To simulate an experiment, provide input files as for a normal analysis, except for the read files, which will be simulated. If you are using the command line, edit the simulation parameters in `simulator/sim_parameters.json` and run the program with the `-sim` flag. Example for linkage analysis simulation:

```
./easymap -sim -w snp -ed ref_bc_parmut -n mySimulation -r
<refseq> -g <gff_file>
```

Acknowledgments

This work was supported by grants from the Ministerio de Ciencia e Innovación of Spain (PGC2018-093445-B-I00 [MCI/AEI/FEDER, UE]) and the Generalitat Valenciana (PROMETEO/2019/117) to JLM. SDL held a predoctoral fellowship (ACIF/2018/005) from the Generalitat Valenciana.

References

1. Schneeberger K, Ossowski S, Lanz C et al (2009) SHOREmap: simultaneous mapping and mutation identification by deep sequencing. Nat Methods 6:550–551
2. Candela H, Casanova-Sáez R, Micol JL (2015) Getting started in mapping-by-sequencing. J Integr Plant Biol 57:606–612
3. Schneeberger K, Weigel D (2011) Fast-forward genetics enabled by new sequencing technologies. Trends Plant Sci 16:282–288
4. Allen RS, Nakasugi K, Doran RL et al (2013) Facile mutant identification via a single parental backcross method and application of whole genome sequencing based mapping pipelines. Front Plant Sci 4:362
5. Wachsman G, Modliszewski JL, Valdes M et al (2017) A SIMPLE pipeline for mapping point mutations. Plant Physiol 174:1307–1313
6. Lup SD, Wilson-Sánchez D, Andreu-Sánchez S et al (2021) Easymap: a user-friendly software

package for rapid mapping-by-sequencing of point mutations and large insertions. Front Plant Sci 12:655286

7. Chen S, Zhou Y, Chen Y et al (2018) Fastp: an ultra-fast all-in-one FASTQ preprocessor. Bioinformatics 34:i884–i890
8. James GV, Patel V, Nordstrom KJ et al (2013) User guide for mapping-by-sequencing in Arabidopsis. Genome Biol 14:R61
9. Wilson-Sánchez D, Lup SD, Sarmiento-Mañús R et al (2019) Next-generation forward genetic screens: using simulated data to improve the design of mapping-by-sequencing experiments in Arabidopsis. Nucleic Acids Res 47:e140

Chapter 24

Characterization of the Different Levels of Variation in 45S rRNA Genes

Fernando A. Rabanal

Abstract

Although commonly regarded as nearly identical in sequence, 45S ribosomal RNA genes represent a massive source of genetic variation at different levels. Unfortunately, due to their repetitive nature and the difficulty to assemble their multiple copies in tandem, these important genomic elements remain largely unexplored in sequencing projects. Here, I describe how to exploit next generation sequencing data to estimate their copy number in an organism and detect true polymorphic sites within and among individuals. Furthermore, for species that carry multiple 45S ribosomal RNA gene clusters, I show how to make use of experimental populations to assign some of these variants to their cluster of origin.

Key words Ribosomes, rRNA genes, 45S rDNA, Copy number variation, Natural variation, Polymorphisms

1 Introduction

In eukaryotes, the RNA component of ribosomes (rRNA) consists of four subunits: the 18S, 5.8S, and 25S rRNAs transcribed via a common 45S rRNA precursor; and the 5S rRNAs often separately encoded [1, 2]. Hundreds to thousands of 45S rRNA genes very similar in sequence identity are tandemly arrayed head-to-tail in large ribosomal DNA clusters [3, 4]. Unfortunately, and despite the increasing contiguity of assembled genomes achieved with long-read sequencing technologies, 45S ribosomal DNA clusters (hereinafter referred to as rDNA clusters) remain largely unassembled [5, 6].

Due to the repetitive nature of the rDNA genes and the limitations of current assemblies, a technical question emerges: how to extract more information about these loci in large next-generation sequencing (NGS) projects? A practical approach consists in align-

Christophe Lambing (ed.), *Plant Gametogenesis: Methods and Protocols*, Methods in Molecular Biology, vol. 2484, https://doi.org/10.1007/978-1-0716-2253-7_24,

ing NGS short-reads to a single reference rDNA unit. For instance, to estimate the rDNA copy number in a given individual and variation in a population, one can make use of the mean read-depth estimate from a representative rDNA gene unit relative to the read-depth genome-wide. Moreover, to detect sequence variation among the multiple rDNA gene copies, genotype calls can be extracted directly from aligned reads paying special consideration to sequencing artifacts such as strand bias. This particular error is observed in NGS data when the genotype at a given position inferred from the positive strand and negative strand are highly inconsistent [7]. Nevertheless, one can implement a strand bias score for all variable sites and exclude those above a certain threshold, similar to the procedure implemented when analyzing heteroplasmy in human mitochondria [8].

For genomes that carry multiple rDNA clusters, once we identify rDNA variants within a genome, we do not know exactly where these variants "sit," in other words, to which rDNA cluster they belong. This information sometimes is of great importance, for instance, if one would like to use these variants as reporters of rDNA cluster-specific expression [9]. To add positioning information to rDNA variants, we need to make use of a segregating population originated from a cross between parents with strain-specific rDNA variants, and only those variants private to either parental genome will have the possibility to be unambiguously assigned to a particular rDNA cluster. We will exemplify this approach with *Arabidopsis thaliana*, a model plant species that embodies the simplest case of a genome having multiple rDNA clusters in different chromosomes. In *A. thaliana*, rDNA clusters are located at the top of chromosomes two (rDNA-2) and four (rDNA-4) [10]. Entire rDNA clusters and flanking genomic regions are usually inherited as haplotype blocks due to the lack of recombination between homologous rDNA clusters [10, 11]. Therefore, in a recombinant inbred line (RIL) population, it is possible to infer the parent-of-origin of each rDNA cluster –and therefore, rDNA variant– by knowing the genotype of the single-nucleotide polymorphisms within the first few hundreds of kilobases right next to them (Fig. 1).

In this chapter, I will focus on the bioinformatics necessary to estimate rDNA copy number, detect rDNA variants, and assign these variants to their corresponding rDNA cluster-of-origin when NGS data for RILs is also available.

The methods described here assume a basic level of familiarity with command line scripting, including the installation of software with the open source package manager conda [12].

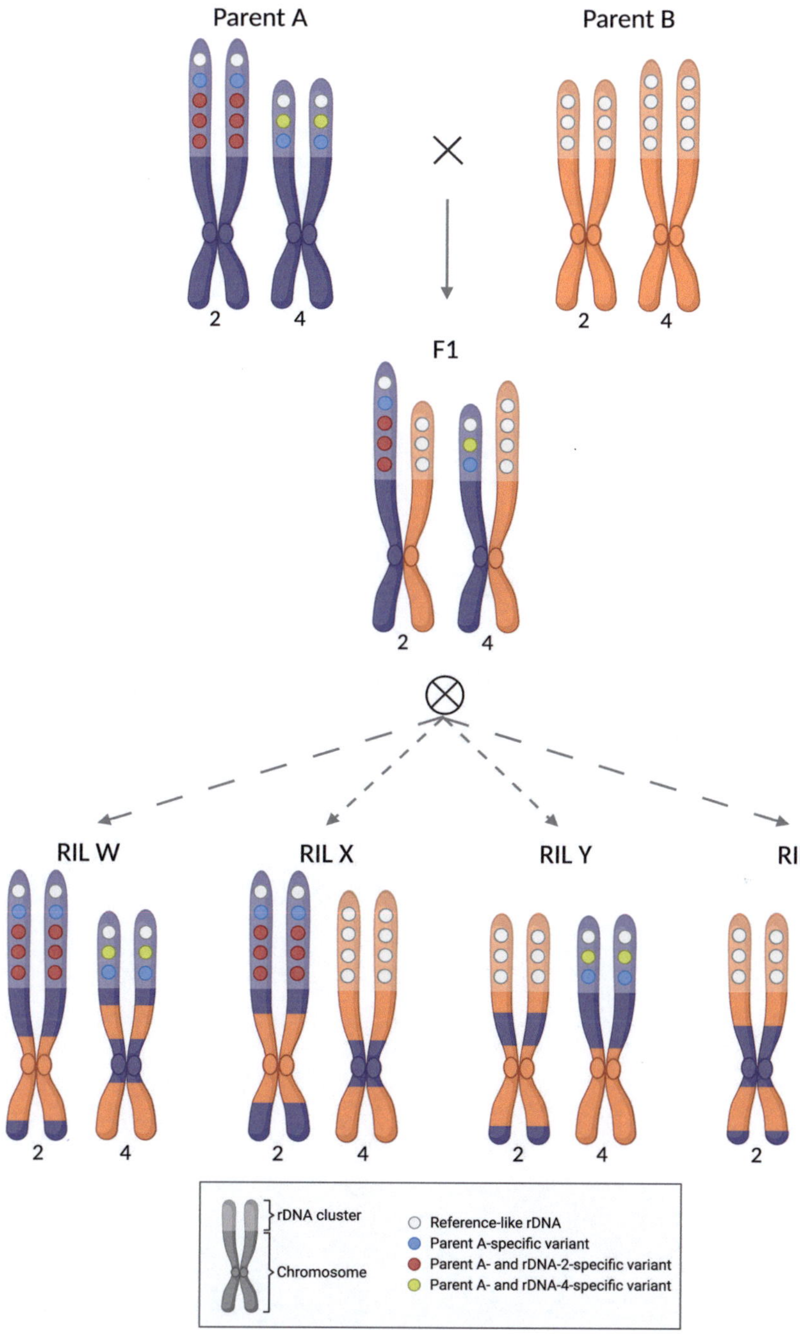

Fig. 1 Schematic representation of the strategy to assign rDNA variants to specific rDNA clusters. Parent A and Parent B strains are crossed, and the resulting F1 is selfed to produce a segregating population that after several generations of single-seed descent propagation produces multiple recombinant inbred lines (RILs). RILs with rDNA clusters of different parental origin are of interest to this approach (RILs X and Y). Shaded portions of the chromosomes represent the rDNA clusters. Circles represent abstract rDNA variants. (Figure created with BioRender.com)

2 Materials

2.1 Software Requirements

This pipeline relies on conda (Anaconda or miniconda) [12]) and R [13] having been installed. Information on how to install conda and add the bioconda channel [14] is available on https://bioconda.github.io/. The methods in this chapter were tested on the software versions listed in Table 1.

2.1.1 Create a Virtual Environment and Install Required Software with Conda

All commands in Subheading 2 must be executed in a shell for Unix/Linux.

```
# Create virtual environment.
conda create --name rDNA_variation

# Activate virtual environment.
source activate rDNA_variation

# Install pysamstats (v1.1.2).
conda install -c bioconda pysamstats

# Install BWA-MEM (v0.7.17).
conda install -c bioconda bwa

# Install samtools (v1.11).
conda install -c bioconda samtools

# Install sra-tools (v2.9.6).
conda install -c bioconda sra-tools

# Install cutadapt (v3.4).
conda install -c bioconda cutadapt

# Install bc (v1.07.1).
conda install -c conda-forge bc

# Install wget (v1.20.1).
conda install -c anaconda wget

# Download R scripts necessary for rDNA variant detection and plotting.
wget https://github.com/frabanal/rDNA_variation/archive/refs/heads/main.zip

unzip main.zip

# Deactivate virtual environment.
conda deactivate
```

Table 1
Required software. Software and the version numbers that were tested for this protocol are listed in the table. For most tools, both earlier and later versions should deliver similar results. For pysamstats, v1.0.0 or higher is recommended

Software	Version	Reference or URL
Conda	4.9.2	[12]
R	4.0.2	[13]
Python	3.8.8	[23]
Sra-tools	2.9.6	[24]
Cutadapt	3.4	[25]
Bwa	0.7.17	[26]
Samtools	1.11	[27]
Pysamstats	1.1.2	https://github.com/alimanfoo/pysamstats
Bc	1.07.1	https://www.gnu.org/software/bc/
Wget	1.20.1	https://www.gnu.org/software/wget/

Table 2
Sample information for SRA files used in this protocol. *Example SRA file used from Subheadings 2.2.2–3.1.5

SRA file	Sample	rDNA-2	rDNA-4
SRR3711864	Cvi-0	Cvi-0	Cvi-0
SRR3712051	Ler-0	Ler-0	Ler-0
SRR3711866*	CVL10	Cvi-0	Cvi-0
SRR3711897	CVL13	Cvi-0	Ler-0
SRR3711976	CVL20	Ler-0	Cvi-0
SRR3712016	CVL5	Ler-0	Ler-0

2.2 Data Requirements

The methods described in this protocol are exemplified in *A. thaliana*. For the first part, consisting of the estimation of rDNA copy number and detection of rDNA variants, we will need the Col-0 TAIR10 reference genome [15, 16], as well as a reference of a single unit of the 45 rRNA gene [17]. Additionally, for the second part, corresponding to the assignment of rDNA variants to specific rDNA cluster, we will make use of a handful of *A. thaliana* lines (Table 2) from a popular RIL population between the natural strains Cvi-0 and Ler-0 [18] that have already been sequenced at low coverage [17].

2.2.1 Download TAIR10 Reference Genome and 45S rRNA Reference Gene

```
# Activate virtual environment.
source activate rDNA_variation

# Download TAIR10 reference genome. (This action has to be performed
only once!)
wget
https://www.arabidopsis.org/download_files/Genes/TAIR10_genome_release/T
AIR10_chromosome_files/TAIR10_chr_all.fas

# Download and decompress 45S rDNA reference. (This action has to be
performed only once!)
wget
https://www.g3journal.org/highwire/filestream/483337/field_highwire_adju
nct_files/0/FileS1.zip

unzip FileS1.zip
```

2.2.2 Download Example Dataset

The following step is meant to be performed for each of the sample files in a given dataset. To properly illustrate Subheading 3.2 in this protocol, it is necessary to download all files listed in Table 2. To that end, substitute the value of variable "SRRID" for the Sequence Read Archive (SRA) ID of the remaining samples.

```
# Declare Sequence Read Archive (SRA) ID for test example.
SRRID=SRR3711866

# Download SRA file and convert it to fastq format.
prefetch -v $SRRID

vdb-validate $HOME/ncbi/public/sra/$SRRID.sra 2>&1 | tee $SRRID.vdb-
validate.txt

fastq-dump $HOME/ncbi/public/sra/$SRRID.sra --split-3 --gzip -O ./

rm $HOME/ncbi/public/sra/$SRRID.sra

# Deactivate virtual environment.
conda deactivate
```

3 Methods

3.1 Estimation of rDNA Copy Number and Detection of rDNA Variants

Commands in Subheadings 3.1.1–3.1.5 must be executed in a shell for Unix/Linux.

3.1.1 Processing Reference Genomes

```
# Activate virtual environment.
source activate rDNA_variation

# Index reference genomes.
bwa index -a bwtsw TAIR10_chr_all.fas

bwa index -a bwtsw FileS1.fa
```

3.1.2 Processing Short-Reads

Similar to Subheading 2.2.2, the following Subheadings 3.1.2–3.1.5 have to be performed for each of the sample files in a given dataset. To properly illustrate Subheading 3.2 in this protocol, it is necessary to download all files listed in Table 2.

```
# Trim adapters from raw paired-end reads.
cutadapt -j 8 -q 20,15 \
  -A TruSeq1=AGATCGGAAGAGC -a TruSeq2=AGATCGGAAGAGC \
  --trim-n \
  -m 90 \
  -o "$SRRID"_1.cutadapt.fastq.gz -p "$SRRID"_2.cutadapt.fastq.gz \
  "$SRRID"_1.fastq.gz "$SRRID"_2.fastq.gz
```

3.1.3 Alignment of Short-Reads to the 45S rDNA Reference Gene, and Generation of Files Containing Statistics Necessary for Variant Calling

```
# Align trimmed reads to the 45S rDNA reference gene.
bwa mem -t 8 -R "@RG\tID:$SRRID\tSM:$SRRID" FileS1.fa \
  "$SRRID"_1.cutadapt.fastq.gz \
  "$SRRID"_2.cutadapt.fastq.gz \
  | samtools view -@ 48 -Sbh - > $SRRID.45S.unsorted.bam

# Sort and index bam file.
samtools sort -@ 8 -o $SRRID.45S.sorted.bam $SRRID.45S.unsorted.bam

samtools index $SRRID.45S.sorted.bam

# Calculate variation per nucleotide with forward/reverse strand counts.
pysamstats --type variation_strand -f FileS1.fa --max-depth 1000000 --
min-baseq 20 --pad $SRRID.45S.sorted.bam >
$SRRID.45S.variation_strand.txt
```

3.1.4 Alignment of Short-Reads to the TAIR10 Reference Genome

```
# Align trimmed reads to the TAIR10 reference genome.
bwa mem -t 8 -R "@RG\tID:$SRRID\tSM:$SRRID" TAIR10_chr_all.fas \
  "$SRRID"_1.cutadapt.fastq.gz \
  "$SRRID"_2.cutadapt.fastq.gz \
  | samtools view -@ 8 -Sbh - > $SRRID.TAIR10.unsorted.bam

# Sort and index bam file.
samtools sort -@ 8 -o $SRRID.TAIR10.sorted.bam
$SRRID.TAIR10.unsorted.bam

samtools index $SRRID.TAIR10.sorted.bam
```

3.1.5 Estimation of rDNA Copy Number

Although overall the multiple rDNA copies are very similar in sequence, the internal (ITS) and external transcribed spacers (ETS) display a substantial amount of variability [6, 9, 19–

22]. Therefore, it is recommended to estimate the mean read-depth of either one of the rRNA subunits, or of the average of all three subunits (18S, 5.8S and 25S). In this case, we have opted for the 18S rRNA gene, but *see* **Note 1** for the coordinates of the other subunits in this reference rRNA gene.

```
# Estimate mean read-depth 18S rRNA gene.
rdna_coverage=$(samtools coverage -r 45S_At3:2195-4002
$SRRID.45S.sorted.bam | grep -v "^#" | awk '{print $7}')

echo $rdna_coverage
```

To obtain the rDNA copy number, the abovementioned estimate has to be divided by the estimate of read-depth genome-wide. However, since the coverage can vary substantially at some genomic regions (e.g., centromeres, rDNAs, organellar insertions), it is better to select a fairly large region that displays a uniform read depth. In this case, we first inspected the coverage along chromosome 3 (Fig. 2) and, based on this observation, we selected the first 9 Mb as our genome mean read-depth.

```
# Identify a sizable portion of the TAIR10 reference genome that does
not contain either deeps or spikes in read-depth.
# In this example, we are displaying coverage along chromosome 3.
samtools coverage -A -w 100 -r 3 $SRRID.TAIR10.sorted.bam
```

Once normalized by the mean read-depth at this region, the rDNA copy number estimate for SRR3711866 should be 415.94.

```
# Estimate mean read-depth on a sizable portion of the TAIR10 reference
genome.
genome_coverage=$(samtools coverage -r 3:1-9000000
$SRRID.TAIR10.sorted.bam | grep -v "^#" | awk '{print $7}')

echo $genome_coverage

# Estimate copy number of 18S rRNA genes.
rdna_copy_number=$(bc <<<"scale=2;$rdna_coverage/$genome_coverage")

echo $rdna_copy_number
```

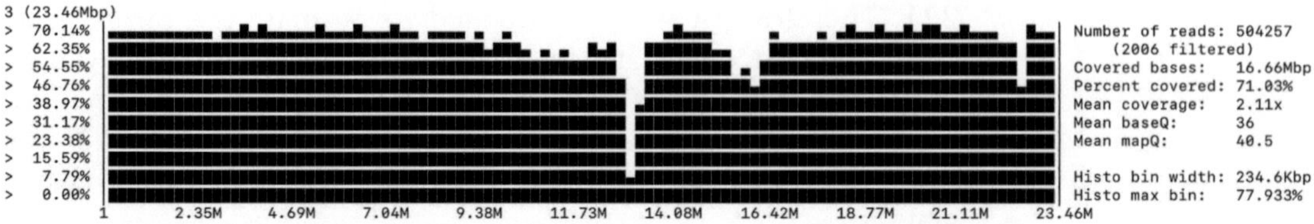

Fig. 2 Command line output of "samtools coverage". Binned read depth along chromosome 3 in TAIR10 for line CVL10 (SRA file SRR3711866). Coverage does not reach values closer to 100% because individuals in this dataset were sequenced at low coverage; mean read depth for SRR3711866 was 1.45×

3.1.6 Detection of rDNA Variants

Commands in Subheadings 3.1.6–3.1.8 must be executed in an R session.

The function "rDNA_variant_detection" reads all the pysamstats outputs (*.45S.variation_strand.txt) generated during Subheading 3.1.3 present in the specified directory (argument: "input_directory").

For each sample, it only processes positions with a minimum read depth (argument: "depth_cutoff") and skips rDNA variants that do not reach a certain allele frequency within an individual (argument: "fraction_cutoff"). For major alleles of variants above the "fraction_cutoff" value, the function calculates a strand bias score. Failing to pass either "depth_cutoff" or the strand bias score threshold (argument: "SB_threshold") sets the allele frequency value to "NA," making this position unfit for downstream analyses. Alleles having within individual frequencies above an additional threshold (argument: "proportion_cutoff") will be considered among the variants to plot in Subheading 3.1.8.

The function returns an R-list with five elements. First, "Variation_output$samples", a character object with the names of all samples analyzed. Second, "Variation_output$Reference", a data frame with all reference alleles in the 45S rRNA reference gene (rows) and metadata associated with each sample (columns) such as read depth, strand bias score, and frequencies of the four nucleotides, deletions and insertions. Third, "Variation_output$proportion_cutoff", a numeric object that simply returns the same input specified by the user "proportion_cutoff". Fourth, "Variation_output$Proportion_variants", a data frame with all reference alleles in the 45S rRNA reference gene (rows) and the number of samples that passed the "proportion_cutoff" for each allele (columns). Fifth, "Variation_output$phen", a data frame with all variant positions (columns) in which at least one individual in the population

```
# Define working directory.
input_directory <- './'

setwd(input_directory)

# Load R-function to detect rDNA variation.
source(paste(input_directory, "/rDNA_variation-
main/R_scripts/FUNCTION_rDNA_variant_detection.R", sep=""))

# Execute "rDNA_variant_detection" function with desired parameters.
Variation_output <- rDNA_variant_detection(depth_cutoff = 100,
SB_threshold = 0.80, fraction_cutoff = 0.02, proportion_cutoff = 0.05,
input_directory = './')
```

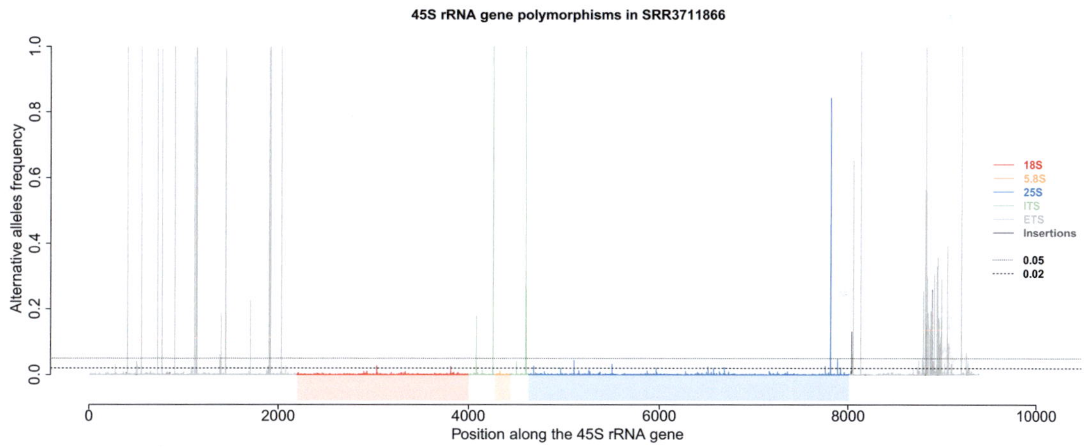

Fig. 3 Output of R function "plot_allele_frequency_indiv". Alternative allele frequencies along the 45S rRNA gene for line CVL10 (SRA file SRR3711866). Vertical lines represent single nucleotide polymorphisms or deletions in the ETS (gray), 18S (red), ITS (green), 5.8S (yellow), and 25S (blue) regions along the 45S rRNA gene. Black vertical lines represent insertions. Dashed and dotted horizontal lines indicate 0.02 and 0.05 allele frequency, respectively

passed the "proportion_cutoff". Each row represents an individual in the population. For every alternative allele, also the reference allele frequency is displayed (in a separate column).

3.1.7 Plot the Frequencies of Alternative Alleles Along the 45S rRNA Gene for Each Sample

The function "plot_allele_frequency_indiv" plots the sum of frequencies of all alternative alleles for each position along the 45S rRNA gene (argument: "Reference") for each sample in the population (argument: "samples") into separate PDF files at a designated output directory (argument: "output_directory") (Fig. 3). Frequencies of insertions are overlaid.

```
# Load R-function to plot rDNA variants allele frequencies per
individual.
source(paste(input_directory, "/rDNA_variation-
main/R_scripts/FUNCTION_plot_allele_frequency_indiv.R", sep=""))

# Execute "plot_allele_frequency_indiv" function giving as input
parameters elements of the list created by the "rDNA_variant_detection"
function.
plot_allele_frequency_indiv(samples = Variation_output$samples,
Reference = Variation_output$Reference, output_directory =
"./allele_frequency_plots")
```

3.1.8 Plot the Proportion of Individuals in a Population with an Alternative Allele Frequency Above a Certain Threshold

The function "plot_allele_proportions_pop" plots the proportion of individuals in a population (argument: "samples") with an alternative allele above a certain threshold (argument: "proportion_cutoff") for each position along the 45S rRNA gene (arguments: "Reference" and "Proportion_variants") into a single PDF file at a designated output directory (argument: "output_directory") (Fig. 4). Frequencies of insertions are overlaid.

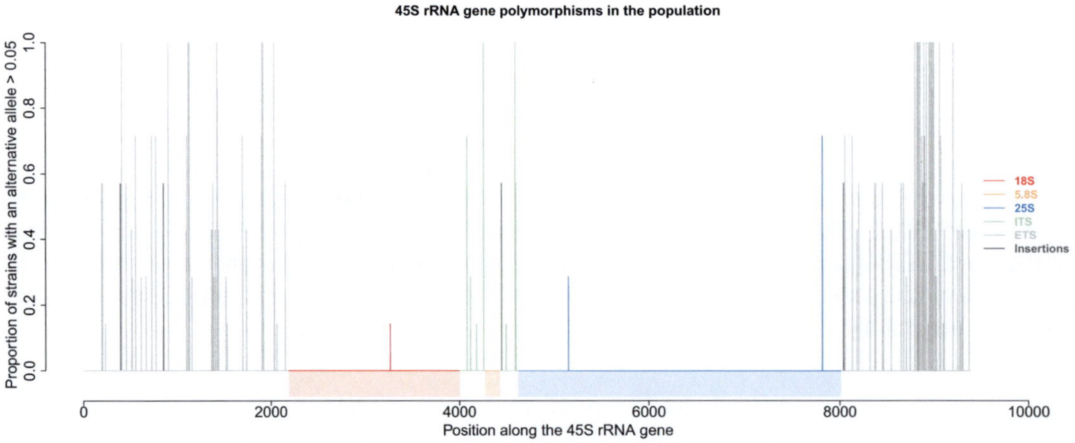

Fig. 4 Output of R function "plot_allele_proportions_pop". Proportion of samples from Table 2 with an alternative allele (defined as >5% within an individual) along the 45S rRNA gene. Vertical lines represent single nucleotide polymorphisms or deletions in the ETS (gray), 18S (red), ITS (green), 5.8S (yellow), and 25S (blue) regions along the 45S rRNA gene. Black vertical lines represent insertions

```
# Load R-function to plot the proportion of individuals in the
population with an alternative rDNA variant allele above a certain
threshold at the individual level.
source(paste(input_directory, "/rDNA_variation-
main/R_scripts/FUNCTION_plot_allele_proportions_pop.R", sep=""))

# Execute "plot_allele_proportions_pop" function giving as input
parameters elements of the list created by the "rDNA_variant_detection"
function.
plot_allele_proportions_pop(samples = Variation_output$samples,
Reference = Variation_output$Reference, proportion_cutoff =
Variation_output$proportion_cutoff, Proportion_variants =
Variation_output$Proportion_variants, output_directory =
"./allele_proportion_plots")
```

3.2 Assignment of rDNA Variants to Specific rDNA Clusters

The logic behind the assignment of rDNA variants to specific rDNA clusters relies on having these clusters effectively unlinked from each other in experimental populations. Knowing the identity of the single-copy chromosomal stretch adjacent to the rDNA cluster, allows us to infer the parental origin of that particular rDNA cluster. Although we exemplify this approach with an *A. thaliana* RIL population that has been previously genotyped [17], the same strategy can be applied to F_2 lines (*see* **Note 2**). With this approach, however, it is only possible to "annotate" rDNA variants specific to either of the two parental strains of the RIL population. In this example, the focus is solely on variants specific to Cvi-0.

Besides the parental strains, in principle (although also *see* **Note 3**), we only require one RIL representative of each of the genotype combinations rDNA-2-AA rDNA-4-BB (RIL X in Fig. 1) and rDNA-2-BB rDNA-4-AA (RIL Y in Fig. 1) to conclude for each rDNA variant whether it is specific to rDNA-2 (red variant in Fig. 1), rDNA-4 (yellow variant in Fig. 1) or shared between both rDNA clusters (blue variant in Fig. 1).

The following commands must be executed in an R session, and the data frame "Variation_output$phen" generated by the function "rDNA_variant_detection" from Subheading 3.1.6 is required. Let us call lines Cvi-0 "ParentA," Ler-0 "ParentB," CVL13 "RILX," and CVL20 "RILY."

```
# Create dataframe with rDNA-origin information for each SRA file.
rDNA_origin <- data.frame(sample_id  = c("SRR3711864", "SRR3712051",
"SRR3711866", "SRR3711897", "SRR3711976", "SRR3712016"), sample_name =
c("Cvi-0", "Ler-0", "CVL10", "CVL13", "CVL20", "CVL5"), rDNA2 = c("Cvi-
0", "Ler-0", "Cvi-0", "Cvi-0", "Ler-0", "Ler-0"), rDNA4 = c("Cvi-0",
"Ler-0", "Cvi-0", "Ler-0", "Cvi-0", "Ler-0"))

# Merge with an object, taken from the list created by the
"rDNA_variant_detection" function, that has rDNA variant frequencies for
each variable site across samples.
rDNA_origin <- merge(rDNA_origin, Variation_output$phen, by="sample_id")

# Define row of Parent A.
ParentA <- 1

# Define row of Parent B.
ParentB <- 6

# Define row of RIL with rDNA-2 from Parent A, and rDNA-4 from Parent B.
RILX <- 3

# Define row of RIL with rDNA-4 from Parent A, and rDNA-2 from Parent B.
RILY <- 4

# Loop through each rDNA variant.
for(i in 5:length(rDNA_origin)){

  # Skip rDNA variant if there is data not available in at least one of
the parents or RILs.
  if(is.na(rDNA_origin[ParentA, i]) | is.na(rDNA_origin[ParentB, i]) |
is.na(rDNA_origin[RILX, i]) | is.na(rDNA_origin[RILY, i])){


    #print(paste("rDNA variant ", colnames(rDNA_origin)[i], " CANNOT be
evaluated due to missing data.", sep = ""))

  # Evaluate whether the rDNA variant is specific to Parent A.
  }else if(rDNA_origin[ParentA, i] > 0.05 & rDNA_origin[ParentB, i] <
0.01){
```

```
    # Determine whether the rDNA variant is specific to rDNA-2.
    if(rDNA_origin[RILX, i] > 0.05 & rDNA_origin[RILY, i] < 0.01){

      print(paste("rDNA variant ", colnames(rDNA_origin)[i], " is
specific to rDNA-2 in Cvi-0", sep = ""))

    # Determine whether the rDNA variant is specific to rDNA-4.
    }else if(rDNA_origin[RILY, i] > 0.03 & rDNA_origin[RILX, i] < 0.01){

      print(paste("rDNA variant ", colnames(rDNA_origin)[i], " is
specific to rDNA-4 in Cvi-0", sep = ""))

    }else{

      #print(paste("rDNA variant ", colnames(rDNA_origin)[i], " does NOT
seem to be specific to either rDNA cluster", sep = ""))

    }

  }else{

    #print(paste("rDNA variant ", colnames(rDNA_origin)[i], " is NOT
specific to Cvi-0", sep = ""))

  }
}
```

The diagnosis for all variable sites in Cvi-0 is printed to the screen, and we learnt that while only rDNA variant “1392.A.G” is specific to rDNA-2, rDNA variants “1378.G.A”, “4497.C.A”, “4601.T.C”, “8037.C.Ins” and “8743.T.A” are specific to rDNA-4.

4 Notes

1. Besides the 18S rRNA subunit (coordinates 2195–4002), the 5.8S (coordinates 4271–4434) and the 25S subunits (coordinates 4623–8009) can also be used for rDNA copy number estimation. It is also recommended to verify that the estimates of all three subunits are consistent with each other.
2. When interrogating F_2 lines for the assignment of rDNA variants to specific rDNA clusters, keep in mind that heterozygous rDNA loci are not useful for this purpose. For species with two rDNA clusters, only four out of sixteen possible combinations

of genotypes at the rDNA loci are homozygous for both rDNA loci, and of these, only two F_2 lines would have inherited homozygous rDNA cluster of different parental origin.

3. Since rDNA copy number is unstably inherited and might vary dramatically within a few generations [17], it is advisable to analyze independent RILs that share the same rDNA cluster configuration as RILs X and Y in Fig. 1.

Acknowledgments

This work has been supported by a Human Frontiers Science Program (HFSP) Long-Term Fellowship (LT000819/2018-L) to FAR and by the Max Planck Society. I thank Sonja Kersten and Ilja Bezrukov for proofreading the manuscript and testing the code.

References

1. Moss T, Langlois F, Gagnon-Kugler T et al (2007) A housekeeper with power of attorney: the rRNA genes in ribosome biogenesis. Cell Mol Life Sci 64:29–49
2. Layat E, Sáez-Vásquez J, Tourmente S (2012) Regulation of pol I-transcribed 45S rDNA and pol III-transcribed 5S rDNA in Arabidopsis. Plant Cell Physiol 53:267–276
3. Ritossa FM, Spiegelman S (1965) Localization of DNA complementary to ribosomal RNA in the nucleolus organizer region of drosophila melanogaster. Proc Natl Acad Sci U S A 53: 737–745
4. Wallace H, Birnstiel ML (1966) Ribosomal cistrons and the nucleolar organizer. Biochim Biophys Acta 114:296–310
5. Michael TP, Jupe F, Bemm F et al (2018) High contiguity Arabidopsis thaliana genome assembly with a single nanopore flow cell. Nat Commun 9:541
6. Sims J, Sestini G, Elgert C et al (2021) Sequencing of the Arabidopsis NOR2 reveals its distinct organization and tissue-specific rRNA ribosomal variants. Nat Commun 12: 387
7. Guo Y, Li J, Li C-I et al (2012) The effect of strand bias in Illumina short-read sequencing data. BMC Genomics 13:666
8. Guo Y, Cai Q, Samuels DC et al (2012) The use of next generation sequencing technology to study the effect of radiation therapy on mitochondrial DNA mutation. Mutat Res 744:154–160
9. Rabanal FA, Mandáková T, Soto-Jiménez LM et al (2017) Epistatic and allelic interactions control expression of ribosomal RNA gene clusters in Arabidopsis thaliana. Genome Biol 18:75
10. Copenhaver GP, Pikaard CS (1996) RFLP and physical mapping with an rDNA-specific endonuclease reveals that nucleolus organizer regions of Arabidopsis thaliana adjoin the telomeres on chromosomes 2 and 4. Plant J 9:259–272
11. Copenhaver GP, Browne WE, Preuss D (1998) Assaying genome-wide recombination and centromere functions with Arabidopsis tetrads. Proc Natl Acad Sci 95:247–252
12. Anaconda Software Distribution (2020) Anaconda Documentation. Anaconda Inc. https://docs.anaconda.com/
13. R Core Team (2020) R: a language and environment for statistical computing. R Foundation for Statistical Computing, Vienna, Austria
14. Grüning B, Dale R, Sjödin A et al (2018) Bioconda: sustainable and comprehensive software distribution for the life sciences. Nat Methods 15:475–476
15. Arabidopsis Genome Initiative (2000) Analysis of the genome sequence of the flowering plant Arabidopsis thaliana. Nature 408:796–815
16. Lamesch P, Berardini TZ, Li D et al (2012) The Arabidopsis information resource (TAIR): improved gene annotation and new tools. Nucleic Acids Res 40:D1202–D1210
17. Rabanal FA, Nizhynska V, Mandáková T et al (2017) Unstable Inheritance of 45S rRNA Genes in Arabidopsis thaliana. G3 (Bethesda) 7:1201–1209

18. Alonso-Blanco C, El-Assal SE, Coupland G et al (1998) Analysis of natural allelic variation at flowering time loci in the Landsberg erecta and Cape Verde Islands ecotypes of Arabidopsis thaliana. Genetics 149:749–764
19. Pontvianne F, Abou-Ellail M, Douet J et al (2010) Nucleolin is required for DNA methylation state and the expression of rRNA gene variants in Arabidopsis thaliana. PLoS Genet 6: e1001225
20. Abou-Ellail M, Cooke R, Sáez-Vásquez J (2011) Variations in a team: major and minor variants of Arabidopsis thaliana rDNA genes. Nucleus 2:294–299
21. Havlová K, Dvořáčková M, Peiro R et al (2016) Variation of 45S rDNA intergenic spacers in Arabidopsis thaliana. Plant Mol Biol 92:457–471
22. Chandrasekhara C, Mohannath G, Blevins T et al (2016) Chromosome-specific NOR inactivation explains selective rRNA gene silencing and dosage control in Arabidopsis. Genes Dev 30:177–190
23. Van Rossum G, Drake FL (2009) Python 3 reference manual. CreateSpace
24. SRA Toolkit Development Team. http://ncbi.github.io/sra-tools/
25. Martin M (2011) Cutadapt removes adapter sequences from high-throughput sequencing reads. EMBnet.J 17:10–12
26. Li H, Durbin R (2009) Fast and accurate short read alignment with Burrows–Wheeler transform. Bioinformatics 25:1754–1760
27. Li H, Handsaker B, Wysoker A et al (2009) The sequence alignment/map format and SAMtools. Bioinformatics 25:2078–2079

Index

Christophe Lambing (ed.), *Plant Gametogenesis: Methods and Protocols*,
Methods in Molecular Biology, vol. 2484, https://doi.org/10.1007/978-1-0716-2253-7,